How to Build a Small Budget Recording Studio from Scratch

About the Authors

Mike Shea has more than 40 years' experience as an independent contractor specializing in the construction of rehearsal and recording studios. He has taught graduate courses at the Institute of Audio Research, served as editor of *Recording World* and technical editor of *International Musician*, and has written extensively about all aspects of professional audio. Mr. Shea resides in New York.

F. Alton Everest was a leading acoustics consultant. He was cofounder and director of the Science Film Production division of the Moody Institute of Science, and was also section chief of the Subsea Sound Research section of the University of California.

How to Build a Small Budget Recording Studio from Scratch

Mike Shea
F. Alton Everest

Fourth Edition

New York Chicago San Francisco Lisbon London Madrid
Mexico City Milan New Delhi San Juan Seoul
Singapore Sydney Toronto

The McGraw-Hill Companies

Cataloging-in-Publication Data is on file with the Library of Congress.

McGraw-Hill books are available at special quantity discounts to use as premiums and sales promotions, or for use in corporate training programs. To contact a representative please e-mail us at bulksales@mcgraw-hill.com.

How to Build a Small Budget Recording Studio from Scratch, Fourth Edition

2 3 4 5 6 7 8 9 0 DOC/DOC 1 9 8 7 6 5 4 3 2 1

ISBN: 978-0-07-178271-5
MHID: 0-07-178271-0

Sponsoring Editor
Judy Bass

Acquisitions Coordinator
Bridget Thoreson

Editing Supervisor
David E. Fogarty

Project Manager
Nancy Dimitry,
Joanna Pomeranz, D&P Editorial Services

Copy Editor
Nancy Dimitry, D&P Editorial Services

Proofreaders
Joe Cavanagh,
Don Dimitry,
Don Pomeranz, D&P Editorial Services

Production Supervisor
Richard Ruzycka

Composition
D&P Editorial Services

Art Director, Cover
Jeff Weeks

Contents

v

Part III   New Remedies to Common Acoustic Problems

21   An Introduction to Today's Premanufactured Acoustic Panels and Systems

Part IV   Music Instruments

Preface

I began my career as a recording studio engineer in the 1960s, at a time when one was not only designated as the audio mixer but also the studio designer/builder along with part of the crew that built the studio's console. By 1970, I was attending lectures at MIT by Leo L. Beranek, a partner with R. H. Bolt of Bolt, Beranek and Newman, and a leading auditorium acoustician. Attending college close to full time, attending seminars and lectures, and working in a recording studio while running my own four-track "portable" recording business, you'd think that all kinds of needed information would be pouring in my direction. While all of the above gave me a great deal of insight into the field of acoustics, it seemed there wasn't anybody talking about the things that were most directly important to pop music recording. I, like many others at this point in time, was among the first to deal acoustically with a whole new situation. Take two electric guitarists, each running through mega-power stacks comprised of "Hi-Watt" amps on top of dual 4×12 inch speaker cabinets, a bass player using dual Ampeg folded horn cabinets with 18-inch speakers driven by a 250-watt Plush amp, a drummer with a giant drum kit played with sticks held backwards so as to have the thick end out, and a keyboardist using a Hammond organ attached to two Leslie speakers (with variable motor speed drives) and a synthesizer capable of producing sub-audible tones plugged directly into the console. Place all of this in a 20×25 foot room, hit record, and you'll have an idea of what I'm referring to.

Then along came F. Alton Everest's groundbreaking book in the field of recording studio acoustics and his generosity in not only making available these findings, but also doing so in an accurate yet easily understood manner. You'd never understand what a big deal this was unless you've tried to explain to a professional acoustician/noise control engineer that it would not be possible to "simply" decrease the electric guitarist's amplifier level or remove one of two 4×12 speaker cabinets that made up his stack, and yes, the electric bass guitarist did feel it was "appropriate" to thump his fingers against the instrument's strings, thereby causing those low frequency shock waves of vibration. My introduction to the first edition of Mr. Everest's book did not adequately put forth the feeling of gratitude that I hold for him. To further illustrate this point, I will explain what it was like back in 1971 when I was a student of the science of acoustics. At that time, one of the better publications in the field was the Soviet acoustician V. S. Mankovsky's *Acoustics of Studios and Auditoria*.

The drill with this book was to read a short amount of text, then labor through a given mathematical equation. Read another line or two of text that varied the acoustical situation slightly, then work your way through the resulting more complex equation.

The section covering Helmholtz resonators is only 15 pages long, however, it contains no fewer than 37 mathematical equations!

Not only did it take me close to two full semesters to work through this book's 360 pages, but being caught, even in the campus library, surrounded by pieces of paper covered with mathematics was a serious blow to one's image—what little of it there was. Now don't get me wrong; I have never regretted this arduous undertaking because at its completion, Mr. Mankovsky's teaching method had been entrenched in my mind for good, providing me with a deep understanding of the fundamental principles of the art and science of acoustics. Yet what a tremendous joy it was for me and a lot of other "recording studio types" when not eight years later in 1979, F. Alton Everest, building upon the work of a great many previous acousticians undertook the simplification of the process of acoustically correct recording studio construction with the first publication of *How to Build a Small Budget Recording Studio*. Mr. Everest gave an unheard of 12 suite designs that he had actually built and thoroughly tested to boot! In his explanation of Helmholtz resonators, he gave all details including construction blueprints, using only one very simple equation with an easily understood explanation, and specific examples of how they are to be used in only six pages.

Later, in 1981, with the publication of *The Master Handbook of Acoustics*, he let everyone in on even more by giving a full description of his experiments along with details of the scientific tests he used to confirm results. Here, he also expanded on his coverage of slot-type Hemholtz resonators and on construction methods with the use of photographs. This material was far more in tune (pun intended) with the specific requirements of a popular music recording facility, all made easily understandable and compiled in a manner that left readers with a much more useful reference work. I would, therefore, like to take this opportunity to say "Thank you very much Mr. F. Alton Everest" for myself and, I'm sure, many others.

Mike Shea

Preface to the Third Edition

This book is about small studios: How to build them and how to treat them acoustically. Details of design, construction, and treatment of twelve actual studio suites are included. Acoustical principles are discussed along the way in the context of real life projects and problems rather than as blue sky theory.

The emphasis of this book is on budget studios which are eminently suited to efficient production of radio, audiovisual, video, film, and television program material on a day-to-day, routine basis and for training students in these fields. These studios operate in a different world from the flamboyant recording studios with gold records on the wall designed to catch the eye of the well-heeled client. The operators of the studios described in this book stress function and economy over glamour yet strive for good (natural, faithful) sound quality.

The word studio is often used in this book in the inclusive sense to include control and monitoring rooms. Acoustical defects in the recording studio quite obviously affect the recorded sound. If the control room acoustics are poor, the processing changes made by the operator may actually degrade the signal because they are correcting for acoustical defects. For this reason the same care is given to control room acoustics as to those of the studio.

The studios described in this book (with the exception of Chapter 7*) follow closely, but not exactly, designs I prepared in my consulting practice for many different types of organizations in various parts of the world. I am greatly indebted to the following for permission to share the design of their studios with the readers of this book: Missionary TECH Team, Longview, Texas (Chapter 3); proposal submitted to the Christian Church in Zaire (Chapter 4); Hong Kong Baptist College, Kowloon, Hong Kong (Chapter 5); The Russ Reid Company, Pasadena, California (Chapter 6); Centro Bautista de Comunicaciones, Montevideo, Uruguay (Chapter 8); Golden West Broadcasting Ltd., Altona, Manitoba, Canada (Chapter 9); The Paraguay Mission, Southern Baptist Convention, Asuncion, Paraguay (Chapter 10); Medios Educativos, A.C., Mexico City, Mexico (Chapter 11); Baptist Caribbean Media Center, Nassau, Bahamas (Chapter 12); Cathedral Films, Westwood Village, California (Chapter 13); and Far East Broadcasting Company, LaMirada, California (Chapter 14).

F. Alton Everest

---

*Chapter numbers from the third edition have changed in the fourth edition, since 5 new chapters were added in the beginning. Therefore, Chapter 7 is actually Chapter 12 in this edition. Add 5 to all other chapter references in this paragraph to find the chapter number in this edition.

Introduction

Since the third edition of this book was published a decade ago, more than a little has changed in the field of small budget recording. The number of studios has multiplied beyond all imagination. Back then, it was not surprising to find full-time professional musicians who had their own small recording studio. Now it seems like everyone involved in music has home recording facilities. This made a new edition of this publication a necessity so that it would retain its original intent of providing the most accurate and up-to-date information to the small recording studio owner/builder dealing with the constraints of a tight budget.

So much had to be added on the subject of acoustic materials that it is now a whole section in Chapter 2 (pages 9–32) that includes new material as well as Chapters 16 "Bits and Pieces of Acoustic Lore" and 19 "Electronics and Acoustics" from the third edition. Another section, Chapter 21, had to be added just to handle the new acoustic panels and systems being manufactured today.

It used to be that the years one spent "coming up" from gopher, to tape op, then assistant engineer, to finally manning the board as a house engineer provided a thorough education in electronic technology, acoustics, and the best microphone placement methods for capturing the sounds of many music instruments. Those days are pretty much gone, and while it is very difficult to explain micing techniques in a book (as opposed to actually watching the process and hearing the results), I can tell you enough about the music instruments themselves so that decisions on microphone placement become second nature. Which instruments? *All of them!*

Part I, Acoustic Huggies, was designed to get you out of the crib and into the acoustic playpen, as it were. We start off with some basic theory just to set up a solid base. Then we get right into practical acoustics with a description of absorptive materials and how the properties of thickness, density, and mounting affects their functioning as per frequency. Next comes the importance of modes (both fundamental and harmonic) and how they can detract from a room's even response. Since these modes occur at lower frequencies in small rooms, they often must be dealt with using active absorbers. Therefore, there is a thorough discussion of Helmholtz resonators as well as panel and diaphragmatic absorbers—*not* just in theory but in actual construction methods. This topic is covered in detail. Here the reader is provided with mathematical equations for tuning Helmholtz absorbers and given an easy-to-use jig for determining center frequency, bandwidth, and peak absorption, along with the required air space behind plywood panel absorbers.

While reverberation time is not generally a factor in small rooms, it must be fully understood before problems such as flutter echoes can be addressed. RT60, reverb time vs. absorption, critical distance, and echoes are all covered in full. Before you can determine which materials, panels, or systems will cure any acoustic problem, an understanding of the specifications and standards that provide their absorption coefficients, sound transmission loss, and noise reduction ratings is required. These standards are also covered in full along with critical frequency, coincidence effect, mounting methods, and facings.

The electronic measurement of a room's acoustic response has always been important in achieving an honest listening environment. We go back in time and explain sequential frequency response analysis, then move into more modern equipment such as real time analyzers. Noise generators (pink, white, and pulse) and measuring reverberation time are also covered. There is a brief history of FFT, TDS, and TEF measurement and their importance. Finally, you are given a detailed description of how to set up a very low-cost but accurate laptop acoustic measurement system. Using this type of system means you'll *know* what you need to make your monitoring system accurate and, thus, eliminate costly purchasing mistakes.

In Part II, Brick and Mortar (Drywall and Stud) Studios, you are given 23 room designs by F. Alton Everest. But first he starts out by covering the basics, including size and shape of the room, then the components that make up a recording studio, such as sound locks, acoustic doors, weatherstripping, HVAC noise, wall construction including resilient mounting, staggered studs, double walls, concrete and masonry walls, floor-ceiling construction, electrical wiring, lighting, observation windows, absorber mountings, reverberation time, and even advice on obtaining a construction permit.

Then within the very detailed descriptions of the rooms he designed, built, tested, and then adjusted for correct frequency response and the required reverberation time according to the room's intended use, he explains: wideband wall units, control room ceiling treatment, control room wideband modules, semicylindrical units, reversible wall panels, low-frequency, midband and wideband treatments, drum booths, and studio floors, walls, and ceilings. He leads you through the practice of "trimming" the acoustics of studios, control rooms, and even talk booths. Nothing is overlooked: air conditioning and duct routing, conduits, power facilities, cyclorama curtains, variable louvered absorbers, wall splaying, floating floors, and using internal and external walls to eliminate noise such as traffic from getting into your studio and stopping sound from getting out and possibly disturbing your neighbors.

While the book gives 12 tested designs, it actually contains more than twice that many rooms counting control rooms, studios, and listening rooms. All the rooms are acoustically correct and can be used as per the builder's application. In fact, one commercial recording studio whose design was taken from the book was changed to match the size, intended use, and layout of the raw space available. Here the studio design was used for the control room, and the control room layout was used for the studio. In other words, since all the individual spaces are "ready to go" as per resulting reverberation times, it's your call as to how they are used, with *no* worry about a decrease in sound performance.

The dimensions range from $11' \times 9'1'' \times 8'\,^3/_{16}''$ high to $27' \times 39'4'' \times 16'11''$ high. That's a span of just under 800 cubic feet to over 18,000 cubic feet! You may want to consider a design that is slightly smaller than your space to allow for the implementation of noise reduction, HVAC duct runs, power, audio wiring, storage and other requirements such as combo amp isolation.

It's hard to think of a situation these designs would *not* cover, whether it be a small private music writing/preproduction studio, church A/V facility, or professional radio or television house. Even the essentials of a great home theater/entertainment center or an honest-sounding production room for a DJ could be handled with ease using one of these designs. All because F. Alton Everest has already done the hard part by calculating the room dimensions so as to avoid those all too common problems such as bass buildup or other anomalies such as resonant frequencies or standing waves. Furthermore, his design principles are based on the desire to achieve a specific room reverberation time as dictated by the use for which the room is intended, be it speech, music and/or critical listening, all with his very experienced eye trained on the practical everyday functionality that a real working environment demands. See Table I.1 for a description of the tested designs found in this book.

Part III, New Remedies to Common Acoustic Problems, is all about correcting the acoustics side of your monitoring system. In other words, how to *honest up* your critical listening/recording environment.

Several types of acoustic products are covered generally, and as examples, we touch on some specific acoustic remedies offered by several manufacturers. Today's laptop recordists are often not capable of or interested in constructing full-fledged recording facilities from the ground up, and they are perfectly willing to make use of any space available to them. This more often than not ends up being a small bedroom.

Small rooms suffer from the dual problems of low-frequency build up caused by their dimensions being about the same as the half wavelengths of the low frequencies being reproduced and high-frequency reflections between parallel walls, floor and ceiling. Laptop recording equipment costs very little so there are millions upon millions of bedroom studios out there, all suffering from the same problems. Because of this, there are also a slew of manufacturers looking to provide what could be very lucrative answers to the bedroom recordist's needs. Research and competition are intense to say the least, but as cutthroat as it can be, no one is outright malicious. I did find a bunch of new products that are not only fairly inexpensive, but also very effective at curing the acoustic problems they were designed to handle. *And* they just keep getting better and better. So thank your lucky stars and dive into the new acoustic landscape where it appears *everyone* is out to help you with your small room acoustic difficulties.

Acoustics becomes even *more* critical in *small, one-room facilities* where the bedroom recordist must utilize a too-small-sized room as both a critical listening environment *and* as a space conducive to aiding the propagation of acoustic instruments. This *will* be a tough go, to say the least, but thirty years ago it was insurmountable. Today it may cost you, but it is now a can-do situation with most any size room.

With today's economic conditions, it is often just not possible to build a new brick and mortar (sheet rock and 2×4) world-class recording studio. However, adjusting an existing room to achieve professional sound quality is now more feasible than ever. Unfortunately in most cases, we are not talking about a studio and a control room, but combining the control room and studio into a one room "facility." The acoustical demands placed on recording as opposed to mixing differ widely. Critical listening requires that there be little to no "acoustic liveness" between the speakers and the listening position, while recording requires an environment conducive to the propagation of music instruments (except when tracking electronic instruments where no acoustics are involved).

Small size means larger acoustic problems; yet in most cases, both critical listening during mixdown as well as live recording while monitoring with headphones is being attempted in less than adequately sized, acoustically untreated rooms. Luckily, every-

Ch	Description	Size, Internal Dimensions
8	Audio Visual Budget Recording Studio	11′5″ × 13′11″ × 10′ h
	And Control Room	9′ × 11′5″ × 13′9″ h
9	Studio Built in a Residence, Studio	18′4″ × 23′1″ × 13′3″ h
	Control Room	11′ × 16′2″ × 13′4″ h
10	Studio For Instruction and Campus Radio	18′10″ × 14′9″ h and 11′5″ × 8′8″ h
11	Ad Agency A/V and Jingle Control Room	13′8″ × 11′5″ × 8′11″ h
	And Studio	10′2″ × 18′4″ × 8′11″ h
12	A Multi Track Studio and Control Room in a Two-Car Garage Measuring	21′7″ × 21′5″ × 9′8″ h
13	A Radio Production Facility Studio	25′5″ and 27′1″ × 18′ and 18′3″ × 11′3″ h
	And Control Room	13′7″ and 13′9″ × 10′8″ and 9′4″ × 9′11″ h
14	A Commercial Radio Station with a Booth	11′ × 9′1″ × 8′³⁄₄″ h
	A Production Room	15′7″ × 12′11″ × 9′11″ h
	And a Control Room	15′6″ × 12′8″ × 9′11″ h
15	Two Studios, One for Music	16′1″ × 23′2″ × 9′5″ h
	One for Speech	11′10″ × 14′9″ × 9′5″ h
	And a Single Control Room	11′10″ × 14′9″ × 9′5″ h
16	A Video Production Facility Studio	20′7″ × 20′4″ × 12′4″ h
	And Control Room	19′2″ × 15′9″ × 12′4″ h
17	A Facility for Both Video (TV) and Multitrack Audio Recording with a Video and Multitrack Studio	27′ × 39′4″ × 16′11″ h
	A Control Room	13′11″ × 17′ × 12′3″ h
	And a Smaller (General Speech Work) Studio	11′5″ × 13′11″ × 10′ h
	With a Second Control Room	11′5″ × 13′11″ × 10′ h
18	Screening Room for Film and Video	35′7″ × 23′2″ × 11′ h
19	Multiple Studio Facility Studio C	22′9″ and 24′7″ × 14′9″ and 17′ × 10′ h
	A Control Room	10′9″ and 12′ × 13′7″ and 14′5¹⁄₂″ × 10′ h
	And Studio A	10′10″ and 12′ × 13′7″ and 14′5¹⁄₂″ × 10′ h

TABLE I.1   23 Tested Room Designs Provided by F. Alton Everest in This Book

thing has changed to the point where you can actually forget about modifying the dimensions of the room to correct modal problems because it is now possible to acoustically treat small rooms by simply adding premanufactured acoustic panels and systems to control unwanted reflections or remove excessive resonance. You may not be surprised by this if you grew up with it, but believe me, forty years ago this was unheard of.

I put a good deal of time into this section. Space was limited so I had to choose products that helped make specific points, such as how important it is for you to dig deeper and deeper into the manufacturer's literature *and* to call them with any remaining questions that need clarifying. I also tried very hard to come up with a worthless acoustic product, kind of like the old sand-filled paint of the 1960s, but could not find a single product that could be called disreputable. On the other hand, I found many products that were very good and several that were stunning.

RPG's Modex Corner™ is a triangular shaped box that stands just under two feet tall (23⅝") and has a front facing that's just 12 inches across. Yet it has a random incident absorption coefficient of 0.54 at 125 Hz, 0.64 at 100 Hz, and 1.12 at 80 Hz. Impedance tube testing showed it to have absorption coefficient averages of: 0.88 at 80 Hz, 0.92 at 70 Hz, and 0.895 at 60 Hz. That's correct; the thing is less than *two* cubic feet in size!

Acoustics First's Art Diffusor® model E is sized 15" × 15" and is 9 inches deep, which extends the bandwidth from to 125 Hz to 16 kHz. These run $66 each, but a box of four costs $210 instead of $264. That's only $52.50 each! So 6 of these panels will run you $345.00 for a little over 9 square feet, and 12 will cost $630.00 for almost 19 square feet. That's very nice but the real kicker is these 15" square panels weigh just a single pound each. This means four small squares of Velcro will hold them up so that now when you are recording, monitoring through headphones, and don't care about side wall reflections, you can spread them all around your recording area for a more dispersed and bigger sounding room!

Three-inch thick UltraTouch™ insulation has the following absorption coefficients:

125	250	500	1000	2000	4000 Hz.
0.95	1.30	1.19	1.08	1.02	1.0

That is very impressive for just three inches in thickness, but this stuff is not just environmentally friendly. It has 85% recycled content, uses natural fibers, contains no carcinogens or formaldehyde, is mold and mildew resistant, is light weight, and is it ever fire retardant. I once saw a demonstration where a propane torch was used to melt a piece of copper lying on top of some UltraTouch. Even though the copper was liquefied, the UltraTorch was merely blackened. They achieved this fire *stopping* capability by adding boron (think borax) into the material. Boron is also used as an insect deterrent so UltraTouch gained some serendipitous insect repellant along with the ability to completely stop fire. The 85% recycled content? Recycled blue jeans!

Auralex puts out a block of foam with wedges cut into it with the following absorption coefficients:

100	125	160	200	250	315	400	500	630	800	1k	1.25k	1.6k	2 k	2.5k	3.15k	4k	5 kHz
1.19	1.6	1.30	1.31	1.34	1.32	1.36	1.29	1.25	1.25	1.26	1.20	1.22	1.25	1.21	1.18	1.20	1.24

It's called the Venus Bass Trap, and it's the proverbial full-frequency open window.

Somewhere along the line (maybe due to the proliferation of "electronic musicians" out there), the art of recording acoustic instruments has been overlooked. For you, that must change because you are going to need as much knowledge as you can gather in order to pull off capturing all the nuances of acoustic instruments played in a less than ideal setting (such as your bedroom). The first step is to understand how music instruments produce and propagate sound.

Part IV, Music Instruments, starts off with some basic facts about sound production by music instruments, such as resonance, note duration, attack, wolf tones, impedance, sound radiation, sound transmission, radiators, and resonators. Then we move into specific sound-producing mechanisms, such as tines, rods, and tongues. We then take on the sound transmission of acoustic guitars and discuss making use of physicists' research experiments. Tube and pipe instruments are dealt with by discussing tubular instruments, the brass and woodwind families, edge tone instruments, ports, pipe organs, reeds and brass, bagpipes, accordions, mouthpiece instruments, trumpets and other cone shaped instruments, as well as the human voice. Vibrating strings are certainly not left out, as we go over bows, hollow-bodied instruments, violins and the making of a violin, bowing, bridges, tuning bridges, middle bridges, violin bodies, electric guitars, some guitar manufacturing theories, setting a guitar's intonation, harps, mechanically plucked strings, the harpsichord, and pianos.

Percussion instruments? How about blocks, marimbas and xylophones, vibraphones, sound actuators, membrane radiators, timpani or kettle drums, bass drums, drum heads, drum bodies or shells, drum tuning, drum setup, pellet drums, tenor drums, snare drums, tube drums, talking drums, tambourines, castanets, friction drums, steel drums, and cymbals. We end up with a bunch of "drummer stories," how to get a bead on an artist before the session starts, and finally a discussion of multi-micing techniques, including some specific examples.

Finally, additional supporting material can be found on an associated Web site at www.mhprofessional.com/shea4. I was not going to touch a word of what F. Alton Everest wrote, so I decided to place two chapters [17 "Acoustic Equations" (Chapter 23 on the Web site) and 20 "Reading Blueprints" (Chapter 24 on the Web site)] and Appendix D "Manufacturers of Acoustic Materials," all of which I had written, up on the Web.

You'll also find chapters on High Reliability (mil-spec) Soldering by Hand, Writing Trouble Reports (with a preprinted blank trouble report page to copy), along with both a broad-band frequency analysis chart and a 31-band RTA chart for you to copy.

There's also a chapter on Audio Equations. This one covers level conversions (VU, peak, average, RMS and peak to peak, as well as dBV to dBu), time, pitch, and tempo equations for vari-speed and delays, and other mathematical formulas, such as changing beats per minute and number of measures to match spot timing, which are handy for a recording engineer to have at the ready.

There are also two updates, one on the ratings of duct silencers and another on metal wall studs. Both are key to achieving low background noise.

Finally an appendix, "Manufacturers of Acoustic Test Equipment," will be placed on the Web site.

With regard to the photos on the front cover, clock-wise from the top left, they are Acoustics First's Model E Art Diffusor; top center, RPG Diffusor System's FlutterFree® above their FlutterFree-T; top right, Auralex's Venus Bass Trap; bottom right, Acoustics First's Double Duty polycylindrical bass trap/diffuser; bottom center, (the one that started it all) RPG Diffusor System's QRD-734 quadratic-residue diffuser; and bottom left, Acoustics First's Model E Art Diffusor.

Oh, and that studio in the center? That's an early 1980s photo of Jimi Hendrix's Electric Lady Studio C, taken just before it was completely revamped. At the time, the trend was to overdub in the control room so musicians could hear themselves on the main monitors, (here the forerunners to the Westlake Reference Series).

This and the need to expand to 48 track capabilities meant upgrading to a larger control room. While they were at it, they added an SSL 56 input console along with dual Studer 24 track recorders. I was told that Jimi's old Neve console went back to the manufacturer who completely refurbished it and resold it to some *very* lucky individual. I'm sure the studio got rid of the Dolby A rack—those things could be a beast.

The half rack of Kepex and Gain Brain modules are probably long gone too, but I'd be *mighty* surprised if they let go of Jimi's 1176 and LA-2A limiters. You can't see it in this photo, but there were *three* LA-2As stacked one above the other.

Just shows how tastes change. Today *any* audio recordist would give their left foot to have that room. You already know what it sounded like!

Mike Shea

How to Build a Small Budget Recording Studio from Scratch

PART I

ACOUSTIC HUGGIES

CHAPTER 1

Theoretical Acoustics

Wave Theory

"Wave theory? We don't need no stinking wave theory!"

Don't worry, we are not going to get too technical right off the bat. Waves are simply a means of transferring energy from one place to another. Sound energy, once initiated, travels in all directions meaning three dimensionally.

Now the wave surface, which indicates where a pressure change has gotten to at a particular time, will, in anechoic or free-field conditions (that is without any acoustics involved), be spherical (see Fig. 1.1).

Imagine two people sitting on the top of two poles a mile up in the sky, as shown in Fig. 1.2. There are virtually no acoustics involved in this situation. Say the poles are 100 feet apart. When person A's voice is heard by person B, the distance it has traveled

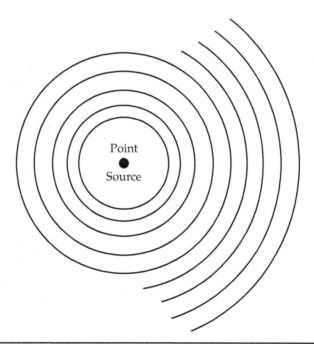

Point
Source

FIGURE 1.1 Sound waves are spherical.

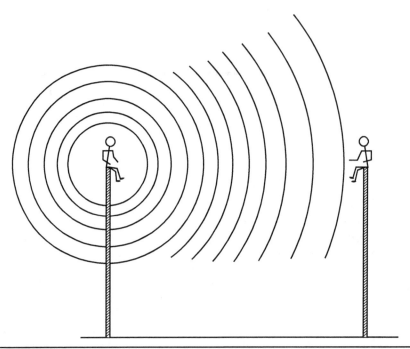

FIGURE 1.2 Two people on poles (with no acoustics involved).

(100 feet) is just the radius of the area that energy has covered. Therefore, the sphere surface would be 4 × pi × 100 feet squared, or 125,000 square feet, upon its arrival at person B. Even if person B's ears were one square foot each, they would still be receiving only 1/62,500 of the total energy given off by person A. And if person B were to move his pole twice the distance away to 200 feet, the energy received would be only one half of that. This is called the inverse square law and it states that, in nonreflective environments, the level will be halved (or drop by 6 dB) with each doubling of the distance between the source and receiver.

Time for Some Reflection!

In practice, even if the surface reflecting sound is hard and smooth, it will not be as perfect as that of light bouncing from a mirror. Therefore, the example we will use of a wall giving a mirror image of the wave to the listener is true to an extent, but cannot be taken at face value because no surface is a perfect sound reflector.

Now let's drop those two people to the ground, and for the sake of this explanation, we'll assume that the ground's surface is as reflective of sound as a mirror is of light. As shown in Fig. 1.3, person B will now receive two sets of waves, one direct as above and another that goes from person A to the ground and is then reflected to person B. The reflected wave would not be as strong as the direct, but the increase in level it causes would be noticeable and helpful.

FIGURE **1.3**   Two people on the ground.

What happens when we add a wall behind the speaker (see Fig. 1.4)? Now three more paths in addition to the direct one are possible, giving B a total of four waves:

1. The direct path from A to B (no acoustics involved)
2. The reflected path from the floor (one surface)
3. The reflected path from the wall (one surface)
4. The reflected path from the wall reflected once again by the floor (two surfaces)

If we add a surface parallel to one of the first ones, the waves would bounce back and forth (as per our theoretical mirror reflections) setting up an infinite number of paths (infinite additional sound sources). Picture two mirrors parallel to each other—if you look at an object placed in between them, the reflections would continue indefinitely. If we completely enclose the two people in a room with everything parallel— walls front–back, left–right and floor–ceiling—we would be setting up three paths of infinite reflections, and those two people would find it quite difficult to communicate.

To illustrate this differently, we'll leave our theoretical mirror image reflections and return to reality. About the closest we will ever get to the above perfectly reflective room is an empty indoor swimming pool. Here, the length of time that all the reflected paths last (reverberation time) might be several seconds. This is enough reverberation to make communication difficult because it takes sound longer to travel along multiple reflective paths to the listener and this will cause each syllable to be overlapped by the ones preceding it, causing speech to become garbled, slurred, or at least confusing.

So where has all this stinking theory gotten us? No reflections make hearing more difficult and so do too many reflections. Obviously good acoustics lies somewhere

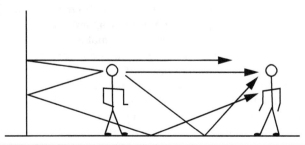

FIGURE **1.4**   Two people on the ground with an added wall.

between the two. Take your living room for example, with its carpet, curtains, furniture and possibly an open window or two; it can make conversation quite pleasing. This is due to the fact that this room has softer and more irregular surfaces and, therefore, is quieter sounding than an enclosed empty swimming pool.

On the other hand, everyone knows about singing in the shower. Here the surfaces are hard and smooth with the tiles causing a good deal of reflection. Therefore, it's easier to reach a higher level of volume; and because the surfaces are fairly close together (showers being relatively small in size), the lengthening of syllables here tends to even out the imperfections of even the most amateurish singing. As compared with our previous examples of either no reflections (lacking of volume or level) or too much reverberation (confusion), showers seem to be closer to ideal, that is, until you put a microphone on that singer and find that there is no control in this particular acoustic situation. That is why every studio that utilizes the bathroom as an additional "reverb chamber" uses the speaker level along with the microphone gain so they, at least, get a handle on the reverberation level and time.

By deciding what materials the walls are made of and where they are placed, we can arrive at a compromise reverberation time that is acoustically suitable to the intended use of a room. As you will see, small rooms tend to amplify some frequencies more than others, while large rooms, or rooms with complex shapes, have more complex reflection paths so they end up having a flatter response. Volume, reverberation time, paths, energy, reflections, standing waves, and all those other mysterious, "scientific" aspects of acoustic theory that cannot be overlooked are actually not that terribly complex.

Molecular Level

Let's take a look at acoustics on the molecular level.

"Molecular level now? Oh Nooooooooooo!"

It's really not all that bad and it will help you to understand why some materials work as acoustic absorbers and others are just an acoustical rip-off. So by diving into the fundamentals you'll end up saving yourself both money and self-esteem. Any discussion about acoustic materials boils down to whether sound coming into contact with it is reflected off, absorbed by, transmitted through, or some combination of all three. That is pretty much the whole ball of wax as far as everyday acoustics goes. However the job of the acoustician not only involves the design stage, but also specifying the materials to be used. It is necessary to know about properties to determine how vibrations or airborne sounds will be transmitted, dampened, or attenuated by them. A material's stiffness or "modulus of elasticity" and resonant frequency are important to its ability to isolate vibration, while its peak Sound Transmission Loss (STL) is important to its ability to stop airborne sound transmission.

But in order to understand how acoustic materials function, you must first understand the physics of sound. Energy that causes a variation in pressure is a result of molecular movement. Sound energy starts at its source, travels along some path, and reaches the receiver. The path must be an elastic medium. Molecular movement is constant for each medium. The magnitude of the movement depends on the pressure and temperature of the medium. In any medium, molecules are constantly in motion, smashing into each other and bouncing around. Sound energy is superimposed on this random motion. To see what effect sound has on molecular motion, you must freeze or stop all the continuous movement for an instant.

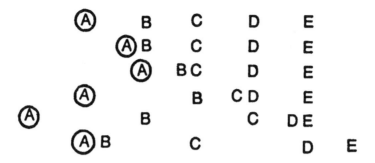

FIGURE 1.5   Molecular motion caused by wave incident or vibration. Molecule A's complete movement, from top to bottom, is one cycle.

Figure 1.5 shows what happens to a molecule when force is exerted upon it by sound. A is displaced to one side and strikes B, setting it in motion. More important than its bouncing back is the fact that A not only returns to its original position, but continues past it to a location opposite the initial direction of movement. Then it returns to its original position. If this movement is repeated at regular intervals, it is a vibration. The areas where A+B, B+C, C+D, and D+E are close together are called *compressions*. The voids in pressure those compressions cause are called *rarefactions*. The distance between two adjacent points of rarefaction or for that matter two adjacent points of compression equals the wavelength of the frequency in that particular medium. This holds true for sound power through air as well as mechanical vibration in solids.

Sticking with the molecular example of Fig. 1.5, the sequence of A's movement from top to bottom is one cycle, or a complete excursion of that molecule. Its frequency is the number of these cycles that occur during the period of one second. The maximum displacement of A is equal to the amplitude of motion, and with audible sound, the excursion or movement is a very small amount.

The important aspects of the path or the medium the sound passes through are elasticity and density. Greater mass causes greater inertia. Elasticity or stiffness is also a factor along with hysteresis (internal friction). Therefore, in any consistent medium, the speed of sound traveling through it will be fixed. The velocity or speed of sound is proportional to the elasticity and inversely proportional to the density of the material through which it is passing.

In other words, the more a material weighs, the slower sound will pass through it; while the stiffer it is, the faster sound will travel through it. These two seemingly simple relationships can become quite complex. As examples, the speed of sound through lead, which is very dense, is approximately 4000 feet per second, while in soft wood it is 11,000 feet per second, and in even less dense air, 1130 feet per second (water: 1,128 / concrete: 10,000 / glass: 12,000 / hard wood: 14,000 / iron and steel: 17,000). These facts would not appear to make sense unless the hardness of the material is taken into consideration.

CHAPTER 2

Practical Acoustics

Sound Absorptive Materials

Absorptive materials come in two basic types: porous and diaphragmatic (resonant or reactive). There are literally hundreds of porous absorbent materials with the most common being mineral or glass wool, molded or felted tiles, perforated panels, sprayed-on fibers with binders, foamed open-celled plastics, elastomers, and even normal room furniture such as cushions, upholstery, carpets, as well as draped fabric curtains. All are fuzzy and/or soft, and when sound pressure is exerted upon them they move. This movement causes friction which converts the pressure to heat and thus depletes the acoustic energy. In order to determine the amount of attenuation they provide, one must delve a bit deeper into the inner workings of sound absorptive materials.

The air inside a porous network of fibers, cells, granules, or particles is pumped or oscillated back and forth within the material when sound pressure is exerted on its facing. This offers some resistance to the flow and causes the network to vibrate which causes frictional losses. In practice sound enters fiberglass or for that matter any porous absorber and causes the internal fibers to move. The fibers then rub against each other and the friction causes heat. More importantly, the fibers retard the progress of sound through the material and therefore deplete its energy. The deeper the sound goes the greater the resistance or more correctly impedance and the greater the loss of energy. The effectiveness of the sound absorber depends on the thickness of the material, the density of its fibers and even the way in which it is mounted.

This energy loss is expressed in coefficients that relate the original energy to the attenuated energy. If a material absorbs 100 percent of the sound (acting like an open window), it would be 100 percent absorptive. The unit of one square foot of totally absorbent surface is given the value of 1.0 and all absorptive materials are compared to this. Therefore, if an acoustical material absorbs 45 percent of sound energy, it would have an absorption coefficient of 0.45. While this rating is not meant for real-world usage, it works well when comparing one material's absorptiveness to that of another.

The internal structure of porous materials can be one or more of several basic types shown in Fig. 2.1, but in actual material the networks are rarely this regular. What these networks do is introduce resistance, or impedance, to the flow of the air inside the material. If the resistance is too low, no frictional loss occurs. If too high, the airflow will be too restricted causing little fiber movement and therefore little frictional loss. In between these two points lies a fairly broad area and an optimum resistance to airflow is not an exact value for most materials. The internal structure of absorptive materials must have openings. As an example, while two foams may look the same, an open-cell

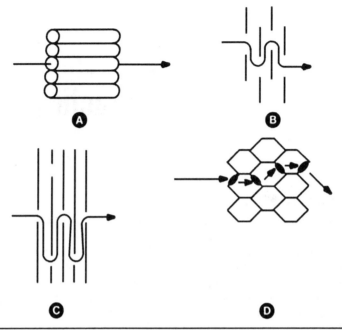

FIGURE 2.1   Internal structure of porous materials: (A) tabular, (B) plate, (C) orifice, and (D) cellular.

foam is sound absorptive, but a closed-cell foam can easily be reflective. The fibers, particles, or cells should be randomly orientated and have irregular size and shape so sound traveling through them takes as complicated a path as possible. If the internal fibers are too large they will reflect and if too small they could end up being too dense and offer too high a resistance to airflow.

Foams

Foam rubber is made by whipping latex, adding a gelling agent, and pouring the mixture into molds. The result is an open-celled material of variable density. Plastics can be foamed by chemicals which create gas bubbles and yield closed-celled structured foams before modification. Foams can transmit, absorb, or reflect sound depending on their permeability. Permeability is measured by the pressure drop of a gas passing through a material. Pressure drops as pore size increases, but it can also be affected by the foam's surface roughness and number of closed cells. Plastic foams normally have a thin skin interconnecting many of the pores, which can block the passage of sound and act as a barrier. To allow for more air movement within the material and increase absorption more open cells are required. This is accomplished by reticulation.

Thermal reticulation is performed by passing a flame through the foam, which produces a fire "polishing." In chemical reticulation, the foam is immersed in a caustic chemical. Both methods remove some of the cell walls, leaving the strands between the walls intact. The air within the foam now moves more freely, and more vibration or friction occurs at the strands so that more energy is dissipated. When most of the cell walls are removed, the foam can become virtually transparent to sound, such as the foam used on microphones and speaker grilles.

Foams are only fair at dampening, and yield some vibration isolation, but unless very thick, do not make such great sound barriers. In addition to wedged and convoluted shapes, some manufacturers provide foams with fuzzy surface coatings that increase absorption at higher frequencies.

It is important that foam be backed by a hard reflective surface. This not only aids in stopping sound transmission, but the sound absorption increases as reflection takes place. The phase of the reflected energy is reversed from that of the energy. Varying the pore size allows for the tuning of absorption to a particular frequency. Therefore, in theory the efficiency of foam can be varied. As an example, one-inch wide foam with 20 pores per linear inch might be only 20 percent efficient in absorbing a particular frequency. But with 60 pores per linear inch it might improve to 40 percent efficiency and with 90 pores to 80 percent.

The permeability of foam and its resistance to airflow is measured in *rayls*. Units of flow resistivity show flow resistance per thickness. If flow resistance is high, permeability will be low and if flow resistance is low, permeability will be high. Even though foams may have the same thickness, density, number of cells, and cell size, their absorption can be different due to the amount of reticulation because this affects permeability. Flow resistance is proportional to the number of membranes reticulated. To a certain degree the following holds true: While less than 100 rayls means almost complete reticulation and more than 400 rayls yields high flow resistance, absorption in between these two points is not exactly predictable. However, lower rayls tend to have better high-frequency absorption, and higher rayls better low.

Foam's Qualities

The load compression of foam denotes its ability to support itself as well as loads.

In general, polyester-based foams are less flammable than polyether. Additives can give both more flame-extinguishing features but also tend to change flow resistance. Polyester-based additives are more durable in highly humid environments, while polyether can survive very dry climates. Polyester is also not as affected by ultraviolet light. Polyether can be produced with greater than 90-percent open cells, while polyester about 60 percent. Here, a higher percentage equals lower airflow resistance and 90 percent is, therefore, better when used as absorptive wedges, while 60 percent makes better sheet foam absorbers. Polyether is also less susceptible to mechanical fatigue. Low cutoff frequency is a function of airflow resistance. Values for polyurethane-polyether should range from 40 to 50 percent and for polyurethane-polyester from 30 to 40 percent.

When used as anechoic wedges, they can yield close to 100 percent absorbency below 100 Hz, when using wedges of 36-inch depth. Thus it can work as well as fiberglass, and actually end up doing a better job in the long run because this material's composition has less variation per unit and remains stable much longer. The shock absorption and compression capabilities of foam are functions of its resiliency. The density of foam ranges from 1 to 2.5 pounds per square foot, and anything over that will not increase the absorption coefficient.

Foam panels are generally real good at absorbing mid and high frequencies (500 Hz and above) but can be quite poor at low frequencies (below 250 Hz). Most are meant to be directly mounted on a surface such as a wall or ceiling and can be fastened with glue. Foam comes in a vast array of densities, pore sizes, thicknesses, and colors. It is also easy to work with. Just have a go at it with one of those electric meat-carving knives that you can pick up for a couple of dollars in second hand stores the day after Thanksgiving.

The only problem I have with foam is its tendency to capture, retain and constantly give off odors, such as those from cigarette smoke and stale beer. You know the scent; it is fairly common in all rehearsal studios that utilize foam for noise abatement. You can try and get around this problem by painting it but brushing on thick wall paint or worse yet using a paint roller (I've seen it done) will just clog the pores and reduce its ability to attenuate sound. You can try to wash it out (it is possible), but foam retains water almost as well as a sponge and the size of most panels makes squeezing out the water close to impossible. And just as with those thin square sheets used in conjunction with whisper fans for air filtering, once wet they collect funk more efficiently. Music sessions can very quickly turn into impromptu parties. Clean-up *must* be immediate in recording studios, but the end results can linger to disturb nasal passages for days. In rehearsal studios where beer and cigarettes are a constant, the smell is permanent. I have always kept an eye open for ways around this problem. For instance you can spray disinfectant on filters to kill some of the germs, but if you do the same to large acoustic panels you'll be stuck with *that* smell for a looooong time.

Washable Foam?

The Dow Chemical company came out with a line of Quash™ *odorless*, sound management, foam planks. These low density polyethylene materials offer uniform cell size along with accurate control over density. They use CFC- and HCFC-free blowing agents, are easily fabricated, are impervious to most chemicals, perform consistently over a wide range of temperatures, handle harsh environments, and are 100% recyclable. This is all very nice, but buried in the literature is the astounding fact that this foam has a rigid, yet soft, structure that does not absorb or retain water, allowing it to be readily cleaned. In fact, Dow's literature claims that it "can be washed without affecting the acoustical performance" and "because these foams do not deteriorate or promote corrosion, they can be removed, cleaned, and then reused." Take it outside, hose it down, then either let it air dry or aim a fan at it. An outfit called Norseman Allfoam (www.norsemanallfoam.ca) manufactures baffles with mounting hardware, wall panels with frames, and complete enclosures with access doors using Quash mounted in anodized aluminum frames that interlock with each other without bolts.

Acoustical Surfaces also markets a "moisture-proof" foam ceiling and wall panel. Called Sound Silencer™ it is an open-cell, Arpro®, porous, expanded polypropylene-bead fiber-free foam. It has a flat, textured surface that looks like it is made up of short lengths of ⅛" tubing mashed together using some form of heat process. The result is a lightweight, structural, "impact, moisture, chemical, mold and mildew resistant, UV stable" material that is "washable, tackable" (think thumb tacks in an office cubical wall), and paintable. Again, I veto the use of thick, brush-on wall paint, as it will surely cost you in the absorption department.

So what do we have here? Along with all the specifications, I received a couple of 6" × 6" × 1" thick samples, so let's find out. Number one in my book is *fire hazard*.

Here ratings can get as complicated as they do with sound, but as opposed to say transportation where there are at least a half dozen different test methods, building and construction materials are usually tested to ASTM E 84 specifications. Both these materials have an ASTM E 84 Class A fire-resistant rating, but Dow Chemical warns that

"Quash Sound Management foam products are combustible and may constitute a fire hazard if improperly used or installed." I take this as a cover-thy-butt line because they also state that it "contains flame retardant additives to meet the fire-test response characteristics."

The fact is both Quash and Sound Silencer proved to be *self extinguishing* using *my* test method—which is holding a cigarette lighter to the material out-of-doors. They both did melt a little, held the flame but did not enlarge it, gave off a burning plastic-type odor, but as soon as the lighter was removed *all* such action ceased immediately.

How about the water proof part? Water pretty much runs right off the stuff, but both materials will take on water if submerged and held under water for awhile. Quash has larger pores so it takes on more. Most of the water can be shaken out of Sound Silencer, and it will air dry in a few hours. You'll be shaking the Quash a long time before getting all the water out, but aiming a fan at it reduces the drying time to several hours. Remember I *submerged* these panels and held them under (both float) for a good length of time, so they both passed my water torture test.

While the Quash weighs a tad less than the Sound Silencer the rest of their specs are pretty much like all flat surfaced 1-inch foam panels meaning not very good below 500 Hz, yet they provide an absorption coefficient of around 0.5 at 500 Hz, 1.0 around 1 kHz and a little better than 0.8 at frequencies above that. But graphs and tables tell the full story.

Table 2.1 showing the Sound Silencer mounting method versus absorption proves to be quite enlightening. Here one- and two-inch thick Sound Silencer is tested at the six standard absorption coefficient frequencies with four mounting methods. The standard A against the wall mounting, D 20 mm in front of a ³⁄₄-inch air space, C 20 mm in front of a 1-inch thick bonded acoustic pad, and E 405 the standard acoustic ceiling mount in front of 16 inches of plenum space. The graphic representations in Fig. 2.2A (one inch) and 2.2B (two inch) sum up the results. Table 2.2, shows sound transmission loss, and the graphic representations in Fig. 2.3 show this material's transmission loss to be effective even down to 125 Hz with a TL of 6 dB.

	Mount	125 Hz	250 Hz	500 Hz	1000 Hz	2000 Hz	4000 Hz	NRC*
Sound Silencer 1"	Wall	0.06	0.01	0.18	0.84	0.63	0.78	0.40
Sound Silencer 1"	Wall with 3/4" air space	0.06	0.10	0.52	0.62	0.51	0.58	0.45
Sound Silencer 1"	Wall with 1" BAP**	0.09	0.59	0.96	0.59	0.59	0.60	0.70
Sound Silencer 2"	Wall	0.05	0.13	0.70	0.63	0.51	0.63	0.50
Sound Silencer 2"	Wall with 3/4" air space	0.09	0.25	0.89	0.59	0.72	0.76	0.60
Sound Silencer 2"	Wall with 1" BAP**	0.19	0.84	0.84	0.73	0.78	0.78	0.80
Sound Silencer 1"	Ceiling	0.63	0.55	0.33	0.34	0.54	0.53	0.45
Sound Silencer 2"	Ceiling	0.50	0.48	0.33	0.56	0.58	1.00	0.50

*Noise Reduction Coefficient   Sound Absorption independently tested to ASTM C-423 standards
**BAP (Fiberglass-free Bonded Acoustical Pad)

TABLE 2.1   Sound Silencer Mounting vs. Absorption/Noise Reduction

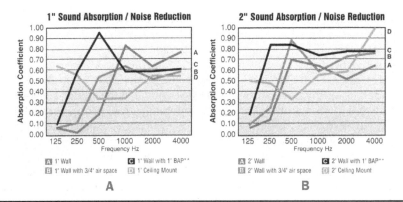

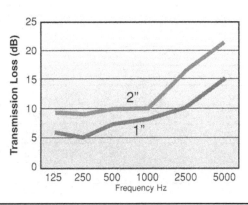

FIGURE 2.2A AND 2.2B   1- and 2-inch-thick Sound Silencer absorption graphs per mounting method.

	125 Hz	250 Hz	500 Hz	1000Hz	2500 Hz	5000 Hz	STC* Rating
Sound Silencer 1"	6	5	7	8	10	15	9
Sound Silencer 2"	9	8	10	10	17	22	13
Sound Silencer with 5/8" gypsum both sides	27	27	29	31	32	45	32

*Sound Transmission Coefficient   Sound Transmission Loss independently tested to ASTM E-90 standards

TABLE 2.2   Sound Silencer Sound Transmission Loss

FIGURE 2.3   Sound Silencer sound transmission loss graph.

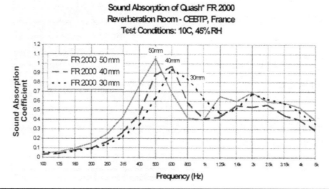

Figure 2.4   Quash absorption coefficient reverberation room graph.

Now with Quash, we are talking about Dow Chemical, which is an International company, so their testing must apply to European as well as U.S. standards. Figure 2.4 shows the absorption coefficient for 30 mm (1" with skin) as tested in a reverberation room. Figure 2.5 (disregarding the "without skin" trace) is interesting in that a 1" (30 mm) thickness has a peak of almost 1, call it 0.975 at 800 Hz while with the reverberant room test the same 1" panel peaks only slightly lower at 0.925 *but* at 630 Hz. This is just about a 25% difference, but in frequency *not* level. The STL graph for Quash in Fig. 2.6 seems to show this material as being better than the Sound Silencer until you note that the sample is not 1" thick but 65 mm or over 2 and a half inches thick!

Dow also provides a single graph (see Fig. 2.7) that displays absorption per frequency for Quash in thicknesses of 30, 40, 50, and 60 mm. In addition to the changes in

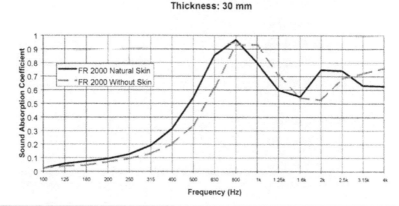

Figure 2.5   Quash absorption impedance tube graph.

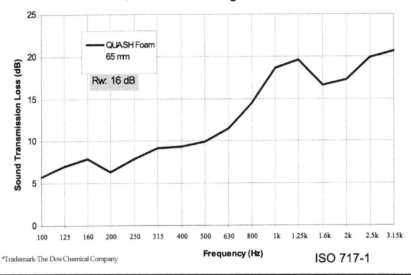

FIGURE 2.6   Quash STL graph.

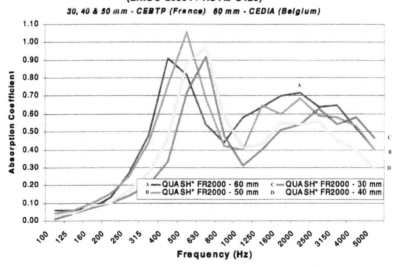

FIGURE 2.7   Quash multithickness trace absorption graph.

peak frequency and absorption per thickness: 0.91 at about 750 Hz for 30 mm (1"), 0.95 also around 750 Hz for 40 mm (1½"), 1.5 at 600 Hz for 50 mm (2") and 0.90 at 450 Hz for 60 mm (2½"), variations per thickness in bandwidth and high-frequency absorption are very revealing. For instance, from 1 to 4 kHz the thickest (60 mm) absorbs the most, but from 4 to 5 kHz where the graph ends, the *thinnest* (30 mm) absorbs the most.

Quash weighs a little less than Sound Silencer, but not by much, and Sound Silencer is *much* more structurally stable. In fact, it can be used for outdoor enclosures without the need of additional supports. It is also impact resistant (think school gymnasiums and rehearsal rooms).

I went through this whole exercise to show the amount of information that can be dug up on most any acoustical product and just how detailed it can get. The punch line? After quite a few days of research, I got down to brass tacks as in how much $$$$ these cost Sound Silencer will run you $572 for 104 square feet, or about $5 per square foot. Quash? It seems it is no longer being made, or at least nobody is selling it. And, it is no longer used by Norseman. So I tried turning them on to Sound Silencer. We'll see if anything develops.

A final note on painting foam: It *can* be done without seriously reducing absorption, but you need to use just a light spray-on coating. Even the ¹⁄₁₆-inch-thick vinyl used to coat and protect foam panels usually has perforations. So get creative and use your imagination. Spread some beads, rice, or other small objects across the foam before you spray paint it, and you will end up with better absorption along with unique and interesting patterns.

Functional Baffles and Shaped Absorbers

Functional absorbers are panels, baffles and hung absorbers in the shape of drums, cones, double cones, and free-standing room dividers. All are easily installed. Panels, for instance, when hung from the ceiling can expose five or all six of its sides to sound. Since most are three dimensional, they give higher absorption than two-dimensional absorbers. They are extremely helpful when the existing surfaces are not suitable for on-wall panels or when the ceilings are blocked by pipes, lights, ducts, etc. Because they are three dimensional and cannot be tested flat against the surface, just as with other three-dimensional absorbers like ASC TubeTraps™, they must be rated in Sabins per frequency as opposed to absorption coefficients.

Space absorbers are pretty much the same as functional absorbers. They are either freestanding or hung where other absorptive materials are not practical to mount. These again can be thick panels of various materials, but more often are of a more spectacular three-dimensional shape such as hollow triangular pyramids of fibrous material, wood fiber cones, or two cones joined at the base and thus enclosing a large amount of air. Others are hollow cylinders of glass fiber with perforated metal or screen facings. Functional and space absorbers can offer a high amount of absorbency.

Curtains and Fabrics

Most curtains, fabrics, and even upholstered seat cushions make good absorbers. Cloth curtains or drapes are certainly porous enough to absorb sound. How well they absorb

depends on the weight of the material, the depth of the folds and the distance it is mounted away from a wall. There have been studies concerning these factors and their effects on absorption. Weight and folds are pretty much self explanatory, but distance from the wall gets a little tricky. I'm not going to get involved in a theoretical dissertation on the subject, but it has been found that the absorption is greatest at distances that equal ¼ and ¾ of the sound's wavelength.

Okay, say you want to get rid of some 100-cycle boominess. 1130 divided by 100 is 11.3 feet. ¼ of that is about three feet and ¾ would be around nine feet. Giving up that much space is generally out of the question. Once again it is obviously going to be easier for curtains to handle higher frequencies that have shorter wavelengths.

The Selection of Materials

We know that energy sets a sound wave in motion and that power is the amount of energy that produces an amount of sound pressure. Sound pressure falls off with distance. Because sound power cannot easily be measured directly, sound pressure at the receiver's location is what is normally measured.

A very important aspect of understanding the way in which sound absorbing materials work is knowing how pressure affects them. Take for example a hollow rubber ball in a pressure chamber. If the outside pressure is increased beyond the internal pressure of the ball, it would collapse. If the pressure in the chamber was decreased below that of normal atmospheric pressure, the ball would expand. Note here that the air that constitutes the pressure did not have to enter or escape from the ball to achieve either effect.

Sound absorbent materials offer resistance to sound pressure variations and this resistance varies with the material. If a material offers low resistance, there is little or no attenuation and sound will easily pass through unless it is very thick or dense. Sound barriers offer high resistance to sound pressure, so most of the energy is reflected back and little or no sound escapes through to the other side. It is precisely for these two reasons that many acoustic products utilize a combination of high and low resistive materials to achieve losses in both sound transmission as well as sound absorption. When choosing materials for sound absorption you have to consider their physical properties in terms of durability, along with the way in which they absorb sound energy and the frequency range they cover. Porous materials have a cellular structure made up of interlocking walls and sound energy is converted to thermal (heat) energy within these cells (see Fig. 2.1). This conversion is not what you would call ideal, but sound waves passing through are met with some resistance and there is additional loss of energy due to the friction caused by the motion of the material's fibers. The absorption coefficient depends on the material's porosity, thickness, density, resistance to airflow, and even fiber orientation.

Thickness

As expected, absorption increases with thickness, but the increase is not a linear 1:1 ratio. First of all, an increase in thickness will aid in low-frequency absorption more than high and a doubling of thickness will not necessarily result in a doubling of absorption. But four inches of 3 lb. per cubic foot density fiberglass will act as an open window (have an absorption coefficient of 1.0) from around 200 to over 4000 Hz. While absorption will increase with thickness, at some point added thickness will yield no additional absorption.

Density

Acoustical properties relate directly to the physical properties of stiffness and density. The porosity of a material is also related to its density, which is unique to each material. But this relationship again is not linear. As an example, fiberglass density, if changed from 1.6 pounds per cubic foot to even up to 6 pounds per cubic foot, can show little difference in absorption. Most porous materials like fiberglass come in different densities from soft, fluffy, cotton-candy-like stuffing inside plastic bags to sheets rigid enough to support their own weight. But on its own density is not an accurate indication of absorption. Fiber, pore, cell, and particle size are important, but so is their orientation. To picture this, think of many layers of ordinary screening. This structure could be absorptive but its ability to absorb would vary with the spacing of the strands, orientation of strands, and the spacing between screens. While density has some effect on sound absorption, it is somewhat less important than thickness and certainly less important than mounting method.

Mounting Methods

Mountings are very important. Manufacturer's ratings are calculated using very carefully followed mounting methods, and thus they should be consulted in regard to the proper mounting for a desired effect. Most materials may be attached directly to a hard, impervious surface but are often suspended (hung ceiling) on runners or furring strips, leaving an air space behind them. This air space has the effect of broadening the absorption bandwidth, but can also decrease high-frequency absorption with a crossover point at around 500 Hz. Depending on the materials used, air space can vary from 2 to 16 inches. Above 16 inches there is little added effect. Due to this, lab tests are commonly performed with mounting method #7 or E 405, which is a mechanical suspension system with 16 inches of air space behind it.

There are many ways to mount absorptive panels and as you will see, they are most effective in reducing low frequencies when placed where low frequencies tend to build up, such as in corners across the intersection of two walls, wall and floor and wall and ceiling. In fact mounting an absorptive panel away from a wall increases its absorption coefficient markedly because as much as twice the surface area is available to absorb sound.

Working with Fiberglass

I would never discourage anyone from building their own sound absorbers, but you need to do the math first before you start. The next thing you should do is *bag* all fiberglass before you bring the stuff into your studio. A great source for fiberglass bags is your local dry cleaner. Those suit, or better yet, the longer overcoat-sized bags should run you about a nickel each, and the plastic used is so thin, it offers no real impedance to any sound passing through it. They come on a roll, which makes the work of sliding the fiberglass into the bottom and taping up that end along with the smaller top hanger hole, go more smoothly. Get yourself a pair of long rubber dish-washing gloves and a cheap nose/mouth filter mask. Take your box of fiberboard outside and place it downwind. Tape the coat bag hanger hole closed, remove a piece of fiberboard, immediately place it into the bag and then fold over the bottom and tape it closed. Once all the fiberboard is encased in plastic, chuck the mask and handle the stuff as much as you want because from now on your fiberboard will no longer give off itchy fibrous debris.

If at some point you need to cut it, untape the bottom, move the bag up so that there is enough plastic to rewrap both ends after the cut. Make your cut and immediately fold over and tape down the excess plastic over all the exposed ends. This is the reason it is better to use overcoat-sized bags. This may seem a little excessive, but I have seen some pretty bad skin reactions to fibers of various types over the years. Believe me, it is best to never find out by surprise that you are susceptible to allergic reaction. I once had the job of testing battleship PA speakers for a government contractor. This manufacturer had their own "anechoic chamber" that, in this case, was a small 6′ × 10′ × 5′ affair in a room whose dimensions were actually three to four feet larger in each dimension. All that extra space was completely taken up (even under the raised screen floor) by raw uncovered fiberglass. The most accurate microphones at the time were still tube driven, making not only the test microphone very hot, but also the interior of that chamber about 120 degrees F. Sound absorption material *does* make great thermal insulation. I spent many days bringing speakers in, hooking them up, testing them, and then carrying stacks of them out of that room. I found out all about exposed fiberglass and its effect on sweaty open pores and it is no joke.

To this day I still specify that all fiberglass be bagged, even when it is to be placed in the air space inside walls. Hey, someone has to put it in there, and somebody might have to drill, cut, pull cable through or remove that wall at some point. I think it is only fair to everyone involved if you act as if *you* are going to be that person.

Here's Mr. Everest's Cheapest Wideband Absorber

"For a truly budget studio there is need for an absorbing panel having respectable characteristics and of very low cost. Glass fiber panels 24 inches by 48 inches by 2 inches and 3 pounds per cubic foot cost something like $3.00 each. The crudest and cheapest approach would be to slap a few gobs of acoustical tile cement on the back of the raw panel and stick it to the wall. By expending a modicum of imagination and effort, the panel can be covered with cloth of some sort, which will not only control the irritating glass fibers but add a touch of class and color.

All budget approaches have their compromises. Stretching cloth over the panel tends to round off the corners and destroy the neat shape. The panels of glass fiber are fragile. If gentle handling can be guaranteed, no problem, but the first elbow blow will be memorialized permanently. All of these disadvantages can be at least minimized by building the panels along the line of that described in Fig. 2.8.

Edge absorption can be largely retained by using edge boards of pegboards as thin as available. Impact insurance is in the form of a wire-screen, perforated-plastic sheet (20 percent perforation or greater), or plastic fly screen. The cloth cover should be lightweight and loosely woven. Burlap is a possibility and it is available in many colors or you can tie-dye your own! The cloth is wrapped tightly around the edge boards and cemented on the back with acoustical tile cement. Mounting the panel to the wall can be done in a rather unrelenting way with cement. The panel is so light, 4 pounds for the glass fiber plus edge boards and cloth, it could be supported easily by a couple of rings attached to the panel by a cemented cloth tape.

The 2-inch depth of the panel carries with it poorer absorption coefficients at 125 Hz (0.18) and 250 Hz (0.76). If panels are built of two 2-inch thicknesses, good absorption could be expected down to and including 125 Hz."

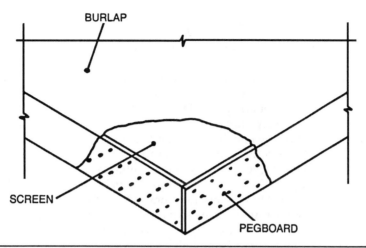

BURLAP

SCREEN

PEGBOARD

FIGURE 2.8   An inexpensive homemade absorbing unit made by reinforcing the edges of a panel of 3 pounds-per-cubic-foot density glass fiber with pegboard and stretching burlap over it, then cementing the cloth on the back of the panel. A screen may be used if impact resistance is required.

Fiberglass and Mineral Wool

So much information exists on fiberglass and mineral absorbers that it is not necessary to rehash it here. Stated as simply as possible, fibrous materials bonded together by resins handle high temperatures better than foams, are about equally absorptive, but are irritating to work and live with. Felts also have good absorptive characteristics. They can be woven, nonwoven, or knitted. Their fibers are often bonded via pressure, chemical action, and/or heat. They demonstrate better wear resistance than glass fibers and are denser than either foam or fiberglass. Wood fibers can also be bonded together. They are much more rigid and are used for wall paneling and roofing. As roofing, wood fibers are bonded together with cement which yields a higher transmission loss especially with the additional layers of tar and stone pebbles that are commonly used.

Spray-Ons

Spray-on absorptive materials can be very convenient for reaching hard to get at areas or when covering irregular surfaces. These combine fibers with bonding materials. If the bond is too tight, they yield a high resistance. If the bond is too loose, not only will the resistance be too low, but the material could fall apart. When mixed properly, these materials can stay up even with a 2-inch thickness. Yet, this will generally only attenuate the higher frequencies and is only a little better than putting carpeting on a ceiling. However, if applied to hung panels these materials can exhibit absorption over a broader bandwidth. Therefore, the absorbency of spray-ons depends not only on the thickness applied but also on the base they are applied to. It is essential to get the manufacturer's specifications on application methods when using these materials.

Sound Absorption per Frequency

If you were to set up a continuous sound source in a room that had no absorptive surfaces and no openings or means for the sound to escape, theoretically the sound pressure within that room would continue to increase until the source was cut off. In the real world of rooms that do have leaks and at least some absorptive surfaces, the sound pressure would only build up to a finite level. So what's the point? Simply that leakage and absorption are very important to the acoustic response of a room.

If you wish to delve into this matter further look to Sabine, as he did extensive studies of the growth and decay of sound in rooms. For our purposes, we are only concerned with how the amount of absorption within the room affects the sound pressure buildup and decay. Theoretical discussions usually assume that the density of the sound energy is equal throughout the room and that the absorption is ideal. But in reality, only some of the sound energy striking an absorptive surface is actually absorbed and the rest is reflected. In large rooms with long reverb times the sound waves will be reflected many times before being fully absorbed.

We need to have some way to determine the effectiveness of absorptive materials. Lucky for us this has already been taken care of. It calls for the use of a room where the acoustical energy is perfectly diffused. The energy in the room is measured (per frequency) with and without the material present and the two results give a ratio that is the random-incidence, sound-absorption coefficient. The uniformity of the results hinges on that random incidence or near perfect diffusion. This means that we can now simply go to a table and look up the absorption coefficient of the material in question. There is also a test for absorption based on the decay of sound in a reverberant room, with and without the material present (after Sabine). The difference in the two resulting coefficients can be as much as 25%, but most absorption tables are simply listed as the absorption coefficient and for our purposes the disparity proves to be minimal.

The absorption coefficients of the materials that make up the room's surfaces can be noted from these tables and used to determine the room's average coefficient per frequency. Frequency is very important to the results, as are room temperature and atmospheric pressure, but the latter two only matter in larger venues.

So everything ends up being pretty straight ahead. As you will see in Part II, you measure the total surface area of each material and average the absorption per frequency to find out the total absorption of the room you are dealing with. But for correcting specific acoustic problems, finding the absorption at a narrow frequency bandwidth will more likely be the main concern.

Take the following as an example:

Material	125	250	500	1000	2000	4000 Hz
Carpet on bare concrete	0.02	0.06	0.14	0.37	0.60	0.65
Carpet on foam rubber	0.08	0.24	0.57	0.69	0.71	0.73

TABLE 2-3   Absorption Coefficients at Octave Center Frequencies

While the absorption of carpeting is minimal at low frequencies, it can be fairly substantial at the high end. As you will see in Part II, this often dictates the use of additional low-frequency absorption as a "contra" carpet cure to balance out the room's frequency response.

Also noteworthy is the difference the rubber underlay makes. It not only increases the absorption of the carpet, but does so across the board, frequency-wise. Even though the absorption at low frequencies remains minimal, this is in fact a rare occurrence, as most materials attenuate only specific bands of frequencies.

Combining Materials

Roof Decks

As absorptive systems, roof decks are structural materials that have absorptive undersides. Whether perforated metal with fibrous material behind it or a strictly sound absorptive panel, roof decks not only reduce reverberation and the level within, but also aid transmission loss. Aside from their use as roofs, many are used as combination ceiling/floor units.

Absorptive materials can be selected to meet environmental conditions. Temperature, moisture, dust, abrasion, high-pressure levels, and force can cause some absorptive materials to become useless in a short period of time. There are however materials or combinations of materials that can be utilized even in the worst conditions. Metal wools, for instance, are unaffected by humidity or temperature, while foamed plastics or ceramics can often handle moisture well. High sound pressure can destroy most materials; yet urethane foams do well here, as do porous masonry blocks, porous rough-faced metals and dampened metal panels.

Facings

Modifications to absorptive materials are also helpful. Various facings can be used as long as their porosity allows the sound wave to cause air movement within the material. Impact and abrasion problems can be alleviated with the use of facings. These protective coverings can also lend strength, hardness, ease of cleaning, moisture resistance, durability, and a more pleasing appearance to absorptive materials.

One modification is to cover absorptive materials with a perforated material such as metal, which is much more easily maintained. These facings can affect absorption by decreasing the high-end absorption and increasing the low. When the reflective area between the perforations is large more high-frequency content will be reflected. On the other hand, low frequencies with their longer wavelengths simply diffract around the reflected areas. Perforated facings generally vary from roughly 10 to 50 percent open area, depending on the thickness of the facing material, hole size, and spacings.

Unperforated films can also be used so long as they are light and flexible enough to offer little resistance to the incident wave. Thin plastic bags, for instance, can completely surround a material. In the case of mineral and glass fiber, plastic bags protect against degradation by moisture, pollutants, and oils, while giving the ability to be cleaned. Additionally, plastic bags allow them to be moved about with less detrimental effects to the materials or those moving them

Painting

Painting absorbers, such as acoustic tiles, can be tricky and the manufacturer should always be consulted. Even if a manufacturer states that a material can be painted, the results on absorption should be known beforehand. Paint is, in effect, a rigid, non-limp film and might cause some materials to reflect rather than absorb sound. Some ceiling tiles are purposely made with large openings to allow for painting. They often have a 15% to 20% open area that prevents the bridging or filling in of the hole by normal brush- or roller-applied paint.

Maintenance of absorptive materials is very important. Spray-ons can crumble and are not easy to clean. Thin perforations can clog from oil and other pollutants. Obviously anywhere that impact or abrasion is possible, the absorptive material should be checked regularly.

Let's Talk about Absorption and Wavelength

We all know that parallel walls can cause nasty flutter echoes and that the old school way of adding copious amounts of absorptive material is unacceptable due to the resultant loss in high-frequency content.

The amount of attenuation caused by any absorptive material depends on its size in square footage, its thickness and the frequency of the sound. It all comes down to the physics of wavelength. When the frequency is high—say in the neighborhood of 10 kHz—most porous materials can easily absorb a good deal of the sound because the wavelength is only slightly over an inch and a quarter. When you drop down to 80 cycles, you are dealing with a wavelength of around 14 feet! No passive absorber is going to be able to attenuate that sound wave anywhere close to the same degree.

Passive Absorber?

Curtains, carpets, and foam are passive in that they take no action nor do they act upon or have any interplay with the sound wave coming in contact with them. At very low frequencies they may actually be moved physically, but for the most part only their fibers are set into motion, rub against each other and generate a faint amount of heat. Passive absorbers must be very thick in order for them to be of any significance when dealing with low frequencies. How thick? At least four to eight inches for glass fiberboard and up to twelve inches for foam. Plus any required air space and protective framing just adds to the overall size.

Modes

It is difficult to achieve quality sound in small rooms simply because the dimensions of the walls, floor and ceiling are very close to the half wavelengths of the low-frequency sounds being reproduced. This in turn sets up numerous harmonics that cause a buildup in level at particular frequencies. This not only distorts the sound, but can also cause peaks and dips in level throughout the space. It can be bad enough so that some locations are not fit for microphone placement.

Low-frequency modes depend on the size of the room. Because the fundamental (lowest mode) and subsequent harmonic modes are determined by the speed of sound divided by twice the length of each dimension, small rooms often need help at the low end. For example, the fundamental mode of a 40-foot wall is 14 Hz, with a 20-foot wall it's 28 Hz, the fundamental of a 10-foot wall is 56 Hz, and you are even more likely to run into a problem with the 71-Hz fundamental of an 8-foot tall ceiling. But using two walls separated by $56\frac{1}{2}$ feet as an example provides a clearer picture of fundamental modes and their harmonics because here the fundamental mode is 10 Hz, so the harmonics end up being 20, 30, 40, 50, 60, and so on.

Figure 2.9 shows a bar coaster hand-out from Acoustic Sciences (the makers of the "TubeTrap"). It is a graphic representation of the fundamental and five upper harmonic modes for various wall spacings.

Accuracy here is ballpark. Why? Well right from the start there's the speed of sound in air (1128 to 1130 feet per second being the general consensus), but that's even debatable because it depends on the room temperature.

But going with 1130 feet per second, let's take a look at the resulting fundamental frequency modes and harmonics for different sized parallel wall spacings in Table 2.4. Again, 1130 divided by twice the distance between two parallel walls results in the fundament mode to which the same amount is added for each subsequent harmonic.

While a bit tedious, this mathematics exercise can point to some problems that may be easily avoided, such as *not* having wall spacings of 12, 16, 20, or 24 feet along with an eight foot ceiling height. That 141 Hz build-up will be a whole lot easier *and* much less expensive to deal with during the design stage. How so?

Well, let's just say you have a room that is 12 by 16 with an 8-foot ceiling height. You could lay the room out so that it is 12-feet wide by 16 feet-deep and give up about a foot of depth for a back-wall bass trap, or start out with a 16-foot wide layout and add a few splayed sections to the side walls to not only break up problems caused by parallel surfaces, but also redirect first order reflections around the listening position. Hey, you may opt for something totally different, but at least you know what your problems are likely to be *before* you make any purchases and start construction. If that is not enough incentive, getting surprised by a design flaw in a "finished" studio may require the

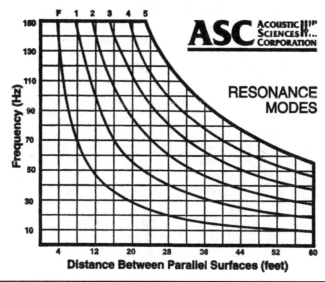

FIGURE 2.9 Mode graph bar coaster from ASC.

Distance between Walls in Feet	Harmonics					
	Fundamental	**First**	**Second**	**Third**	**Fourth**	**Fifth Harmonic**
4	**141.25**	282.5	432.5	565.0	706.25	847.5
8	70.63	**141.5**	211.87	282.5	353.13	423.75
12	47.083	94.16	**141.19**	188.33	235.15	282.49
16	35.312	70.624	105.96	**141.25**	176.56	211.87
20	28.25	56.5	84.75	113.0	**141.25**	169.5
24	23.54	47.08	70.62	94.16	117.7	**141.24**
28	20.17	40.34	60.51	80.68	100.85	121.02

TABLE 2.4   Fundamental Frequency Modes and Harmonics for Different Wall Spacings

dismantling of some of your previous work, which means more costs and time delays. Now think of the mess moving a wall creates. All electronic equipment may have to be removed from the room! The best bet *is* to do the math first.

Active Absorbers

Diaphragmatic absorbers are considered to be active because they resonate at certain frequencies. Low frequencies will pass right through them, but they will also set the panel in motion, which depletes their energy somewhat.

Polycylindrical Diffusers

The facing of polycylindrical panels is convex. They are used for dual purposes: to help control low frequencies and to diffuse high frequencies. Due to their convex curved surface they reflect high frequencies at many different angles. They can also act as a diaphragmatic panel, and are moved by low frequencies. In order to accomplish both of these, the surface of plywood panels must be finished. This means a lacquer or high gloss enamel-like surface to promote high-frequency reflection. The area behind the polycylindrical panel is lined with absorptive material to aid low-frequency absorption. The half circle spacers and end caps are very important components. They securely brace the panel, which is necessary, especially with thin facings and they help prevent any internal resonance. Furthermore, having several attachment points makes bending the wet plywood much easier and also helps hold everything in place while drying. Remember the thicker the facing, the lower the resonant frequency absorbed. Due to this fact, "Masonite" is often *not* used because it is not stiff enough. Most panels use ¼"-thick ply; you can start off with ³⁄₁₆-inch, but save the ⅛- and ¹⁄₁₆-inch ply for smaller more radical radiuses.

The best bet with polycylindrical diffusers is having the dimensions, in terms of width, depth, radius and arch, all figured out before you start to cut wood. I always start with the arch because that gives you the width of the plywood needed to complete the project.

As an example let's take an arch that allows us to get two four-foot long panels out of a single 4×8 sheet.

Well, the width of the panel will end up being $43\frac{1}{4}$ inches if the depth is 9 inches. So mark the 9-inch depth off at the center point of $21\frac{5}{8}$ inches on your one-inch-thick cap that has been sized to $43\frac{1}{4}$ inches in length. Wrap a string around a pencil at the length of 9 inches, hold or pin the pivot end to the zero ($21\frac{5}{8}$ inch) center point of your one-inch-thick cap/divider and draw your arch. It may take a couple of swipes, but you'll get it. I find it easier to do the drawing of the arch on cardboard first and then cut the curve out and trace it onto the wood. Everything seems to come out a little more even this way. The number of these curved braces you mark off and jigsaw out depends on the overall length of the panel. If you are using a $4' \times 8'$ sheet of ply, meaning your polycylindrical diffuser is going to be 96 inches long, you'll need 5 to 7 braces and 3 to 4 for half that, and remember to offset the center ones.

Diaphragmatic Absorbers

When a sound wave falls upon a panel, the panel vibrates at its natural period. Panels are moved more by low frequencies than high, and in fact, most high frequencies are reflected off the panel. Because no panel is perfectly elastic some friction takes place. In addition, energy is also lost due to internal damping in the panel as well as damping due to the means of panel support. All panels have a resonance frequency and in a resonant condition the amplitude at which they vibrate is greatest. Depending on the panel's size and other characteristics, it will transmit most of the energy hitting it at its resonant frequency and higher order harmonics and they are usually most effective between 50 and 500 Hz.

Panel absorbers can include sheets of gypsum board, wood paneling, windows, suspended ceilings, ceiling reflectors, and even wood platforms (such as staging). Thin sheets of plywood, plastic, metal, or even paper can be used as diaphragmatic absorbers. There are vacuum-formed ceiling panels of vinyl, damped sheet metal, and plywood assemblies in use today. One unit is of molded fiberglass no more than $\frac{1}{8}$ inch thick. It is shaped into pyramids approximately 2 feet by 2 feet and used as a hung ceiling with air space behind it. They have good broadband absorption with equal coefficients from 125 to 4000 Hz. Because diaphragmatic absorbers are more absorptive at lower frequencies they are used to supplement other materials or to address specific low-frequency problems. Since it is often not practical to utilize very thick amounts of fibrous materials for low-frequency absorption, they are highly cost-effective. Furthermore if the walls and other surfaces behind the panel absorber are lined with porous material, you end up with what is termed a bass trap. Think of this from the perspective of the propagation of a low-frequency wave. The wave hits the panel and sets it in motion. It passes through but loses some energy. It then encounters the porous material and causes its fibers to rub against each other, which converts some of the wave's energy into heat. It then hits the wall, reflects off of it and has to pass through the porous material again for more energy loss. Finally it passes back through the panel again setting it in motion for more energy loss. In this case some of the low-frequency energy is returned to the room but at an attenuated level.

However in the case of a "tuned" diaphragmatic resonator, almost all of the low-frequency energy can be completely "trapped" with nothing returned to the room. This wonderful design has a single flaw. It requires that you give up some space. How much depends on the thickness and density of the panel material and the frequency you are looking to eliminate. As you will see, panels can be combined with other materials and facings into systems such as Helmholtz resonators and bass traps that make it possible to "tune" the absorption to a range of frequencies and even for peak absorption at a specific frequency. Panel absorbers are fairly inexpensive and easy to make. The lower the resonant frequency and the thinner the panel, the greater the volume of air needed. However absorbers with thicker panels can be used as well.

How thick and how thin? If you use ½-inch plywood for the panel, the air space required for it to have a peak resonance at 100 Hz would only be two inches deep (one of which will often be taken up by 1 inch of fiberboard). With ¼" ply the air space required for a 100 Hz peak resonance doubles to 4 inches.

A Jig for Designing Resonant Plywood Absorbers

Okay, this is a *big* gift. There was some thought of putting this on the web, but I wanted you to be able to go to it and look up the correct panel thickness and air space at a moment's notice. I can't remember exactly where I got this, but I do know it came from a society made up of plywood manufacturers. It was a very smart move on their part and I have no idea why it seems to be no longer available. Look at Figs. 2.10 and 2.11. I know they look kind of weird, but once you know how to use them, they'll provide near instant answers to your panel absorber bass trap needs.

Your task starts with the selection of the "Q" or bandwidth of the absorption desired. Go to Fig. 2.10, and eyeball curve A. If you take the center frequency to literally

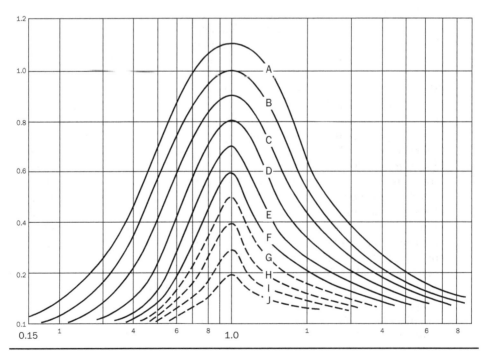

FIGURE 2.10 Plywood panel jig Q selector.

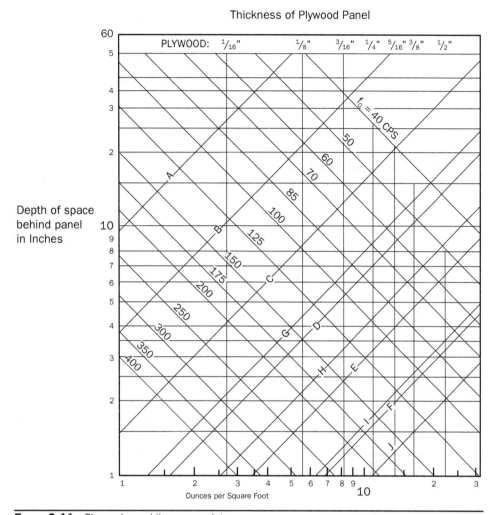

FIGURE 2.11 Plywood panel jig cross points.

be 1000 Hz, you can see that in addition to getting complete absorption at 1 kHz, you will end up losing half the sound power down to 450 Hz and also out to 2.5 kHz. That's a huge hole in your room's audio spectrum. On the other hand, while curve J is nice and narrow, it renders mere pip-squeak sound reduction at the resonant frequency. Curves B, C, D, and E are often the more practical choices.

Next you decide on the resonant frequency of the panel. Now go to Fig. 2.11, and find the intersection of your curve and frequency. Let's say you want to attenuate 125 Hz. And you'd like a bandwidth that yields a moderate amount of loss in the low-mid range of the spectrum, something in the order of 0.2 at 400 Hz as would occur with curve C. Now we come to the part where you have to wiggle the outcome because frankly put, plywood only comes in set thicknesses.

There are rules for wiggling (seems there are rules for *everything*).

NOTE *Figure 2.11 was designed to work with your normal commercial plywood, that is, panels with a density in pounds per square foot in the mid-thirties at most. Now if you substitute heavier woods like ash and oak, which come in at the mid-forties, you'll need to multiply the center frequency on the chart by 0.85. And if you want to use other materials such as plastics and metal, you'll need to move across the bottom to the correct density in ounces per square foot.*

If the most important factor in your decision is maintaining the exact resonant frequency, move along the diagonal frequency line and find the nearest standard plywood thickness. The resonant frequency will remain the same, but the bandwidth and the amount of absorption will change.

If the amount of absorption or bandwidth is more important, then move along the letter diagonal line to the nearest standard thickness. The bandwidth and amount of absorption will remain the same, but the resonant frequency will change.

You can also move horizontally from the frequency and bandwidth crossing point to the nearest standard thickness. Here the resonant frequency, the amount of absorption and the bandwidth will all be changed, but only slightly.

All right, it's best we get this down once and for all, using the example of 125 Hz and a bandwidth of curve C.

> To keep the resonant frequency unchanged, move along the frequency diagonal line to the nearest standard thickness.
>
> If you choose $\frac{1}{16}$-inch ply the Q or bandwidth will widen to curve B and the amount of absorption will increase by 0.1 Sabins.
>
> If you choose $\frac{1}{8}$-inch ply the Q becomes almost as narrow as curve D and the amount of absorption will decrease by 0.1 Sabins.
>
> To retain the same amount of absorption and bandwidth, move along the letter line to the nearest standard thickness.
>
> If you choose $\frac{1}{16}$-inch ply the resonant frequency will change to 200 Hz.
>
> If you choose $\frac{1}{8}$-inch ply the resonant frequency will change to 100Hz.
>
> For little change in both resonant frequency and absorption move horizontally to the nearest standard thickness.
>
> If you choose $\frac{1}{16}$-inch ply the absorption will remain the same, but the resonant frequency will be 150 Hz.
>
> If you choose $\frac{1}{8}$-inch ply the absorption will again remain about the same, the bandwidth will be a touch narrower and the resonant frequency will now be half way between 125 and 100 Hz.

It might take a bit of head scratching, but generally, the choice will pretty much hit you right between the eyes. For instance, if you are dealing with a resonant frequency that is slightly lower than 125 Hz the last example might be the best choice because only a little absorption is lost, the bandwidth remains close to the same and the resonant frequency decreases.

On the other hand with a resonant frequency of 100 Hz and a bandwidth of C no wiggling is needed as the two diagonal lines cross almost exactly on $\frac{1}{8}$-inch ply.

In most cases you'll find that the thinner plywood ends up being the way to go. Why? Because thinner ply is less expensive, easier to work with and it can be bent into a half circle to additionally benefit your acoustics by helping to disperse high frequencies. This is a win in terms of price, performance, *and* labor. Wow, that does *not* happen very often!

How about the *depth* of the air space behind the panel? Simply look at the left column. In our case, 112.5 Hz, curve B, and $\frac{1}{8}$" ply, the air space behind the panel should be about 6 inches. If we were going after 100 Hz and using $\frac{1}{16}$" ply, we would need 7 inches. At 85 Hz and $\frac{1}{16}$", 23 inches of air space would be needed. Want to kill 60 cycles using $\frac{1}{16}$" ply? Now you're talking about *four feet* of air space. However, sticking with 60 cycles, if we use $\frac{1}{8}$" plywood, the air space required drops to 22 inches. $\frac{3}{16}$" ply only needs 15 inches of air space, and $\frac{1}{4}$" ply only 12 inches, while at $\frac{5}{16}$" it's just 9 inches, with $\frac{3}{8}$" ply 7.5 inches, and with $\frac{1}{2}$", just $5\frac{3}{4}$ inches.

This not only spells out the relationship between panel density, frequency, and air space but also shows how *not* considering this panel effect on any sheet rock walls or suspended ceilings at the design stage can cause a room to end up with gaping holes in its frequency response. So how does adding absorptive material behind the panel figure in? Go back to Fig. 2.10, which shows the bandwidth or Q curves. The ones with the dashed lines do not have any absorptive material behind the panel. Compare the amount of absorption of these curves with those using absorptive material behind the panel. The answer is in your face obvious with a doubling or more of absorption using porous material to cover the surfaces behind the panel. However the absorbent material, can be very close to the panel, but should not actually touch it because that could restrict its vibration, and thus, reduce the absorption.

In fact the way in which the panel is mounted is very important. The size is not critical but the support spacing should be no less than 16 inches apart with no bracing in between. Think of the panel as a drum skin, readily vibrated and attached only at the edges. Furthermore there should be an elastic (cork, felt, foam) separator between the panel and the mount. Hey, some designs even get into dividing the space behind the panel into different-sized sections to absorb a wider frequency content.

Plywood Panel Mountings

Many years ago the Hardwood Plywood Manufacturers Association published a four page reprint of a 1948 *Journal of the Acoustical Society of America* article titled "Absorption-Frequency Characteristics of Plywood Panels" by Paul E. Sabine and L. G. Ramer.

This paper discusses absorption coefficient measurements of panels made of birch plywood "using the standard procedure of the Acoustical Materials Association." The paper also discusses the effects of panel thickness, air space depth and absorbent backing on absorption coefficients. That's right; more than 60 years ago, they had it all nailed down right to the resilient mounting. Figures 2.12 and 2.13 show both the mounting diagram as well as the results of $\frac{1}{8}$-inch plywood mounted firmly (curve 1) and with the edge mounted in sponge rubber (curve 2). These results not only show a 10 dB increase in peak absorption but also the center of the peak shifting downward from around 200 to 160 Hz with the use of resilient mounting

My method was to use thin adhesive-backed rubber strips. Those handheld paper hole punchers work real well for making pass-through holes for the panel mounting screws. (Save the little circles that are punched

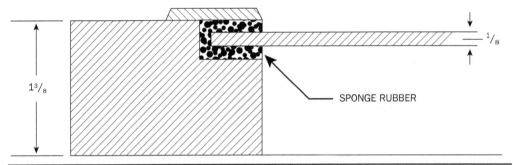

FIGURE 2.12 Plywood resiliently mounted.

Plywood Panel Mountings *(continued)*

out, as they make nice non-abrasive feet for your lighter-weight gear). This strip can be mounted to the furring underneath the panel or to the underside of the panel itself. I also drill holes in the plywood panel that are larger than the diameter of the screw's threaded area. A second adhesive rubber strip is placed on top of the panel edge.

The top furring strip gets starter screw holes and the end result becomes a *sandwich* of material that ends up being: furring, rubber strip, plywood, rubber strip and furring. Now you can adjust the tightness of the mount via screw tensioning and the panel will still not rattle.

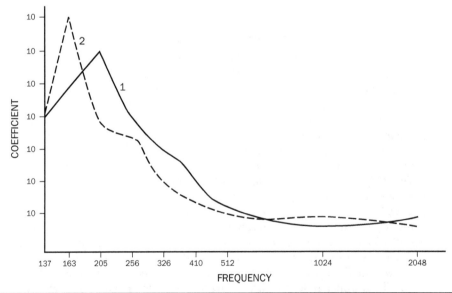

FIGURE 2.13 Plywood panel graph. This shows the effect of mounting on panels of $^1/_8$-inch plywood: curve 1, plywood mounted firmly at edges, curve 2, with edge mounted in sponge rubber.

DIY Bass Trap

So how's a regular non-scientist Joe going to use this info? How about a 17–inch-deep *corner bass trap*

If you are trying to reduce low frequencies, the best location for absorptive materials is in the corners of the room because low frequencies tend to "build up" in these areas. With this plan, you are also starting off with three (top and sides) of the five boundaries already completed. First you have to cut and mount three wood furring strips. Rip a two-foot length of 2×4 right down the middle with the blade set at a 45 degree angle. These are mounted on the walls 17 inches out from the ceiling corner with the wide side against the wall. On the ceiling also distanced out 17 inches out from the corner to match the size of the desired air space, mount a 2×2 (one and a half inch square) furring strip between the two inside angles of the side wall strips. It's a good idea to face these with a strip of cork, felt, or rubber before attaching the panel, to do away with any possible vibration noise and help the mounted panel to remain flexible. Add another length of furring strip in the lower inside corner and one across the bottom between and attached to the two side mounting strips to support the corners of the triangular bottom piece

Mounting a square of 1-inch, *bagged*, glass fiberboard behind the furring strips leaves a bit of space between it and the back of the panel. Personally, I prefer to add *bagged* fiberboard to the bottom piece, the two walls and the ceiling as well. Any 100-cycle sound that makes it through the front panel and the 1 inch of glass fiberboard directly behind it *will* be further attenuated. You end up with a Q that's between A and B. The peak absorption (1.1) will be at 100 Hz, and the two (0.5) points will be 45 and 250 Hz. While working with $\frac{1}{16}$-inch plywood is fairly easy, you can save yourself a good deal of trouble by making it out of cardboard first. .

Helmholtz Resonators

Helmholtz resonators, on the other hand, connect the room air to the air inside the panel through small air-filled tubular openings or slots often compared to bottle necks. In this case, the air moves in and out of the openings much like a spring. Again, the low frequencies pass through and are absorbed by the fiberglass within.

As with diaphragmatic panels they are more effective at low frequencies. They are basically perforated panels with an enclosed air space behind them, which often includes absorptive materials. Describing the function of a single opening (see Fig. 2.14) helps to understand how Helmholtz panels work. When a sound with a large wavelength hits the resonator, it causes the air in the neck of the enclosed cavity to vibrate back and forth. The air in the cavity acts like a spring. The cavity/channel combination tunes to a specific frequency (much like a bottle sounds when air is blown across the top). This resonance effect has a very narrow bandwidth. There are many ways to vary the resonant frequency and broaden the bandwidth. The cavity size and hole diameter as well as length of the hole affect the resonant frequency. The bandwidth can be effectively broadened by adding absorptive material inside the cavity, just behind the panel. Panels have even been made with varying thickness, resulting in varying channel lengths, and thus, tunings. This can also be accomplished by using different hole sizes on the same panel. Here the larger holes and thicker panel sections tune to lower frequencies. Even with these variations, the frequency range of resonant absorbers will not extend much above 400 Hz. In situations requiring a large amount of narrow-band, low-frequency absorption or when utilized to supplement other absorptive materials, Helmholtz resonators are extremely helpful.

They can be attached to ducts or other structures where some low-frequency emission requires attenuation. Here sound reduction can also be augmented by adding a layer of porous materials. Again, combinations of materials can both increase absorption and broaden the affected bandwidth.

Tuning the Helmholtz Resonators

When a peak of sound absorption is needed at low audio frequencies to compensate for other absorbers deficient at those frequencies (carpet, drapes, acoustical tile, etc.), or a sharp peak is needed to tame a troublesome modal frequency, attention naturally turns to Herr Doktor Professor Helmholtz. Panel type absorbers generally peak in the lows, but the absorption coefficient rarely exceeds 0.3 at the peak while Helmholtz resonators commonly come close to perfect absorption (1.0). There is really no theoretical difference between the slot type and the perforated type; consider a long slot nothing more than a row of holes. In the acoustical design of a room it is very helpful to know where the peak of absorption occurs. One way to estimate this is to calculate the resonance frequency. Let us take the perforated resonator of Part II, Fig. 19-12 as an example:

Face panel .... 3/16-inch masonite
Hole diameter .... 3/16 inch
Hole spacing .... 3 inches on centers
Depth of airspace .... 7 5/8 inches

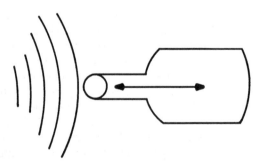

Figure 2.14 Resonant or reactive (Helmholtz) absorber. When wave coincidence occurs, air in the enclosed cavity vibrates back and forth moving in and out of the channel and tuning to a specific frequency much like air blowing across the top of a bottle.

The frequency of resonance of a perforated panel absorber backed by a subdivided air space is given approximately by[5]

$$f_0 = 200 \sqrt{\frac{p}{(d)\,(t)}}$$

where,

f_0 = frequency of resonance, Hz
p = perforation percentage
t = effective hole length, inches
  = (panel thickness) + (0.8) (hole diameter)
d = air space depth, inches

Figuring perforation percentage is easily accomplished by reference to the sketch of Fig. 2-15 and is about 0.31 percent. The effective hole length is $t = \frac{3}{16} + (0.8)\,(\frac{3}{16}) = 0.34$ inches, approximately. With these numbers and the air space depth of $d = 7.6$ inches we can estimate the resonance frequency as follows:

$$f_0 = 200 \sqrt{\frac{0.31}{(7.6)\,(0.34)}} = 69 \text{ Hz}$$

The effective hole length is quite uncertain, being dependent upon the geometry, and the 0.8 (hole diameter) correction is only an approximation. The indication of a frequency of resonance near 70 Hz is where the peak of absorption is expected to appear and the magnitude of that peak is close to $a = 1.0$.

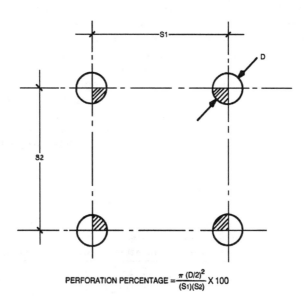

PERFORATION PERCENTAGE = $\dfrac{\pi\,(D/2)^2}{(S1)(S2)}$ X 100

FIGURE 2.15   The perforation percentage of an extended area is easily calculated from its smallest repeated pattern.

Now, what is the shape of the absorption curve so that absorption calculations can be made at other frequencies? This information is best obtained from actual measurements on perforated absorbers, but few such measurements have been reported in this country. Some of the most complete measurements of this type have been reported by the Russian, V. S. Mankovsky[*]. From his excellent book are selected three Helmholtz type resonators with perforated faces. The physical dimensions (translated from metric to closest English equivalent) and other pertinent data are given in Table 2.5. The measured absorption coefficients are plotted in Fig. 2.16, and it is tacitly assumed that Russian measurement techniques are at least roughly equivalent to those in other parts of the world.

A comparison of graphs A and B dramatically shows the effect of absorbent in the air cavity. The peak of graph B is so sharp it fell in the cracks between the 250- and 500-Hz measurement points but the broken line indicates a reasonable guess as to its shape. Without absorbent, this type of resonator has a tendency to "ring," i.e., to die away slowly when exciting sound ceases. A great widening of the absorption peak and slight shift in resonance frequency results from the use of absorbent in the air space. Although the absorbent used (called PP-80) has not been identified by translator or editor, it is reasonably sure that it is common rock wool or glass fiber. The calculated resonance frequency applicable to the A and B graphs is about 467 Hz, which is not very close to the peaks determined experimentally. Graph C is included to show the effect of a high perforation percentage, 8.5 percent. The depth is the same as A and B, panel thickness is the same, but hole diameter and spacing are different. Graph C in Fig. 2.16 is very broad. The calculated frequency of resonance in this case is 466 Hz which is in good agreement with the measured values of graph C. At such high perforation percentages the point is approached at which the cover has no Helmholtz effect. For example, 15 percent perforation (or open space) is used for covers for wideband absorbers (see Fig. 2.15 for the formula for calculating the perforation percentage).

To apply data, such as shown in Fig. 2.16, to any problem at hand is quite difficult in this form. Shapes vary as well as location of resonance peaks. These perforation percentages could well apply to absorbers tuned to other frequencies by using different airspace depths. For example, the perforation percentage of 0.785 percent applied to a box made of $1 \times 8s$ ($7\frac{5}{8}$-inch depth) would resonate in the vicinity of 120 Hz.

If this is where a peak is desired, how can graph A data be shifted down to this resonance frequency to give an estimate of coefficients to use based on the measured A values? In Fig. 2.17 the measured curves of Fig. 2.16 are brought together in what is called *normalized form*. These, in turn, can be applied to systems of other resonance frequencies. For example, graph A actually peaks at 270 Hz. At $2 f_0$ (2) (270) = 540 Hz, the

Graph	Depth of Air Space	Thickness of Perforated Plywood	Hole Diameter	Hole Spacing	Perforation Percentage	Filled with Rockwool ?
A	2"	5/32"	5/32"	1-9/16"	0.785%	Yes
B	2"	5/32"	5/32"	1-9/16"	0.785%	No
C	2"	5/32"	25/32"	2-3/8"	8.5%	Yes

TABLE 2.5   Data on Graphs of Figure 2.16

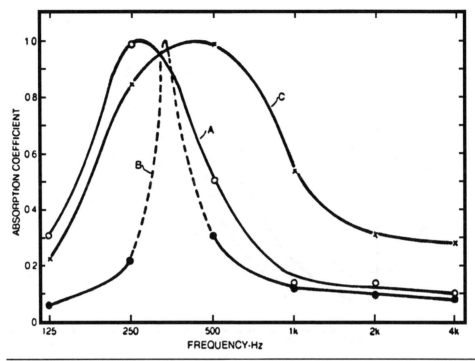

FIGURE 2.16   The absorption characteristics as measured by Mankovsky of three Helmholtz perforated face absorbers as specified in Table 2.5. Curve B is for a unit with no absorbent in the cavity. Curve A is for the identical unit with absorbent. Note the shift in resonance frequency and broadening of the peak resulting from introduction of the absorbent. Curve C is for a similar unit with much greater perforation percentage.

absorption coefficient is read from graph A, in Fig. 2.16 and found to be 0.42. This is plotted in Fig. 2.17 at $2f_0$ as part of the normalized representation of graph A. The same is done to complete all three graphs in Fig. 2.17. An example of the use of the normalized graph A would be to build up an absorption graph using, for example, the f_0 = 120 Hz obtained for the boxes 8-inches deep and perforation percentage of 0.785 percent. At 120 Hz, a = 1.0. At $2f_0$ = 240 Hz, a = 0.42. At $4f_0$ = 480 Hz, a = 0.15.

The same can be done for $1.5f_0$, $2.8f_0$, and other points in between, as well as for frequencies below the resonance peak. Another approach would be to trace graph A of Fig. 2.16 on a sheet of tracing paper and then slide the paper to the left until the peak coincides with 120 Hz. The absorption coefficients with the new tuning can then be read off on the assumption that the shape of the curve would not change for such a modest shift in frequency. These two approaches give, at least, something to use, even though the accuracy leaves something to be desired. Some cases have been noted in which the calculated frequency of resonance does not agree very well with measured peaks of absorption. A relatively simple method of measuring the frequency of the absorption peak of a Helmholtz resonator is diagramed in Fig. 2.18. A loudspeaker is driven by a sine-wave oscillator. The resonator under test is placed 6 to 10 feet from the loudspeaker

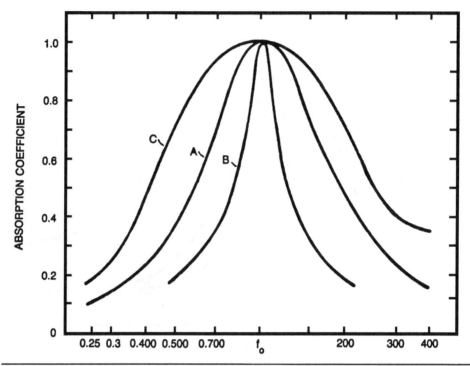

FIGURE 2.17   The three curves of Fig. 2.16 are repeated here in normalized form to assist in applying the curve shapes to other design tasks in order to estimate absorption coefficients to meet specific needs.

with the perforated face toward the loudspeaker. A microphone, placed inside the box, drives an indicator calibrated in decibels, preferably a sound-level meter. As the frequency of the oscillator is swept past the resonance point, a significant increase in reading will correspond to the great increase in sound pressure in the box at resonance.

If the oscillator output, the amplifier response, and the loudspeaker response are all constant with frequency, the box pressure indications are accurate. A second sound-

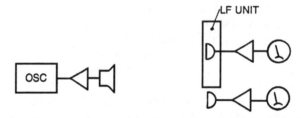

FIGURE 2.18   The resonant frequency of a Helmholtz type absorber may be obtained experimentally outdoors as shown. The rise of pressure inside the box is detected by a microphone placed in the box, as the frequency of the sound from the loudspeaker is swept through the range of interest. A second system measuring the sound pressure outside the box is a possible refinement to separate out instrumental variations in sound pressure.

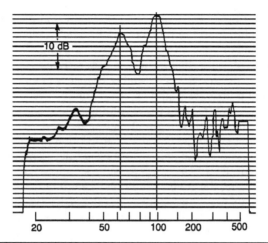

FIGURE 2.19   A graphic representation of the sound pressure per frequency of the Helmholtz resonator as built in Part II Figure 14.2 as the frequency is varied. The peak at 100 Hz is probably the Helmholtz peak, and the one at 63 Hz, a resonance of the back or face panel.

pressure measuring system with its microphone the same distance from the loudspeaker as the one inside the box could be used to measure the pressure outside the box and yield corrections for the other. However, if only the frequency of the peak is desired, just note the frequency at which the microphone in the box indicates the peak. Such a test, made on the Helmholtz resonator of Part II, Fig. 19-12 with a graphic-level recorder making a hard copy of the peak, is shown in Fig. 2.19. Surprisingly, a double peak was found. At first, transverse modes in the air space of the box were suspected, but this theory was rejected when the use of corrugated cardboard dividers in the air space had no effect on the double peaks. Panel resonance of the face and back of the box were then studied. The $\frac{3}{16}$-inch Masonite cover resonates at a calculated frequency of about 69 Hz and the $\frac{1}{2}$-inch-plywood back at about 51 Hz. The calculated resonance peak of the resonator action is 69 Hz. Checking the accuracy of such calculated resonance points is what this section is all about, but pinpointing just what causes which peaks in Fig. 2-19 is somewhat doubtful, although the higher peak at 100 Hz is probably the Helmholtz action and the peak at about 63 Hz is due to panel action, possibly of both front and back acting together. The sound pressure within the resonator box is increased about 22 dB at resonance. This corresponds to a 13-fold increase in pressure. The double peak characteristic may actually be an advantage in that high absorption prevails over close to an octave.

Helmholtz Concrete Blocks and Low Frequency Compensation in the Floor

This is an idea whose time may not yet have come, at least for general application. However, it seems good enough to consider here as it stretches the imagination and points to a possible wider use in the future. There is no denying that it is nice to have carpets in studios, control rooms, and other listening rooms. Nor is there any way to deny their acoustical effect: high absorption in the highs, low absorption in the low frequencies.

If only a special carpet pad were available that would absorb well in the lows and very little in the highs. Just compensating for the carpet, the problem would be solved.

The Japan Victor Company expended great effort in one of their small Tokyo studios to sweep the solution to the problem under the carpet, so to speak.[41] The novel way they did this was to lay the carpet on top of a Helmholtz resonator array. Under the carpet and pad of 1.6-inch thickness is a perforated board and cemented excelsior board of 3.5-inch thickness. Under that is 5.5 inches of air space. A neat solution, to say the least. This eliminated unsightly boxes, or other special ceiling or wall treatments to gain the required low-frequency absorption.

The JVC solution is effective, although rather costly for general use. Is there a less expensive way of placing low-frequency absorption under the carpet? Recent improvements in the Proudfoot Soundblox® (The Proudfoot Company, Inc., P. O. Box 9, Greenwich, Connecticut 06830) suggest a possible solution. Soundblox are proprietary concrete blocks utilizing the cavity inside plus carefully designed slots, metal septa, and fibrous filler as Helmholtz resonators. Figure 2.20 shows the appearance and cross section of one of the many types available, the 8-inch Type R block. All the elements of a Helmholtz resonator are there along with the compartmentation effect of the metal septum and the fibrous filler to improve and widen the absorption effect.

In Fig. 2.21 is a plot of the measured absorption coefficients of the 8-inch Type R Soundblox. The broken line graph shows the absorption coefficients of a typical heavy carpet to show that the carpet absorbs well where the Soundblox is deficient, and vice versa.

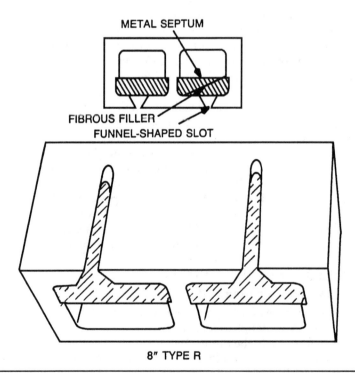

FIGURE 2.20   The proprietary 8-inch Type R Soundblox concrete block, which is a Helmholtz type sound absorber having the absorption characteristics shown in Fig. 2.19. The sound absorption is increased by dividing the cavity with a metal septum and the use of funnel-shaped slots.

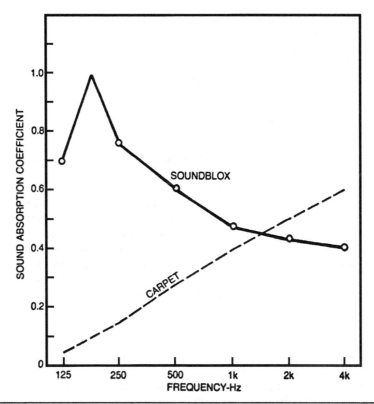

Figure 2.21 Sound absorption characteristics of 8-inch Type R Soundblox tends to compensate for carpet absorption. This suggests the possibility of mounting Soundblox on the floor under the carpet.

If 100 square feet of carpet and 100 square feet of Soundblox were placed in separate locations in a room, the coefficients would be proportional to the Sabins of each patch, and the problem would be one of simple addition to get their combined effect. If carpet is used over the Soundblox, only the sound not absorbed by the carpet reaches the Soundblox. The Soundblox can act only on the sound energy reaching it, which is less than that falling on the carpet by the amount the carpet absorbs. However, at the low frequencies, carpet is almost transparent to sound where the Soundblox have their peak of absorption. In this case, the blocks absorb what the carpet does not. At 4 kHz, the reverse tends to be true and in less definite form. The carpet allows 40 percent of the energy to pass through it and the Soundblox absorb only 40 percent of this. Thus the combined effect is a bit complicated. Certainly adding coefficients to obtain the combined effect is not legal.

Placing a layer of Soundblox, slots up, on the studio floor under a carpet is then a possibility to compensate for the acoustical effect of the carpet. There is the problem of mechanical strength of the Soundblox to withstand a concentrated load such as a grand piano leg. Certainly some load-distributing layer is necessary between the carpet and the Soundblox. The JVC people solved this by using a perforated board and cemented excelsior board. The perforated board, of course, was part of their Helmholtz system. A perforated plywood layer having a high percentage of its area in holes might provide adequate load distribution.

Their cemented excelsior board apparently has its counterpart in this country in such products as National Gypsum's Tectum which might serve satisfactorily as a mechanical protection for the Soundblox. Here are the elements of a possible studio floor with acoustical advantages, but in strong need of further study and experimental verification.

Reverberation

Critical distance is the distance from the sound source where the direct sound energy is equal in level to the reverberant sound energy. Unless you are very close to the sound source, much of the level heard is reverberation. Once you understand the difference between the direct sound, initial early reflections, and later arriving group delay clusters, understanding reverberation becomes a given. It's all so straight ahead it can be represented by a single graph (see Fig. 2.22). Without amplification, the first sound heard will come directly from the source. "Early reflections" are the next to arrive at the listener. These are normally single reflections from the stage boundaries (side walls, ceiling, floor, and back wall) around the source instrument. "Later reflections" start to arrive as clusters of delayed signals. At some point all these reflections start to build up into one continuous sound with reflections from as far off as the venue's rear boundary. This overall level then begins to decay. Reverberation time is calculated from when this decay falls by a specific level. Normally this is 60 dB down (RT 60) from the steady state level, but can also be calculated from level decreases of less than 20 dB.

All of these time periods (early and later reflections) as well as reverberation and its decay are used by our ear/brain system to determine intricate details about the acoustic environment. The direct sound tells us the distance and position of the instrument in relation to ourselves. While this gives us a good deal of information about the sound source, the early reflections further augment this perception. An acoustic guitar, like most music instruments, radiates sound in many directions. Therefore, we can judge its

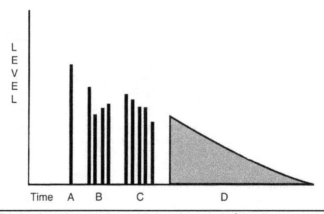

FIGURE 2.22   This level vs. time graph establishes the arrival of the direct sound (A). The first cluster of reflections then arrives from boundaries close to the source (B). The next cluster of reflections (C) arrives from more distant boundaries. As the reflections build, they are no longer perceived individually, but become a wash of reverberation (D) that fades over time.

timbre more thoroughly when we hear its full depth as reflected from the boundaries that closely surround it. These reflections also tell us about the size of the stage area (delay time) and the acoustic properties (frequency response) of the surface coverings (as in wood vs. draped fabric). Later reflections start to come to us from the sides, thus rendering details about the width and size of the acoustic space (delay time). Again this information includes cues to surface hardness (frequency content) and irregularities (number of reflections). Finally the reverberation decay as per time and frequency content fills us in on the overall size, shape and surface makeup of the hall or any other type of space we are in.

If you place a speaker in a room and play a constant tone through it, the sound would spread out and fill that room. It would bounce off of all the surfaces (walls, floor, ceiling, chairs, lamps, books, etc.) not once, but hundreds of times.

The level of the reflected signal would be lower, but because the signal is still coming from the speaker, the reflected signal would add to the overall sound level in the room. At some point this rise in SPL or intensity of sound would reach a peak, level off and remain constant. At this point, the full body of air within the room has been set into motion. If the sound from the speaker is discontinued, the sound in the room will not immediately end, but will fade away over a period of time. Common sense, right? This decay time, also called the reverberation period, is dependent on the size of the room, the room's surface area and the absorption coefficients of those surfaces. Here the logarithms and time factoring can get fairly complicated. For our purposes, it is enough to state that reverberation is the continuation of sound within an enclosure after the original sound has ceased.

Echoes, on the other hand, are distinct repeats of a single sound, as when yelling into a canyon. But a series of echoes or repeats spaced tightly together (flutter echo) can be perceived as reverberation.

However, because an echo is a repeat, in order for it to sound like reverb, the original signal must have little or no attack, in the nature of a legato string line or melodic backup vocals. This is due to the fact that echoes copy the attack of the original signal, while reverberation is a grouping of delayed sound images that are smeared together with no attack sound being discreetly discernible. Echoes are not perceived as repeats unless they occur more than 50 milliseconds after the original. Helmut Hass had this "precedence effect" all figured out back in 1951. This timing also depends on the attack of the source. With a drumbeat, delays as short as 30 milliseconds can be perceived as a separate echo, while the sound of an organ can be delayed by 80 milliseconds or more before a separate echo is discernible.

Noise Control

So far, we've only discussed the adjustment or acoustical treatment of sound within a confined area. Dealing with unwanted sound entering or escaping from a room is considered noise control. Again, we are dealing with the same source-path-receiver situation, but here, since a source is causing unwanted sound, first try to control it at the source, either by enclosing, decoupling, or isolating it. Next, should this not prove effective enough, move on to the path and deal with the sound by using barriers and any reflected sound with absorptive materials.

This process involves a strict step-by-step procedure. Octave band measurements of the area are made to determine the present amount of noise per frequency. Next the source of the offending sound must be located. It is then determined whether the sound can be controlled at the source, along the path, or at the receiver. Normally sound takes many paths, and therefore, may require many solutions. Airborne sound can be direct or reflected and can pass through or around boundaries fairly easily. It can even sneak through pipes and air ducts. Vibratory sound can readily pass through floors and walls. Thus before beginning to build a studio, the degree of noise in the proposed area must be known. In addition other factors about the proposed location, such as outside noise, the building's structural limitations, all piping, duct, and electrical paths, as well as any nearby mechanical equipment must also be assessed.

Sound barriers are acoustic materials that reflect, contain, or isolate. Their function is to stop or block acoustic energy and the transmission of airborne sound from one place to another. Barriers should have high density, be impervious, and limp. Common barrier materials include heavily loaded vinyl, lead sheet, steel sheet, concrete, gypsum board, rubber synthetics, and sheet asphalt.

Barriers have natural vibration modes or resonances, which are determined by the structure's size, mass, elasticity, and the way that they are mounted at their edges. While all resonant modes are important, the fundamental frequency is the point of least isolation because as frequency increases above this so does transmission loss.

More force is needed to accelerate a wall at higher frequencies than at lower. The mass law predicts that transmission loss increases 6 dB per each doubling of mass or frequency. In real life, this increase is closer to 4 dB and is primarily effective at high frequencies. Limpness or stiffness on the other hand is also important, as the resonance of a material is determined by mass and limpness in that transmission loss increases with mass and decreases with stiffness. Walls, ceilings, floors, or partitions must provide enough isolation to achieve the desired level of background noise. Because control rooms should have background noise levels of only 20 to 30 dB and if there are expected to be exterior noise levels of 110 dB, the amount of transmission loss provided by the structures in between the two must reach 70 to 80 dB at the very least.

A good sound barrier must not only be heavy but also airtight. If only 10 percent of a wall is open, it will make absolutely no difference what the rest of the wall is made of. Thus sealing is most important in order for a barrier to reach its full capability. Noise will flow out of every opening, and a great deal of noise can flow out of a very small opening. Barriers should be overlapped at the seams, which should then be caulked and filled. Remember, partitions are seldom just walls. Electrical outlets, television and telephone cables, as well as access doors can undermine how airtight a wall is. All caulking used should be nonshrinking, have good bond strength and a long elastic life.

There are definite frequency limits to the applicability of the mass law to real-life materials. At low frequencies, a wall cannot be considered on the basis of mass alone, its rigidity also has to be taken into account. In addition, most walls are constrained at the edges with the most movement occurring in the center. Not only do stiffness and resonant modes of vibration cause the wall to depart from the mass law but so does the coincidence effect.

Critical Frequency

Critical frequency is a function of the speed of sound, the thickness and mass of the wall, as well as something called the tensile modulus. Below the critical frequency, sound is reflected but above, it is transmitted. In general, it lies above the frequencies controlled by mass. For ½-inch plywood, it is 2 kHz. And for the same thickness in glass, it is 1300 Hz.

Coincidence Effect

Above the critical frequency, there are certain angles of incidence where plane waves can excite the wall, causing a reduction in the transmission loss so that more sound is able to pass through. A wall's movement, though slight, is very complicated. While in motion, a shear wave occurs. When this shear wave is a component of the wavelength of the incident sound wave, transmission loss decreases. Additionally, at every frequency above the critical frequency, there is a particular phase angle of incidence where a panel vibrates as if it were in resonance. Therefore, walls have a drop in transmission loss at the critical frequency when the wavelength in air equals the bending wave of the wall set in motion. In other words, if a wall is driven at its resonance, it transmits more sound.

This all sounds very scary, and you can get all bogged down in it, but it is not a major factor in small studio design because it is easy to compensate for this by adding more material or other materials that have a different density. Double walls avoid this coincidence effect, and their transmission loss approaches the sum of both if well separated. These walls should not be exactly the same, or there can be resonance between them.

Sound Isolation and Dual Panel Wall Partitions

In 1950, Mr. A. London published in the *Journal of The Acoustic Society of America* findings derived from his scientific studies of sound transmission through double-wall partitions (meaning walls made up of two layers of gypsum board with an air gap in between). The results showed that depending on the resonant frequencies involved, the two panels along with the air space between them could in fact act much like a spring. This is now referred to as the mass-spring-mass effect, and it was found to be not as pronounced at high and mid frequencies. Yet the results of his tests did show that dual-wall partitions could actually be less effective in providing low-frequency sound transmission loss than if only one of the two panels were to be used independently. This does not seem logical until you stop to think of that mass-spring-mass model. At and below their resonant frequency, two similar panels, although separated by an air space, will still move in response to each other, and this vibration-like movement is actually augmented by the spring-like effect of the air space between them. While not actually amplified by this effect, it certainly could cause frequencies at and below resonance to pass through almost as if unimpeded.

In short, low-frequency transmission through walls can be due to the mass law, mass/spring resonances, critical frequency, and/or the coincidence effect. Again this happens when the wavelength of the flexing or bending wall itself is equal to the wavelength of the sound wave that is being projected onto it). This ends up tightly coupling the two so as to make them act almost as one. Wave coincidence can cause wall movement. Therefore, the sound transmission will be much greater than through a wall of that particular mass, and results in a transmission loss lower than the mass law would normally indicate. For every frequency above the critical frequency, there is a certain angle of incidence at which this occurs, but that's enough of the theoretical. For more on this subject you should seek out a set of curves given by Magrab, which plots the thickness of common building materials versus critical frequencies. Here you'll find that the critical frequency of ½-inch common plywood is about 2 kHz, while with glass of the same thickness it's closer to 1250 Hz. Of course, you can also look up the equation for critical frequency and do all the math yourself—most enlightening, but very lengthy in terms of time consumption.

High frequencies are normally pretty easily impeded by partitions, but it is more difficult to stop low-frequency sound waves. Yet, even lightweight double partitions can yield high sound isolation if properly designed, and many now are designed this way, thanks in large part to Mr. London's work. This is exactly why I stress gathering as much information on past acoustical work as you can because 99 percent of all the research you need has already been done and it's all mostly free for the asking. You'll find that sound absorptive materials can be installed in the air gap to lessen the resonance. Hey, you can even lower the resonant frequency by making the air space wider. What you're accomplishing is lowering the mass-spring-mass resonant frequency, which in turn reduces sound transmission.

You'll discover that increasing the density of the panels will also yield this same effect and the same kind of results will occur even if only one of the two panels is changed by making it thicker or heavier. If, in fact, you have two panels of different densities, the partition will perform more effectively, as sympathetic vibrations between the panels will decrease.

You'll discover that increasing the density of a panel can be accomplished in many ways, by simply using a denser material or by the building up or adding of additional layers of the same material. Simply making sure each layer's seams do not line up with the ones under it will also cut down on the amount of sound transmitted. If, between those built-up layers, you use a viscoelastic core decoupling, such as NDS damping sheets from Noise Suppression Technologies, Inc. and similar products from many other companies, the partition's low-frequency isolation capabilities will be further increased. This product even comes with pressure-sensitive backing and is easily cut, making installation uncomplicated. You do not have to cover the complete area between the layers of paneling with elastomeric damping sheeting, so adding this type of material in between panels in a kind of patchwork design provides more TL and is more cost-effective. Do the research and pretty soon you'll see structures that leak sound with new eyes that easily lead you to note design and construction flaws.

As with some thermal insulated window glazing in which the panes' edges are glued to the frame around the whole perimeter, rigidly affixed boundaries create a sound bridge that allows mid- and high-frequency sound waves to pass through almost unimpeded, pretty much the way mass-spring-mass resonances do low frequencies. Other paths can occur at structural links, such as the wall joists and studs. The "bridging" transmission due to these structural links can be reduced by using panels made up of multiple layers, as this makes the coincidence frequency higher. By using premade elastomeric stud mounts like those from Sorbothane, Inc., Auralex, and others, you'll isolate the sound energy by lowering the resonant frequency of the whole wall system, and this damping occurs over a broad frequency range.

Laminates, or two layers coupled with a visoelastic core material between them, provide higher transmission isolation than the same material of equivalent thickness. Again, not only are you using two layers instead of one (added mass), but the addition of the core material provides considerably more isolation by reducing structural linkage which helps prevent the passage of vibration from one panel to another. Vinyl material added to the outside or right on top of your walls can yield an improvement. For example, a vinyl barrier product called Prospec from Illbruck, Inc., when installed over a wall, can add up to 26 dB in increased noise transmission loss.

Don't forget any of the other sound leakage pathways often found in studio walls either, including internal wire-ways, connector panels, electrical outlet boxes, windows, door frames, doors, through-wall hanger lags, and bolts, not to mention any flanking paths. With flanking, airborne sound does not have to pass through a wall because it has been left a nice easy pathway above (ceiling), below (floor), or around (adjacent walls) the wall in question.

TL and STC

The sound transmission class rating is usually reasonably accurate and is based upon careful measurements in controlled lab environments. Yet, like absorption coefficients, dependence on these values in the field can lead to problems because buildings are a collection of systems which are susceptible to the weakest link. For an STC of 40, not only do the walls have to meet this rating but also the ceiling, or sound will simply travel up and over the wall. Partitions should be carried from the structural floor all the way to the structural ceiling. Where this cannot be achieved, the plenum space must have equal sound-blocking integrity. Most important in these ceiling passages is the framing around pipes, conduits, and ducts that pass through from one area to another.

Predicting transmission loss is very difficult. Actual partitions must be tested in reverberant rooms, and this is even more straightforward than the tests for absorption coefficient. ASTM Standard E 90 "Test Method for Measurement of Airborne Sound Transmission Loss of Building Partitions and Elements" dictates the test procedures. The specimen wall is mounted between two reverberant rooms with a great deal of precision so that the only sound path between the two is through the specimen. One reverberant room is excited with sound and the sound pressure level in both rooms is measured at sixteen $\frac{1}{3}$-octave bands from 125 to 4000 Hz. The transmission loss is then calculated.

It is much faster to compare a single number as opposed to a graph of 16. Here the STC comes in handy. According to ASTM Standard E 413, "Standard for Determining of Sound Transmission Class," the resultant TL data is compared to a standard curve from which a single number rating is computed. However transmission loss results from ASTM E 90 can be more accurate because it calls for the specimen to be a minimum of 8 feet in size with all doors, windows, etc. of normal sizing. In other words, the specimen should be exactly the same as that used in the field. While most manufacturers publish reliable data, it still is difficult to achieve these ratings in the field.

Ratings change when different mountings are used. For example, even using different metal studs can cause different ratings.

Barrier Materials

Lead is primarily a barrier. Lead sheet and composites are dense, limp, and impermeable, which gives them the ability to block sound. Sheet lead is malleable, readily bent, folded, and formed. It easily conforms to irregular surfaces, does not spring back once formed, and is easily cut without special tools. Joints can be folded by hand, taped, stapled, or crimped. Lead readily accepts adhesives and pressure-sensitive tapes. In cramped plenum spaces, it can be difficult to fit conventional barrier materials around pipes, ducts, conduits, etc. The ease of use in terms of cutting and forming sheet lead make it very simple to install. Thus, a high quality installation becomes less expensive labor-wise. Because all openings are detrimental, lead's formability greatly helps block even the smallest leaks around the smallest of wires, which often must travel through plenums.

More rigid barrier materials are not as conducive to this work. Sheet lead is specified in pounds per square foot and ranges from $\frac{1}{2}$ pound ($\frac{1}{128}$ inch thick) to 8 pounds ($\frac{1}{8}$ inch thick). Lead-loaded vinyl or lead-powder-loaded vinyl can be coated with glass fiber, nylon, cotton, or other fabric backings. They are also specified in pounds per square foot ($\frac{1}{2}$, $\frac{3}{4}$, 1, 1.5, 3, or more) and are acoustically completely limp. Hung as curtains, used for folding doors, blanketing, or pipe and duct wrap, they can be more expensive than plain sheet lead. Their advantage in terms of colors, textures, decorative patterns, and more durable finishes make them very appealing to designers. Lead-loaded plastic materials make good barriers and are easy to install because they are generally very flexible. They can be made into curtains by simply building a frame on which hang them. They can even be transparent. Lead is not expensive in sheet form, and due to its limpness and natural damping qualities, it is not affected by coincidence and follows the mass law relatively closely.

Sheet lead or loaded vinyls, whether lead-loaded or loaded with other heavy materials, can also have foams attached to add sound absorption. Additional absorption material in the receiving area is very beneficial to transmission loss. As an example, Carrol George, Inc. manufactures a loaded vinyl with foam adhered to both sides. While only 0.6 inches thick, it yields impressive TL ratings of:

14 dB at 100 Hz

14 dB at 125 Hz

14 dB at 160 Hz

13 dB at 200 Hz

18 dB at 250 Hz

17 dB at 315 Hz

21 dB at 400 Hz

20 dB at 500 Hz

22 dB at 630 Hz

26 dB at 800 Hz

27 dB at 1000 Hz

29 dB at 1250 Hz

32 dB at 1600 Hz

35 dB at 2000 Hz

38 dB at 2500 Hz

38 dB at 3150 Hz

41 dB at 4000 Hz

44 dB at 5000 Hz

The above ratings were measured by Riverbank Acoustical Laboratories, and the depth of information provided by the lab reports is shown by the following quote.

Description of the Specimen 4′ × 8′ = 32 square feet

The test specimen, 48 inches (1.22 m) wide and 96 inches (2.44 m) high was mounted directly into the laboratory test opening, attached to a frame, and then sealed with a mastic over the entire perimeter to the test opening. The specimen was constructed as follows: a 1 pound per ft.² (4.9 kg/m²) filled vinyl noise barrier septum faced on both sides with a ¼-inch (6.4 mm) thick 2 pounds per ft.³ (32.0 kg/m³) density urethane foam adhered to the septum. The specimen had an overall thickness of 0.6 inches (15.2 mm) and weighed an average of 1.09 pounds per ft.² (5.32 kg/m²). The transmission area, S, used in the computations was 32 ft.² (3.0 m²).

When using lead or loaded composites one should avoid rigidly fastening them directly to stiff surfaces. Visoelastic adhesives or intermittent fastening should be used. In other words, it is more effective to hang lead sheet limply between two walls as opposed to rigidly fastening it to one.

Other barrier materials include brick, gypsum board, wood, concrete, steel, and glass. In all barriers, the more porous the material, the less transmission loss yielded. Mastic cements are heavy, dense, flexible, asphalted materials with elasticity ranging from semirigid to fairly flexible. They are used for barriers and damping in hollow doors, appliances, walls, etc.

Spray-on materials can be applied into a cavity, reducing noise by damping the structure and filling in any leaks. Adhesives can bond any material, as long as the manufacturer's recommendations are met. Sealants like adhesives come in room temperature cure, hot melt elastomers, and tapes with pre-applied pressure-sensitive adhesives.

Window glass also offers barrier properties that vary according to thickness, number of panes, and construction methods. As an example, single-pane windows offer an STC of 19; double pane, 21; ¼-inch, 26; ½-inch, 31; and 1-inch, 34 to 37.

Along with these properties, one must also consider edge restraints. Resilient edge restraints can provide as much as a 6 dB improvement over a stiff-edge attachment.

As noted in the section covering observation windows (Part II, Chapter 6), the strips of rubber or neoprene underneath and bearing the weight of the glass should not compress more than 20 percent under the load, but the side and top resilient mounts can be more pliant. After witnessing many window-seating experiences that turned out badly, I can offer the following tips. Beveling the edges of the glass can save a lot of trouble. Since it often takes several individuals to hoist the control room window into place (and this often includes any musicians who are available at the time), beveling is a good idea as it helps avoid cut hands. Those same sharp edges can also slice and possibly defeat the advantages of using resilient isolation strips. The best bet is to install edge work, such as those used for decorative protective trims, sealing channels, weather stripping, or even the type of U-shaped edge capping normally found surrounding the edges of helmets, onto those glass panels the moment you accept delivery. These are simply pushed on, right over the raw edges. They come in *dual durometer* sponge or solid rubber and have flexible-wire or steel-segmented cores, which make them strong and resilient. Additionally, they'll grip and hold to any contour without your needing to pull or stretch them around those dangerous edges. They render the added benefit, upon installation, of helping to avoid the "crowding" of the glass within the framework, which can cause cracks and actual breakage. This usually occurs some time after the installation when the system has shifted or settled into place. More than once I've witnessed a studio owner in the midst of showing off a newly installed control room or isolation booth window only to sadly find the glass panel cracked. It's important to note that latex foam takes less of a compression set, and for that reason, it is more forgiving than sponge rubber seals, especially if your framework is a little distorted, that is to say, not perfectly plumb.

NOTE *There can be some confusion with durometer measurement results because there are 12 different scales, but all soft rubber, foam, and sponge-like materials are measured using the "A" scale. The results can also be given as "Shore hardness," but there is no difference; it is just a reference to the person who invented the durometer measurement tool back in the 1920s, Alfred F. Shore. With that said you should easily pick up on the beneficial qualities of "dual durometer" sponge.*

Obtaining these seals poses no challenge, as they are readily available from most automotive parts suppliers and outlets, like Stan Pro, or by getting in touch with manufacturers, such as Cooper-Standard Automotive or Lauren Manufacturing Company (both are listed in the manufacturers of acoustic materials appendix on this book's web site). From them you will also get some very educational material. The latter company also manufactures complete doors and windows, and both offer door seals, gaskets, and various elastomeric products, such as strips and flexible cellular materials that can aid in solving vibration isolation problems.

Normal glass may not have the best transmission loss ratings, but when double or triple glazing is utilized, TL increases. Here an air space of 1 inch or more is used, but any more than 4 inches of air space provides negligible results. Laminated "acoustic glass" provides good isolation, as does safety glass, which has multiple layers of thin glass laminated with thick soft layers of polyvinyl-butyrate plastic.

For example, 1-inch glass with $\frac{1}{2}$-inch airspace provides an STC rating of 32; $\frac{1}{4}$-inch safety glass, 31; $\frac{1}{4}$-inch laminated (with plastic) 38 to 40. Two panels of different thicknesses can prevent resonant and coincidence effects. Sealing is also important. Spacings of greater than 1 inch or laminated glass can provide good TL ratings; however, if the window is openable, TL could decrease as much as 10 dB, even when closed. If windows must be openable, use dual cam or catch methods, apply two to three edge gaskets, and close them tightly with as much pressure as possible. In fact, if windows can be opened with the use of a special wrench or tool, it's even better.

If a barrier or wall is made up of different sections with different TL ratings, such as doors and windows, it is important to have an idea of its average transmission loss. This averaging takes the transmission coefficient of each section multiplied by the area of that section and divides the sum by the total area. This method can lack some accuracy due to both real-life deviations from lab conditions and the effect of the very poor TL of the weakest link.

Curtains that are often loaded plastics or vinyls offer good strength and barrier performance while remaining flexible. This flexibility aids sound TL by reducing coincidence and causing them to behave close to the limp mass law. However, since their tops are often open, transmission loss is not maximized. Furthermore, during lab testing, their edges are sealed while in actual usage this may be impractical because access is often important. Therefore, even with curtain edges overlapping, TL ratings will be lower in the field.

Partition systems are used to divide a room temporarily or semipermanently. They are rigid and come in all types of materials, sizes, and finishes. They are most often used in open-plan offices. Operable rigid partitions ride on rollers in a track mounted on the floor and ceiling. Thus they can be easily moved, folded against a wall, or extended out to divide a room. Semipermanent panels can be erected or disassembled on site but generally require tools. They are made of all types of absorptive panel materials, often with air space within the panels and interlocking edges to prevent sound leakage. They come in decorative finishes and allow for various changes in work-space usage. Some sound panels for noise control have heavy barrier material in back of a porous material faced with perforated metal. Thus, they fulfill two tasks by reducing reverberation on one side along with the adding TL from the other. Increasing thickness generally means more absorption and TL. Often these panels are 2 to 8 inches thick. To further increase TL, some use heavily loaded vinyl or lead in between but spaced away from the absorptive and backing layers. To further aid in TL the rigid backing side can also have a damping layer added to it to reduce radiation from the back side, which also helps with coincidence problems.

Enclosures are used to completely or partially surround an offending noise source, such as machinery. They are made up of interlocking panels and can provide protection from dirt, oil, and water by adding a thin facing or by totally enclosing their porous material in a plastic bag. Enclosures have rigid backing and can include windows and access doors for machine operation and maintenance. They are usually rated in TL per machine, as most are custom made. Due to heat buildup (low sound transmission means low heat transmission), ventilation with attenuated ducting is also often provided.

Quiet rooms are like enclosures, but here an isolation area is provided for operators, which effectively separates them from the noise of (for example) a factory floor. Thus, they can also be used for audiometric (hearing) tests or as recording isolation booths, and are similar to telephone booths. However they are generally not very effective in shielding against low-frequency vibrations.

Since sound can easily travel along a plenum above ceilings from one room to another, ceiling systems, while mostly absorptive, must also incorporate barrier materials. Therefore, many are rated for transmission loss. Acoustical ceilings may not be highly thought of, but they offer a wide range of uses. They can be used for anything from a simple decorative covering to industrial or heavy-duty use. They vary in density and thickness and can be made of anything from mineral or glass fiber to wood fiber. Their absorption coefficient ranges from 0.3 to 0.9 with an average around 0.6. Most absorption tests and field mountings utilize # 7 (E 405) mounting, which has a 405 mm or 16 inch air space plenum. Therefore, they can also act as diaphragmatic absorbers increasing low-frequency absorption as opposed to when mounted directly to existing ceilings.

If lead sheet or other barrier materials are included, acoustical ceilings can cover the full range of acoustical properties, if chosen correctly. To aid in this choice, several test methods are utilized. ASTM C 635 and C 636 standards for "Composition and Installation of Metal Ceiling Suspension Systems for Acoustical Tile and Lay-In Panels" as well as AMA 1-11 Ceiling Sound Ratings, which gives the attenuation of sound from one room to another through the ceiling and via the plenum path above the ceiling. These are extremely useful when used as decision-making tools.

Standards

To predict the results to be expected from acoustical treatment, one must understand the difference between the two main absorption coefficients used.

Most acoustical tests of absorptive materials are performed using either random incidence or normal incidence tests. Incidence is the term used to describe the direction of a sound wave striking a surface. Normal is perpendicular and random is from all angles. Both test methods are strictly defined and adhered to. Random is governed by ASTM Standard C 423 "Standard Method of Test for Sound Absorption of Acoustical Materials in Reverberation Rooms." Here the absorption coefficient of a material is determined by the change its introduction causes in the rate of decay of a sound in a reverberant room. First the total absorption of the room is measured without the test specimen. A sound source is turned on until its level reaches a steady state. The sound source is stopped and the rate of reverberation decay measured. The test specimen is brought into the room, and the test repeated. The difference is noted and the total amount of absorption caused by the specimen is divided by the area of the specimen in square feet. The resulting value gives the absorption coefficient. The room must not only be hard, reverberant, and diffused, but also large enough so that the diffused area is not broken up by the specimen. While the specimen must be small enough so as not to break up the diffusion of the reverberant field, it must also be large enough to give accurate results.

Small specimens can give higher absorption values because diffraction around the specimen can increase absorption. Here sound waves that strike the material are bent or diffracted at the edges so they *see* an effective area that is greater than the area of the test specimen. This can often yield coefficients higher than 1, even though this value cannot be exceeded in theory. This however poses no problem for comparing the absorption coefficients of two materials, if the test standards are adhered to.

Should a lab round off or adjust the test results so that the peak of 1 is not exceeded they must according to C 423 specify exactly how this was done. Additionally, due to the fact that an absorber's low-frequency absorption can be increased due to air space behind the specimen, standard repeatable mountings are utilized and that mounting will always be specified in the lab report.

Coefficients based on normal incidence utilize an impedance tube as per ASTM Standard C 384 "Test for Impedance and Absorption of Acoustical Materials by the Tube Method." Here a small sample of the material is placed at one end of a sealed tube and a tone is generated within. A small movable microphone measures the minimum and maximum sound pressure levels and from these the absorption coefficient is found. The impedance tube method uses a sound source that radiates pure tones usually at octave intervals. On the other hand the random method is determined by testing at six different $\frac{1}{3}$ octave bands centered at 125, 250, 500, 1 k, 2 k, and 4 kHz.

Comparing normal incidence test results to random incidence is not easy. One rule of thumb is that normal is $\frac{1}{2}$ random, if random is a small value, but in general normal incidence results are usually lower than random results by $\frac{1}{4}$ to $\frac{1}{3}$.

Absorption is also related to frequency and that is why tests are performed across several frequency ranges. Furthermore, a material's dimensions as compared to the wavelength of sound falling upon it are also important. Theory states that maximum absorption occurs when the size of the material is $\frac{1}{4}$ wavelength of the lowest frequency to be absorbed.

The effective amount of absorption is often lower than calculated. But by distributing the absorption widely and randomly, a closer to calculated result can be reached. This is due to the edge effect or the diffraction of sound at the edges of the material.

More On Standards

Acoustical properties are discussed in terms of ratios and coefficients and the two are very helpful for comparing acoustic material properties. The range of human hearing covers a huge variation in pressure from 0 to 1,000,000. Instead of dealing with all those numbers, the decibel, a logarithm of a power ratio is utilized. The resulting range of 1 to 120 dB is a bit easier to deal with. However as a log of a ratio [10 log (value a/value b)], the decibel has to be referenced to another value. With sound pressure levels, the dB is referenced in dynes per square centimeter, with sound power, in watts. Thus, single number comparisons between the two cannot be made. To further complicate matters, since it is a log, the addition of decibels requires complicated equations. The human ear's response to sound is also not uniform with frequency. It is more sensitive to higher pitched sounds than to lower. The crossover point is around 1000 Hz. Below this point the ear becomes increasingly less sensitive. This fact necessitates the use of weighing scales when measuring sound pressure. Therefore to accurately compare test results it must be known if, and which, scales were utilized during testing. As a matter of interest, the A scale heavily discriminates against low-frequency sounds and closely approximates the characteristics of human hearing, the B scale moderately discriminates against sounds below 500 Hz, and the C scale is essentially flat.

Sound pressure level comparisons may be made as long as the same weighing scale is used. To help avoid confusion, many values will be given followed by the scale used as in dB(A), dB(B), or dB(C). If only dB is given you should check out the test methods used with the manufacturer.

Transmission Loss (TL)

While sound absorptive materials are light weight and porous, sound isolation materials are massive and airtight. The way acoustical materials are rated for their ability to isolate sound starts with measuring the level in decibels of sound that passes through them per frequency.

TL is the ratio of the sound power coming in contact with a structure to that which passes through and it is not unusual to find ratings of 30 to 40 for ordinary heavy building materials. Some examples of this are shown in Table 2.6. From looking at this table it is immediately obvious that transmission loss varies with frequency. It turns out that this variation is fairly predictable. At low frequencies the stiffer a wall is, the better. As we

Material	125	250	500	1000	2000	4000 Hz
4" Brick	30	36	37	37	37	43
$7^5/_8$" hollow Cinder block	33	33	33	39	45	51
$^3/_4$" 2 lb./ft plywood	24	22	27	28	25	27

TABLE 2.6   Transmission Loss (TL) of Common Building Materials

WALL STRUCTURE	125	250	500	1000	2000	4000 Hz
Metal channel stud wall with ½"-gypsum board on both sides 5 lbs. per square foot	25	30	38	47	48	44
2 × 4 stud with ½"-gypsum board on both sides 6 lbs. per square foot	22	30	35	40	41	40
2 × 4 staggered stud wall with ½"-gypsum board on both sides 7 lbs. per square foot	36	37	40	47	52	45

TABLE 2.7   Noise Reduction (NR) of Standard Wall Configurations

approach frequencies that are in resonance with the structure, transmission loss depends on the dampening of the wall. Above the resonant frequency the transmission loss is dependant on the mass or density of the structure. Density relates to thickness and theoretically the doubling of either the frequency or the density increases transmission loss by 6 dB. But in actual practice the increase in transmission loss is a bit less than in theory.

Sound Transmission Class (STC)

There is also a single number for transmission loss called the *sound transmission class,* but it is not something that the average person is going to test because it calls for superimposing a standard contour on actual measurements to keep the STC within 8 dB of the TL rating.

Noise Reduction Ratings

You can also readily find tables that give the noise reduction (NR) of standard wall configurations. These are a major help.

Examples of these are given in Table 2.7

I chose these three examples because they exemplify how both mass and structure affect transmission loss. Because of its superior structure, the first example betters the second even though its mass is lighter. Yet the third example, which has both a beefed-up structure and additional weight shows the best STL at all frequencies tested.

Noise Reduction Coefficient (NRC)

The NRC is an index of a material's absorption capabilities. It is merely an average of the six octave absorption coefficients with 125, 250, 500, 1000, 2000 and 4000 Hz center frequencies. As such it is of limited help due to the fact that it gives no indication of the material's absorption capabilities per frequency.

Mounting

If you do not take the mounting method utilized during the testing process into account, absorption coefficient specifications become a meaningless list of numbers. Check out these two sets of absorption coefficients in Table 2.8.

125	160	200	250	315	400	500	800	1 k	1.25 k	1.6 k	2 k	2.5 k	3.15 k	4 kHz.
0.45	0.48	0.57	0.69	0.98	0.93	1.02	1.15	1.10	1.09	1.10	1.07	1.11	1.09	1.16
0.82	0.82	0.78	0.90	0.90	0.96	1.07	1.09	1.04	1.06	1.04	1.05	1.04	1.02	1.04

TABLE 2.8   Absorption Coefficients Specifications

It seems that the top panel is superior at absorbing mid to high frequencies but below 250 Hz, the bottom panel blows the top panel away. The fact is, both of these sets of panel specifications are for the same RPG Abffusor® panel.

How can this be? It is due to the mounting method.

The top absorption coefficient specifications are from an Abffusor flush mounted on the surface (A mounting). The bottom are from an Abffusor mounted with a inch of semirigid glass fiber and a 16 inch depth air space behind it in the standard E 405 or hung-ceiling mounting. This shows that the way in which absorptive materials are mounted is critically important. Porous material mounted out away from a wall instead of directly on the surface can provide as much as double the absorption at low frequencies. This should not be surprising because the sound is now is exposed to twice as much material. In fact, absorption panels that have exposed edges can often achieve absorption coefficient ratings greater than 1.0.

This is the reason "space absorbers" are so popular. These are usually seen hanging from the ceiling in factories and school gymnasiums and can be in the shape of cones or cylinders or flat baffles. In recording studios acoustic "clouds" are often suspended over the mixing console. Here in addition to absorption they help reduce ceiling reflections into the mix position.

When absorption ratings are derived from different mountings methods it negates any and all comparisons between two materials. You can be led to accept some pretty wild advertising hokum if you do not investigate the standards used to test the material or product. It's actually fairly easy to do. Manufacturers spend a *lot* of money to have certified test facilities rate their products. In most cases the test reports are open to the public. So all you have to do is dig a little deeper into the test report to find the mounting method. But this should help you to understand the different mountings and how critical they are to the outcome of the test.

A—directly against a hard surface

B—cemented to plaster board

C-20—for materials with perforated or expanded facings spaced out ¾″ (20 mm) from a hard surface

C-40—same but spaced out 1½″ (40 mm) from a hard surface

D-20—materials with nonperforated or expanded facings spaced ¾″ (20 mm) from a hard surface

E-405—spaced out 16″ (405 mm) from a hard surface (as with hung ceilings)

But the one that really can cause major confusion is the J mount. This was originally devised for baffles that are placed around machinery to curb noise. These are often sheets of perforated vinyl or some other easily cleaned material that are used to encase absorptive and barrier materials. The powers that be allowed these products to be mounted away from the wall because that is how they are used in practice.

The J mount is also known as the "manufacturer's suggested" mounting method. This is a legitimate test mounting wherein the manufacturer is trying to determine or show the best mounting method for their product.

Test results using the J mounting method are given in Sabins per square foot or Sabins per unit and should never be compared with the absorption coefficients. However, it is a very simple task to convert Sabins per unit into what can be thought of as false absorption coefficients.

NOTE   *No test results using the J mount are ever given in absorption coefficients by the test facilities.*

Facings

The facing of a material is also very important. They are used to protect the more delicate materials from abrasion and also help to prevent fibers from escaping into the room. The opening area of these facings can range from as little as 10% to more than 50%, and this variation has a significant effect on the panel's high-frequency absorption properties. In many cases, a thin sheet of plastic, will do the job and have little effect on the inner material's high-frequency absorption.

Measuring Acoustics

If truth in your listening environment is what you are after, you've got to pay close attention to all those not so small details such as smeared stereo imaging, standing waves, flutter echoes, low-frequency build-up, disconcerting ambient noise and reverb times that differ with frequency. You will have to put proven acoustical correction methods to use, such as corner bass traps, diffusers, Helmholtz resonators or maybe a diaphragmatic absorber utilizing a closet sized space behind it. But how do you know exactly what it is that you need? By getting involved in testing acoustics.

Evaluating Audio (A/B Comparisons)

The audio recordist's job entails both art and technology. The technology side is not as exciting, sexy or fun, but when put to use it opens the door to more expressive forms of the art and points out limitations that can detract from it. Either way, knowing the technology enhances the art. Knowing the facts ends up being much more productive than being lead astray by half-truths, folklore, fairy tales, voodoo, black magic, dogma, generalized suppositions and electro-rhetoric.

Although we can discern improvements in sound such as a reduction in noise content, most judgments about quality end up being purely subjective since it is seldom possible to correctly compare A/B actual before and after conditions. Our ear/brain hearing system has a short memory. Therefore, we cannot attempt to perform A/B comparisons at different times. Otherwise, we may find improvements that do not exist.

Scientists spend a great deal of time making sure they avoid "contaminating" the objectivity of their test results especially by subjective influences. But the field of professional audio is rampant with subjective opinions, which are often readily accepted as fact without any subjective testing. Some may think it's all about: "Let's not rock the boat, but go right along with the program and keep the gravy train of profits rolling."

But for the most part I believe it is due more to a lack of understanding of basic audio principles. It has become almost as bad as with consumer electronics where a cable that cost $400 is falsely believed to be superior because the audiophile who bought it needs to feel post-purchase gratification with their expenditure.

Hobbyist's purchases are no business of mine, but professionals are supposed to be grounded in the real world and not blinded by subjective illusions. Of course, the economic realities of our business do not allow us to ignore client demand and the "wow 'em with delight" factor of having the newest craze in pro audio equipment right there to entice clients to shed extra money. As an example, while I honestly liked the sound of EMT's digital reverb device, it was its R2-D2-like, space-age remote unit that led most studios to spring for its hefty cost.

Big-time client thrill factors require no blind listening tests that compare the device under test (DUT) via A/B switching with a straight length of wire. One of my favorite terms in audio is *transparent*. Its use does not require any testing for extended frequency response, ultra-low distortion, slew rate, damping factor, speed, dynamic range, S/N ratio, or headroom, but "all of those factored in." Yeah, sure.

A good example is cabling. Capacitance, resistance, inductance, attenuation, filtering and induced delay can all be measured. But I once offered to test and review several brands of high-end audio cable for a professional recording magazine and was told by several *specialized* cable manufacturers: "You won't find any difference using tests, but you can hear the difference." Time to take out the wading boots. A/B testing, when properly conducted, bypasses all the hokum and presents us with information that may improve the end product's audio quality. And when you get right down to it, isn't that really what our clients expect from us?

Okay, as an example of proper A/B comparative objective testing, take two equalizers, set them up on the studio's shop-test bench and feed both of them the same exact output from a signal generator. Using a frequency counter and a decibel meter, adjust their processing (decibels of boost and frequency settings) to be exactly the same, and also set their output levels so that they match. Put those outputs into an A/B switcher, and feed its output to a dual-trace oscilloscope that is also displaying the original signal. Now you can actually see any difference in timing caused by equalizer-induced phase shift. Do I expect the average recording studio client to care about the results of this test? Of course not, but I would expect the studio owner to want to know about any differences that may cause sonic inferiority and I certainly expect every audio engineer on staff to be highly interested in those results.

To be adequately revealing, this test should include all pertinent frequencies with varying amounts of boost and cut. This may seem like I'm asking a lot, but after using the Neutrik-strip chart recorder in phase-response mode, I came to appreciate the usefulness of this knowledge in achieving quality sound. When it comes to acoustics, especially control room acoustics, test results are critical. You need to have them before you make any changes or additions to the surface area of your room. Don't have a spectrum analyzer handy? Then use the old time one frequency at a time "sequential" method of room analysis.

Sequential Frequency Analysis

Step into my *way-back* machine and check out "Shooting a Room" 1960s style. Ziiiiiip....
It's 1968, and you and I are going to shoot the control room. While they were available,

we had no access to $20,000.00 spectrum analyzers, so we'll just make a point by point plot. Ready?

Okay, here's our test bench equipment list:

1 studio quality cardioid, pickup-pattern microphone

1 studio quality omnidirectional, pickup-pattern microphone

1 dual concentric (15″ minimum bass cone) speaker

1 sheet of foam at least 4′ × 4′

1 sine wave signal generator

1 frequency counter

1 mono power amplifier

2 preamplifiers

2 multimeters

A bunch of graph or chart paper

In our case, this means a couple of capacitor microphones, an Altec silver colored medium- to large-sized dual concentric speaker, some flat-surfaced (not convoluted, pyramid, or textured in any way) foam, but textured if it were available would be preferable, a Hewlett-Packard sine-wave generator that had its frequency adjusted by a 6-inch diameter, oversized dial; and a Hewlett-Packard four-digit, seven-segment frequency counter where seven-segment displays were generated by filaments inside of vacuum tubes. Any studio quality amp will do, but we will only be using one side if it is stereo. The two preamps can be outboard, but it is certainly more convenient to use two of the preamps in the mixing console.

NOTE *The preamp handling the output of the signal generator that will be fed to the first multimeter and the power amplifier should be panned opposite the preamp handling the microphone output, which will be fed to the second multimeter. Even if the two multimeters have decibel scales, it may be wiser for the readout to be in millivolts. Get a bunch of graph paper, and label every page as to what it pertains to before noting any levels.*

Now we must "calibrate" our test equipment. What we are attempting to accomplish here is to make note of the complete system response without any influence from the room acoustics (see Figure 2.23).

1. Form your sheet of foam into a cone. You can use gaffer's tape to hold it in place.

2. Position the wide end directly in front of the dual concentric speaker cone.

3. Place the omnidirectional pickup microphone into the small end of the cone facing directly into the center of the speaker.

4. Turn on every piece of the test bench equipment and let it warm up for several minutes.

5. Decide beforehand which spot frequencies will be checked:

    As few as only the six absorption coefficient bands: 125, 250, 500, 1000, 2000 and 4000 Hz

    Or ISO octave center frequencies: 16, 31.5, 62.5, 125, 250, 500, 1000, 2000, 4000, 8000 and 16000 Hz,

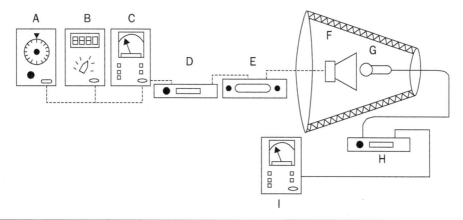

FIGURE 2.23 Sequential frequency analysis calibration setup. A is the signal generator, B is the frequency counter, C is the output multimeter, D is the output preamplifier, E is the output amplifier, F is the coaxial speaker, G is the microphone surrounded by a foam funnel, H is the input preamplifier, and I is the input multimeter.

Or ISO one-half octave center frequencies: 16, 22.4, 31.5, 45, 63, 90, 125, 180, 250, 355, 500, 710, 1000, 1400, 2000, 2800, 4000, 5600, 8000, 11200 and 16000 Hz

Or ISO one-third octave center frequencies: 16, 20, 25, 31.5, 40, 50, 63, 80, 100, 125. 160, 200, 250, 315, 400, 500, 630, 800, 1000, 1250, 1600, 2000, 2500, 3150, 4000, 5000, 6300, 8000, 10000, 12500, and 16000 Hz.

This could add up to a whole lot of time, even if the testing will be at only one location, so first we are going to perform a slow sweep across the complete bandwidth to check for any gross anomalies

6. Choose the frequencies to be tested and begin the testing and plotting sequence.

Dial in 1 kHz as the reference center frequency while looking at the frequency counter readout.

Set the generator level to read at a midway point on the initial multimeters' scale.

Set the signal generator preamp output level for the highest level before distortion using only your ears or a scope connected post preamp (in our case a scope was not used and thus not shown).

Set the amp level to drive the speaker at the normal listening level.

Set the preamp being fed by the microphone to give a midway reading on the second multimeter.

Note this reading in millivolts on the graph paper and mark it as the reference level or 0 dB point. Okay now label the sheet of graph paper as "Omni Cal" and add the levels that are read out for every center frequency. For example: 5 millivolts above the 1 kHz level would be plus 5 and 3 millivolts below the reference level would be noted as minus 3. Return to step # 3, and replace the omnidirectional microphone with the cardioid microphone. Then repeat the calibration procedure, marking the level of every center frequency on a sheet of graph paper that has been labeled "Cardioid Cal." Put the foam aside, place the speaker midway between the left and right monitors and position the omnidirectional microphone in the listening position. Start with 1 kHz, and adjust the levels to match the *calibrated* 1 kHz reading, while at the same time checking for distortion either by carefully listening or by watching a scope. Now repeat the test for every center frequency noting the level as referenced to the original *calibrated level*. This means that, if when the microphone was directly in front of the speaker, you had a reading at 125 Hz that was minus 3 millivolts below the 1 kHz reference level and now with the level as "influenced by the room" you obtain minus 1 millivolt from the microphone at 125 Hz, the graph will be marked *Plus 2*. (You will find that this low-frequency bump is not unusual in smaller rooms).

With this stage completed you may think that you would have a pretty good idea of what steps will be needed to adjust the room for a more even or honest response. Think again! You may be able to make general statements such as: "Okay, we have a low-frequency buildup at 125 Hz so we'll need to add a bass trap" or "There is a bit of excessive high end around 4 kHz so we should add some absorption or diffusion." But how much and where? This is why we went through the trouble to calibrate a cardioid microphone. All right, now we set the signal generator to 125 Hz, and I walk around the room pointing the cardioid microphone to the walls, floor areas and corners while you watch the reading on the second multimeter. You call out "There!" every time the level peaks and I place a piece of painter's blue masking tape on each spot. This process is repeated (using different colored tape) for every frequency whose level is significantly different from the 1 kHz reference level. Don't pack up the test equipment just yet! You realize that the channel the microphone was plugged into (which was last used on a snare drum during a rock song mixdown) still has the equalizer engaged with a setting of plus 6 at 100 Hz and plus 4 at 4 kHz. Ziiiiiip....

Just kidding, but crazy things like that often seemed to occur prior to microprocessor control. However, even if this exercise ended with successful room analysis, you are still talking about several hours of work and less than pinpoint accuracy.

CHAPTER 3

Real Time Analysis

Noise Generators

White noise, (the sound heard when a radio or television is tuned between channels) consists of all audible frequencies at equal level. Pink noise is similar but has been put through a filter that rolls of 3 dB per octave. Here the higher the frequency, the greater the attenuation. This is done because pink noise is used with ⅓-octave analyzers whose bandwidths become wider as the frequency increases. For example, between the first two center frequencies, 16 and 20 Hz, there are but 4 cycles. On the top end, between 12.5 and 16 kHz, there are 3,500 cycles! In other words the level of power of the bottom bandwidths of ⅓-octave spectrum analyzers is much lower than it is at the higher frequency bandwidths. By using a filter that progressively attenuates more level as the frequency increases, we end up with a display that appears flat on ⅓-octave analyzers. Everything becomes all matched up perfectly, when set at ISO center frequencies and ANSI bandwidths.

As an aside, the names white and pink relate to the light spectrum, wherein red equates to lower frequencies and a mixture of the whole spectrum produces white light. Due to its high-frequency roll-off (and resulting low-frequency emphasis), it adds a little red to the white and is therefore "pink" in color. (Crazy world, right?) Like white noise, pink noise is a sequence of all frequencies generated randomly or at least pseudo-randomly. It is because of this randomness that we are required to average multiple readings to achieve accurate results. The Alan Parsons and Stephen Court Sound Check CD has individual ⅓-octave bands at ISO center frequencies. These are more stable than a pink noise generator. In addition the "one band at a time" approach to system adjustment is easier on the eyes, ears, and equipment, as both white and pink noise are not enjoyable listening, especially when levels must be high to overcome background noise.

Using ⅓-octave analyzers has become very simple because of the addition of microprocessors, but care must still be taken so as not to over-EQ. The operator must proceed in a slow, orderly, and thorough manner. If any cutoff filtering (low or high pass) is being used, it must be introduced before the processing because it can have an effect on the other settings. After making any adjustment to a ⅓-octave EQ, the system should be given a bit of time to "settle in" or stabilize before proceeding to the next band to be adjusted. After the EQ adjustment has been completed, it is critical that the system be given a good "listening to," using very familiar programming to make sure the end result is acceptable to all intended users.

Real-time analyzers (RTAs) are capable of obtaining a reading of the whole spectrum in one shot. They are able to follow changing signals and also to provide a graphic readout. This is *way* faster than sequential and even swept analysis. This is all possible because RTAs have a set of filters that match ISO center frequency and ANSI bandwidth standards. In short, a signal is injected into the system, amplified, picked up by a calibrated microphone, and then put through a bank of filters that divide the signal up into multiple frequency bands. These are then scanned, and their levels detected, averaged, and presented on some form of display for ease of viewing and adjustment. This is done very quickly, just as it happens in real time, hence the name.

The speed at which these results are generated made it necessary to have a display that was capable of constant updating while providing ease of viewing. In the early days, this screen was a cathode ray tube much like the ones used in old television sets. These were bulky, delicate, and expensive, but compared to sequential analysis, they were a joy to use and these early RTAs gave very accurate readings of the audio spectrum. By the 1980s, with the use of smaller and less expensive components, such as integrated circuits and less bulky and cheaper LED matrixes, the cost of RTAs dropped radically, even though many of these included microprocessor control and memory functions!

Today RTAs have many different features and capabilities. Most folks consider these as use-once devices unless acoustical parameters in the environment have been changed. Therefore, many studios use one just for a day or two to take measurements and make adjustments. However, as a continuing display tool, a spectrum analyzer can be remarkably helpful. This may be obvious when performing maintenance and calibration routines, but I loved to have one all wired in while I was recording and mixing too. Look, I can walk over to a 31-band graphic EQ during a live show and pull down the exact fader required to reduce even a slight amount of ring. My ears have been trained for over forty years to discern frequencies, but in the studio, when my attention may be focused on one particular instrument, some other sound, or worse yet, a lack of sound may slip by unnoticed.

After having done a good deal of recording with a Klark Teknik DN60 RTA sitting right up on the console's meter bridge, I can understand why some manufacturers outfit their consoles with spectrum analysis capabilities. I can honestly say that its use not only improved the overall sound, but also increased my ability to perceive spaces in the content. I also found that while clients were at first amused by the "light show" movement of the display, they soon started adjusting the frequency content of their instruments' sound and toned down the excessive use of effects devices. The DN60 featured memories, peak hold, a built-in switchable "A" weighting network, microprocessor control, variable response time, and much more. It provided its own signal output in the form of a quality pink noise generator. Output was via an XLR panel connector, and its level was continuously variable (via a rear panel pot) from −6 dBm to +2 dBm.

Gold Line makes a complete range of test equipment for sound contractors including: real-time analyzers, SPL (sound power level) meters, frequency counters, sinewave generators, gated pink-noise generators with timed pulses, decibel meters, impedance meters, a distortion analyzer, and systems that have 30 memories, a printer port as well as RS232 interface. This test system made up of individual components allows you to go with only what you need, but its extensive range also includes a RT 60 reverb timer, loudspeaker phasing, as well as frequency-response measurement tools.

They even give you the option of measurement microphones at two price levels. It can take a lot of test gear to get and keep a monitoring system honest.

I'll never forget the first time I used a Gold Line portable 31-band, third-octave, real-time analyzer. You draw an outline of the walls, ceiling and floor; pump pink noise through the monitors; and walk around the room pointing the microphone into suspected areas and mark down all problem positions with frequency notations on the drawings. I was then able to say to the studio owner: "You have a 4 kHz-flutter problem between these two surfaces, a low-frequency peak at these locations, and a phase dropout at the listening position due to reflections from these locations on those two surfaces."

Later with the Neutrik strip chart recorder, I could hand them hard copy. Then with the Klark Teknik, averages of multiple reverberation decay curves were added! A newer DN600 model RTA from Klark Teknik offered both $\frac{1}{6}$- and stereo or dual $\frac{1}{3}$-octave analysis plus displayed sum and difference and had built-in A and C scale filtering. The display could be set for peak or average, and it had a peak hold function. Its internal signal generator produced sine and pink noise signals, and had tone burst (for reverb timing) and sweep capabilities. The DN600 handled reverb timing without the need of an optional add-on. It could store 32 memories and also had a printer port for direct printer output, but to me, the most amazing addition was this unit's ability to directly interface with the DN3600 programmable dual-channel $\frac{1}{3}$-octave equalizer. Hold onto your hats, because the combination of these two devices equated to: automatic correction of equalization settings!

But everyone should shoot a room the old-fashioned sequential way at least once! The insight gained as far as the benefits of microprocessor control, real-time averaging, and synchronized sweeps will clue you in on why we were happy to spend thousands on test equipment that had these capabilities back in the 1980s. The fact is, that today, some inexpensive software along with the price of a sound card and a microphone can turn your laptop into a real-time analyzer, which can put you in the game. However, attention to detail still separates mere amateurish attempts from the kind of precision measurement that provides conclusive and reliable results.

Measuring Reverberation Time

Measuring reverb is pretty much the same as measuring frequency response, but now your speaker output levels for each frequency tested not only have to be equal, they also have to fit into a fairly narrow dynamic range because too low an output may not trigger the measurement and too high can cause false triggering. These days this is not usually a major problem because most reverberation timers automatically turn the noise source on and off and provide the correct timing synchronization.

Reverberation timers are usually composed of a signal generator that outputs a pulse, spark, pink-noise burst or continuous sign wave (which, once it has built up in the room, is then cut off), timing circuitry, and a synchronizer that locks the two together. A microphone input and preamp completes the package. Here you are checking to see if the decay is unusually long or short at a particular frequency. Some instruments also provide reverberation decay curves across the complete audio bandwidth.

A controllable electronic noise source is used as a signal for these measurements. This signal is fed into a power amplifier and loudspeaker to bring it up to a level above the room's noise floor. Impulse noise as from a starters' pistol or a punctured balloon can be used as the test signal, but the noise they produce does not contain equal energy at all frequencies. Furthermore, the impulses generated by these sources can be variable to the point where they may not trigger the test instrument to capture the data reliably, which is obviously essential to accurate measurement. Because reverberation time varies with frequency, these measurements are usually taken in one-third or at least octave bands. These tests are performed in real time simply because a large number of measurements must be taken at a number of different locations in the room to provide an average result that is representative of the whole room. At least a half dozen measurement positions are required, and this can end up meaning dozens upon dozens of tests. Some real-time analyzers not only turn the noise source on and off to provide the correct timing, but their internal microprocessors also gather, analyze, and store the data. Hey, studio time = $ money, so interruptions to the work schedule must be kept to a minimum.

The Klark Teknik Optional RT-60 Reverberation Timer

The pink-noise source in the DN60 RTA was driven by a line amp that was gatable. The gating was controlled by a microprocessor, and it enabled very accurate timing of the output on and off switching. Reverberation decay time is determined by first removing a signal and then measuring the time it takes for that signal to drop 60 dB. To do this accurately, the sound source level might have to be set very high level (depending on the level of background noise). To ease this measurement process, here the decay time is calculated from a smaller decibel drop. Through an internal "communications" card, the RT-60 optional reverb timer could control the DN60's microprocessor.

Once the RT-60 was hooked up, it was included in the DN60's self test, and this dual unit was capable of measuring reverberation time. It sampled the decay every 550 microseconds, averaged 16 of these samples, and stored them until the cycle was finished, at which point it had 790 8-millisecond samples. The DN60/RT-60 could measure a decay from 0 dB down to −30 dB, or any range in between, with as little as a 2-dB drop and calculate (extrapolate) a measurement for a 60-dB drop. The parameters of the decay curve could be changed after the measurement without affecting the original measured value.

It is normally necessary to take many tests with any reverb timer and average these together for more precise readings. This dual unit did your averaging (accumulating) for you up to 32 readings and could measure and accumulate either the full bandwidth or any individual band. Since the RT-60 controlled the microprocessor, some of the functional parameters of the DN60's switches were changed. Here the peak/hold switch now showed at which frequency band the measurements were taken. The peak/average switch controlled the pink-noise turn-on. This was important to saving your ears, as it allowed you to start the noise just before taking a measurement. The RT-60 was also very user-friendly in that it told the operator via the readout when the unit was properly set up for measurement. It indicated running and calculating, as well as accumulating modes, and more. Once measurement was completed it showed a graph of the decay curve. Here the ⅓-octave frequency bands were converted into time columns. The display's time parameters could be changed after the measurement to one of three settings: 16, 64 or 208 milliseconds per column. This decay curve could also be stored in the DN60's memory. When recalled, it would show if accumulation was used, how many accumulations were made and the frequency band measured; yet the window and time parameters could still be changed. Furthermore the RT-60 accumulator could be used in the spectrum analysis mode to average up to 32 frequency response measurements. All this gave the operator a very powerful tool to work with. I was able to check studios, control rooms, as well as time digital reverbs, tape reverbs, and chambers, all with ease.

However if all we need is a clue to give us insight into possible room anomalies, simple reverberation time only tests can often do the trick. A good example of how easy reverb measurement can be is provided by using Gold Line's GL60 reverberation meter. Here the microphone, timer, and display are built into one small 5″ × 5″ × 2¼″ box, and because it runs on two 9-volt batteries, it is completely portable. This unit mates with Gold Lines' PN3A pink-noise generator which is only 2¼″ × 4″ × 1½″ and runs on a single 9-volt battery. In "continuous" operating mode its output is 95–97 dB (ref 110 dB SPL). In RT60 mode it puts out 11-second bursts with the time off in between each burst being variable from 10 to 150 seconds. These two units will run you only about $600.00.

Okay, say you have excessive levels at 125 Hz. Now with close micing, we can see if the reverb time at 125 Hz differs in various parts of the room. In other words, where are the best places to add acoustic corrections? The GL60 takes measurements at six different frequencies: 125, 250, 500, 1000, 2000, and 4000 Hz. Sound familiar? That's right, the same frequencies at which absorption coefficients are rated. This ends up being pretty handy in that, should you have a problem at 250 Hz (or at any of those frequencies for that matter), you simply look for a material or acoustic panel that has a narrow band peak in absorption at that frequency.

Acoustilog IMPulser

Folks used to input pulses into their system, mic the room, and monitor the output on an oscilloscope. They would vary the frequency or timing of the pulse train until the resonant frequency caused a flutter echo. When they measured the distance between the flutter peaks, they had the timing. Multiplying that by the speed of sound and dividing the result in half gave them the distance. Sounds a bit complicated? Well, it could be quite cumbersome with all the wiring, microphone phantom power supply, preamp, amp, speaker, and scope, but a company called Acoustilog, which was run by an acoustician named Al Fierstein, used to make a small portable box called an IMPulser. Most people purchased it because it rendered *absolute* confirmation of correct: system, speaker, individual driver, and even microphone phasing, but outside of a mic and a scope, it covered everything needed for these tests all by itself! It generated a positive pulse, which had variable width (40 Hz to 10 kHz), rate (0.3 to 10 pulses per second), and level. You simply injected the pulsed signal into the system, and used a mic to pick up the speaker/room output. The IMPulser not only provided the phantom power to the mic but also amplified its signal. This was then sent to the scopes' trigger input, and the direct pulse output was connected to the vertical input. The result was an irrefutable display, showing you either a positive in-phase (above center line) or negative out-of-phase (below center line) pulse. But when used to test the acoustics of a room, you could actually *see* flutter echoes, slap-back delay pulses and ringing waveforms (Fig. 3.1).

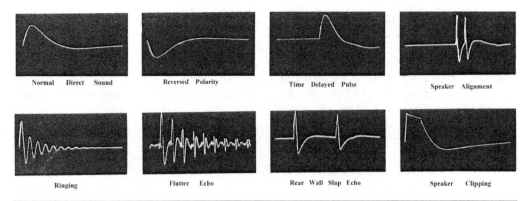

FIGURE 3.1 Examples of Acoustilog IMPulser oscilloscope signal traces.

Using the scopes' screen graticules (divisions) and varying the time base, you could get a pretty good idea of the timing and then do the math, but Acoustilog also made a fairly inexpensive reverberation timer that saved you the trouble. Finding the boundary that is causing a flutter, slapback delay, or phase cancellation is a lot easier when you know its distance from the microphone. The whole idea is to know exactly what you are dealing with before you start to make adjustments to your room or purchase acoustic materials or premade panels. Getting everything right from the start is crucial to keeping cost-inflating mistakes to a minimum. This is why I like using both reverberation time and frequency response measurements and also why we go through all the trouble of working out the mathematics to get an idea of the *expected* reverb time by using the average absorption coefficient of all of the room surfaces, which is well covered in Part II. So now you will have a very good idea of how to perform all three checks.

Equipment Rental

Unfortunately I cannot tell you where to rent real-time analyzers and reverberation timers on the cheap because these devices have to be calibrated every time they are sent out. I put a good deal of time into researching this possibility, and except for the following two came up blank. If you reside in England, you can "hire" a top of the line Nortronics NOR 118 handheld RTA/reverberation meter for 32 pounds ($51.38) per day, minimum of four days plus shipping and, what I am sure are, some hefty insurance fees for this multi-thousand dollar device. States-Side, Acoustical Surfaces, who sell the Phonic PAA3 RTA/reverb timer for about $500 will also rent one to you with the USB interface for $250. The interface is important here because it is much easier to view 31 bands of information on a computer monitor instead of the small LCD screen. Bottom line? I might not recommend these to the small studio owner simply because the cost, while fair, equals the price of a premade acoustic panel which might out and out solve the problem. But do go online and check out what is arguably the "state of the art" Nortronics NOR 118 just so that you have some "dream" device to compare to prospective measurement equipment at noisemeters.com. The Phonic PAA3 is all inclusive and you can check out the full range of its capabilities at http://www.acousticalsurfaces.com/soundlevel_meter/phonic_paa3.htm?d=30.

Acoustic Measurement with a Laptop

Now let's look at a very simple, and thus, commonly used room acoustics measurement setup that's a combination of a microphone, preamp, sound card, laptop, and software. Let's start at the software end. How about using freeware?

First download a free version of TrueRTA™ software. I have used this, and it is a very good spectrum analysis tool. The free version provides a pretty comprehensive generator, analyzer, and display combination. The generator can output sine, triangle and sawtooth waves. You select the frequency (5 Hz to 24 kHz) and level (minus 150 dBu to plus 20 dBu). The generator also outputs impulse and pink noise (up-graded versions, which you have to pay for, add white noise and sweep modes). On the analyzer side, the free version provides only an octave or 11-band readout, but you can still select the top-level reading to be anything between positive 10 dBu to negative 90 dBu. And the bottom of the scale can be set to anything between positive 10 dBu and minus 160 dBu. The high frequency limit of the display can be set from 50 Hz to 50 kHz and the low from 10 Hz to 10 kHz. Finally the speed can range from slow 20 Hz (as in twenty times per second), medium 40 Hz, or fast 80 Hz. Straight out I do not expect you to understand all of the above technical information, but hey, it's a free download so jump on the chance to check it all out. I guarantee you that in a fairly short amount of time you will have it all *down*.

Next is the computer. Most everyone today has a laptop with at least a headphone jack output and a microphone jack input (usually ⅛ inch). The generator output appears from the headphone jack, and the analyzer input is on the microphone jack.

Here we run into our first problem: the headphone output is stereo and the microphone input is mono. You *could* simply patch a stereo jack-to-jack cable from the headphone out to the microphone input. However, this would short circuit one side of the stereo output, and I am flat out against shorting *anything* out in audio. There are ⅛" stereo to mono adaptors, but they *combine* the left and right sides into the mono jack. The real deal here is to cut the end off of a ⅛" stereo to ⅛" stereo cable (leave enough cable attached to the "cutoff" end so that you can use that side of the cable for another adaptor). Now solder a ⅛" mono jack to the cable, omitting the "ring" conductor. You can then plug the stereo plug into the headphone and the mono plug directly into the microphone jack, and work the input and output variables of this software to the bone until you have it all figured out.

Then you get to take the signal generator output from the headphone jack, add an adaptor if needed to the ⅛" mono plug and feed the signal into anything you want. An equalizer is a great place to start. Feed the output of the EQ into the analyzer (microphone) input, and you will be able to *see* exactly what that EQ is doing in terms of an octave or 11-band spectrum. Right quick you'll pick up on the need for more bands, but you'll have gained a huge insight into spectrum analysis, and also understand the need for more professional analogue-to-digital conversion that can be had from a dedicated sound card.

But in order to perform room analysis, you are going to need a microphone, a preamp, and a power supply. So go online and check out a bunch of acoustic test microphones just so you are aware of the differences. Most important here are a flat as possible frequency response, low noise, and the most sensitivity you can afford. Flat frequency response and low noise are straight ahead, but sensitivity should be explained.

Although you are simply measuring levels above and below a reference level, you are *always* dealing with background noise especially when it comes to reverb time so let's get it straight once and for all. Okay, typical analog studios had to have a noise floor that was at least 80 dB below reference level because the audio equipment was only 70 dB down. Fat chance! They were more like minus 30 dB. In today's digital world, the equipment is at least 110 dB down, but studios are seldom any quieter than minus 35 or 40 dB. In order to properly take an RT 60 reverb measurement the signal must be at least 60 dB above the noise floor, and practical levels should be closer to 80 dB. That's quite a range minus 30 to positive 80 and not all microphones have a range of 120 dB. But that is not really necessary. 80 dB may be a little tough, but a 100 dB range will usually work out just fine. The following are just a few of the many sources of microphones used for acoustic testing. Some of these products are way out of your price range, but it's always a good idea to check out the best to get a handle on what you're giving up by paying less.

ACO Pacific, Inc., acopac@acopacific.com

Audix Corp., www.audixusa.com

Behringer, www.behringer.com

Beyerdynamic North America, www.beyerdynamic.com

Bruel & Kjaer, www.bkhome.com.

Crown International, Inc., www.crownaudio.com

Earthworks, www.earthwks.com

Gold Line, www.gold-line.com

Josephson Engineering, www.josephson.com

Rane, www.rane.com

When you are measuring sound pressure, there is always some ambient sound so what you are actually measuring is the difference between the ambient and the introduced sound pressure in terms of Pascals or Pa (1 Pa = 94 dB). Sound pressure level (SPL) is compared to 20 microPascals (one millionth of a Pa) root mean square (RMS). RMS is like the ultimate average. Here multiple samples are taken, each is squared and the square root of the sum of all of the samples is divided by the number of samples. Normally these tests are carried out in a free field environment, as in an anechoic chamber. Free field means *no* reflections, but you will be at best performing point source (close micing) tests where the room has less of an impact on the results.

Test microphones are usually pressure sensing. They have an electrical power input that is affected by the displacement of the diaphragm which then outputs an electrical value. At high frequencies, the size of the microphone diaphragm can equal the wavelength of the sound being measured. That is why microphones with $\frac{1}{4}$" diaphragms have a more even high-frequency response than microphones with $\frac{1}{2}$" diaphragms. Finally you need to power these microphones, and those that utilize common phantom power make things nice and easy, if your console provides 48 volts at the microphone inputs. If not, you can opt for a dedicated phantom power supply, or just pick up a $50

Behringer mixer and get the phantom and the preamps in a package that will come in handy for a slew of other audio chores. The point is that you can start off by jumping in and getting your feet wet for free, and *this* is the best way to learn how to perform frequency-response tests on rooms as well as on just about every piece of electronic audio equipment that enters your life. (These tests do not apply to transducers because microphone testing requires specialized calibrators and speaker testing requires an anechoic chamber.) Believe me when I say this whole proposition wasn't even a *dream* possibility when I was coming up.

FFT, TDS, and TEF Measurement

While ⅓-octave analysis provides a wealth of information, it leaves out the very important ingredient of time. It will tell you what the frequency is and how much of it there is, but cannot tell you when it occurs. Temporal distribution or a sound's place in time is important psychoacoustically. Time-delay spectrometry (TDS) can give you the levels of *only* the reflections, thus providing mathematical hints as to where they originated. Fast Fourier transform (FFT) transforms the time domain to frequency domain.

Let's start with a bit of history. In the early 1960s, a scientist named Richard C. Heyser was working at the California Institute of Technology's Jet Propulsion Laboratories. Jets were still new then; just ask any old-timer about the experimental aircraft like the X-15 and the test pilots who became our heroes as kids. Mr. Heyser was calibrating the frequency and phase response of an open-reel tape recorder and the time lag between the outputs of the two heads (record and reproduce) was giving him some trouble with his measurements. So he just went and assembled a bench-load of equipment to compensate for this offset in time. He found that his new test method not only corrected the time offset, but also canceled out the harmonic distortion. He realized he had something there and kept working on it, and described it in a 1967 AES paper.

Here's my brief synopsis of a time energy frequency (TEF) test procedure as used when testing speakers. A sine-wave generator is used to produce the test signal that is fed through the monitor system, and a mic picks up the output from the speaker. In addition, the test system utilizes a tracking filter in sync with the generator and a delay module, which offsets the input timing, to correct for any delays incurred by the signal, such as propagation through air. The bandwidth and the time delay can be varied to allow for a test of only the direct sound, just the early reflections, or both. Even the earliest of these devices were quite informative to say the least. To say that they completely changed acoustic measurement would *not* be an exaggeration.

Due to the inherent noise rejection of TDS measurements, reverberation decay time can be calculated and displayed even when a high amount of background noise is present. TEF measurements can provide close to anechoic results in any room. A short time segment of a periodic signal can be sampled for a complete or whole number of periods, and from TDF the FFT can be extrapolated using one of the many "windowing" techniques available. This tiny little area of measurement has acronyms for days, so here's a partial list that will help keep you somewhat afloat.

FFT Fast Fourier Transform

TDS Time-Delay Spectrometry

TEF Time Energy Frequency

FTC   Frequency-Time Curves
ETC   Energy-Time Curve
EFC   Energy-Frequency Curves
ALC   Articulation Loss of Consonants
STI   Standard Speech Transmission Index
RASTI   Rapid Speech Transmission Index
RTA   Real-Time Analysis
RT-60   Reverberation Time Decay of 60 dB
NC   Noise Criteria
NLA   Noise-Level Analysis

Back to some more history.

Mr. Cecil Cable picked up on the potential of this measurement procedure and in his 1977 AES paper, he described his research into comb filtering caused by early reflections. In 1979 Crown International (the amplifier manufacturer) began manufacturing their TEF system (a TDS analyzer in combination with FFT) that analyzed time, energy, and frequency. By 1980, Mr. Don Davis was not only teaching "Syn-Aud-Con" (synergetic audio concepts) students how to use a TEF device, but along with Chips Davis, designing and building control rooms based on the early reflection/comb-filtering information compiled by this system.

The resulting live end, dead end (LEDE) control room design had the psychoacoustic ability to make the control room boundaries seem to disappear from the listener's perspective. The only way I can describe the one "true" LEDE control room I've had the opportunity to work in is that your perspective is not one of being in a control room listening to what is going on in the studio, but one of actually *being* in the studio listening to what is going on. The acoustics of the control room did not come between the listener and the sound. The maximum length sequence (MLS) of signals is used with FFT to provide frequency response, speech intelligibility, and time analysis. With random signals such as white or pink noise, the signal must be sampled over a period of time in order for the analyzer to generate accurate readings. A pseudorandom signal can be made up of a group of sounds that display any appropriate parameters needed to perform a specific test. This signal has a set length and is sequentially repeated.

Level versus frequency measurements are known to be an important indicator of a system's capability to produce sound adequately, but time or phase versus frequency measurements are not given the same importance. In fact, many people feel that phase inconsistencies are inaudible. Mr. Heyser's answer to this was to point out the fact that with a system's low-frequency driver positioned relatively close to the listener and the high frequency driver placed a mile back, this system could actually achieve a flat frequency response if the high frequencies were driven with enough level. In this case, the phase response would give a better indication as to how bad it actually sounded. Beautiful response, no?

In source-independent measurement (SIM) FFT analysis, the input is fed to the analyzer and the system. The system's output is then compared to the input signal. Thus, the test signal is "independent " of any restrictions except for being broadband in nature. Therefore, like the dbx RTA-1, it is possible to use music program for the test signal. Here the FFT's "windowing" capabilities allow the operator to focus in on the test results down to a single cycle. By gating the input signal, different sections can be

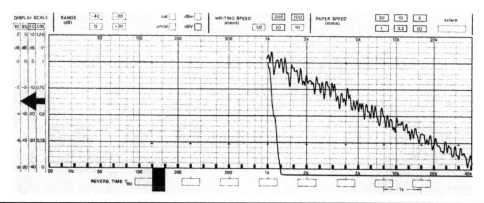

Figure 3.2   Neutrik Audiograph 3000 strip chart recording of the shortest and longest plate presets. Each vertical division equals ⅓ of a second, which is also the shortest reverberation time. The longest is 5 full seconds!

analyzed, and the display can show how frequencies change with time. Short gating times provide superior time resolution, and longer gating times give good frequency resolution. While a wealth of information is provided by the Neutrik 3000 strip chart recordings (see Fig. 3.2), the 3-D plots produced by these tools are no less than spectacular (see Fig. 3.3).

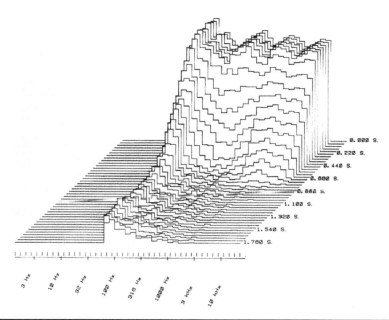

Figure 3.3   3-D level vs. frequency vs. time display. The only thing I remember about this 3-D plot is that I used it as an example when I taught at the Institute of Audio Research. But I *can* tell you that the horizontal and vertical planes are just the common level and frequency scales, but the "depth" is the time starting from 0 seconds at the back and ending at 1.750 seconds at the front. As you can see these plots provide a ton of information.

CHAPTER 4
Vibration

Resonant Frequency Vibrations

Vibrations that are at resonance produce a very high amount of movement, which can often cause damage to very expensive machinery, possibly leading to a complete system failure. Should this happen, the result could be manufacturing downtime. Therefore, a great deal of research has been devoted to this problem, and it behooves us in the field of audio (where large-scale funding for these types of studies is usually not available) to take advantage of all this outside work. In this case, it was found that the effective mass of a vibrating object as well as its stiffness determines its resonant frequency. Now stiffness can be changed by the use of damping and vibration isolators, but should this prove impractical cost-wise, you now have a second, often less-expensive and easier-to-facilitate method of reducing the vibrations simply by experimentally adding mass to the offending part of the system until the vibrations have subsided sufficiently. Thank you very much!

Ambient control room noise is not only due to your HVAC (heating, ventilating, and air conditioning) system, but can also emanate from your electronic equipment as well. These problems can often be cured with a little damping material. Here, a foam of a non-silicone formulation will avoid off-gassing and breakdown (where-in foam turns to sticky dust). Once, during a control room checkout, I hunted down an ambient noise problem to a vacuum fluorescent display that was vibrating up against the back of an outboard processors front panel. The addition of four small pieces of a product called CONFOR® from EAR Specialty Composites to the display's corners completely ended a problem that everyone assumed was an electrical ground loop hum. Another time the same type of problem ended up being caused by an internal "whisper" fan that was rattling from a combination of a missing mounting bolt and a snapped-off blade. Same cure, but this time I cut small pieces from a sample of a closed-cell, Poron® foam from Rogers Corp. Besides having low outgassing and being resistant to temperature changes, chemicals, and abrasion, these urethane foams are no problem to cut and they have bondable surfaces, making them easy to glue in place. Like most closed-cell foams, they display good resistance to compression set, high energy absorption, and excellent all-around dimensional stability, making them suitable for use as gaskets, seals, shock and vibration mounts, isolation pads, and as in this case, vibration isolators for PC boards or other internal electronic parts that might be causing vibrations.

You can also cast your own small rubber vibration isolator parts. A manufacturer named Devcon produces an easily pourable, castable liquid urethane that cures to a rubber with a varying rigidity (durometer) depending on the additives you choose.

A company named Raxxess manufactures a sound isolation equipment rack. It has a clear plastic front door that allows you to view metering, power-on status lights, etc.; blower-aided intake air vents; vibration-isolated mounting rails; and an interior lined with 1-inch wedge-shaped foam. I haven't personally heard the results of parking offensive equipment inside, but it's a great idea for noisemakers that cannot be removed from the studio.

Vibration Isolation

A detailed discussion of vibration isolation is beyond the scope of this publication; yet since the idea here is to prepare the small studio builder/owner for handling the adjustments needed to alleviate problems as they arise, a general knowledge of the theory and terminology on the subject of vibration isolation and a basic understanding of the mechanics involved should prove helpful. There are only two ways of treating vibrating structures: vibration isolation or damping. Isolation separates a vibrating structure from the receiver, while damping reduces the intensity of vibration, thereby quieting down the noise problem. The qualities of damping include internal friction and viscosity. Once the energy is dissipated, it can no longer radiate. Thus, damping reduces the amplitude or the impact energy of the noise.

Damping materials are generally visoelastic, such as rubber or elastomeric plastics. To help in understanding the effects of damping, imagine hitting a piece of metal with a hammer. Now cover all or part of that piece of metal with a thick, heavy piece of rubber, and then hit it again. The sound produced will not only last a shorter length of time, but will have less amplitude. Damping is primarily effective only at or below the dampener's resonance so that a vibration correction plan will depend on whether the excitation frequency is less than, greater than, or equal to that resonant frequency.

Damping ratio is also a key factor since, should the system be overdamped, the response will be slow, causing its return to equilibrium to be sluggish. Yet if underdamped, the system response will be fast, but would continually overshoot the equilibrium position until friction finally caused the oscillation to die out. In between these two amounts lies what is called *critical damping*, and here the system returns to equilibrium as fast as possible without the repeated overshooting. It's all fairly simple, as most often you are dealing with something that is vibrating at its resonant frequency (meaning it is loud). You change its resonant frequency by adding material to it, just as you do when adding a second layer of sheet rock to a wall with a patchwork of loaded vinyl in between. While *any* material can be used to "dampen," those designed specifically for the task are usually "viscoelastic," meaning that they are capable of storing energy when deformed and depleting some of this energy through internal friction (hysteresis). These materials come in sheet form often with preapplied contact cement. Alternatively the vibrating part to be dampened can be dipped into a heated coating that solidifies on cooling, or material can be troweled onto its surface. Increasing the thickness of the material increases the damping.

As stated, vibration isolation separates noise sources from structures, thus reducing the vibrations radiating from the source. Basically all you are doing here is placing some form of flexible material between whatever is causing the vibration and whatever is receiving and transmitting the vibration. Simply put, something is causing the disturbing vibration, and its frequency is one of the most important factors in determining what kind of vibration isolator to use. The source of the vibration is generally easy to locate as the resonant frequency of the vibration is directly related to it. For instance, if the cause of the vibration is a running fan's eccentrics, then the vibration's frequency will be the same as that of the fan's rotating speed. This is called harmonic excitation. It is extremely important to insure that the lowest vibrating frequency is used when determining the correct isolator. Another important factor is the natural frequency of the isolator itself, which is determined by its own spring rate along with the weight of the load that's being supported. No need to be alarmed here, as most manufacturers of vibration isolators list the natural frequencies of their products. You should look for an isolator that has a natural frequency that is at least one-third that of the vibration you are trying to isolate. For that matter, the lower the natural frequency of the isolator, the more pronounced becomes that isolator's effectiveness.

Your first step in dealing with correcting a vibration problem (just as with airborne noise control) is starting your approach by first considering the three main points of attack: right at the source, along the path, or at the receiving end. Unlike a situation on a noisy factory floor, isolating the receiver, meaning a studio control room, may not be practical. The path may still be a possibility, but if it is through the building's foundation that may be way out of the question and usually not the most cost-effective approach. So as always, it's often best to start right at the source of the problem. First, see if you can alter the source by making it more structurally rigid. Try balancing it, possibly changing its mass, or "detuning" it in any other way. For instance, a small vibrating motor that causes a maddening amount of noise when it's placed on a wobbly, wooden stand, produces an almost tolerable amount of noise simply by affixing it to a thick concrete floor. The improvement is obvious even before utilizing any vibration isolation mounting. Think of a washing machine whose foot lengths are adjusted incorrectly. "Hello, spin cycle!"

Vibration Isolators

Decoupling the noise source from the path via the use of vibration isolators is the usual method used for correcting a vibration problem. In general, vibration isolators can be broken down into three categories: metal springs, elastomeric mounts, and resilient pads. The advantages of metal springs are as follows: they are resistant to environmental factors, such as temperature, corrosion, and solvents; they do not creep or drift out of place; they are designed to handle maximum deflection; and they work well at low frequencies. The disadvantages include their transmission of high frequency vibrations in addition to those at their resonant frequency, and that they can be most troublesome if allowed to take on a rocking motion. You can get around these problems fairly easily. The lack of damping in the metal springs can be corrected by using dampers along with them. Rocking can be avoided by the use of an inertia block that weighs one to two times the amount the springs are isolating. High frequency transmission can be blocked by simply selecting and adding rubber pads in series with the springs.

Elastomeric mountings are most often made of natural rubber or synthetic rubber-like materials such as neoprene, which also has the advantage of being more desirable in harsher environments. In a controlled environment, natural rubber is one of the best and most economical isolators, as it has inherent damping and is perfect for machines that operate near resonance or pass through resonance when they start and stop. Rubber used for isolation is rated in durometers (30 being soft while 80 is considered hard). Often, when dealing with very specific or complicated problems, multiple isolators of differing hardness can be used to take advantage of the desirable properties of each.

Isolation pads include materials like cork, felt, and compressed fiberglass. These can be obtained in sheet form, cut to size, and stacked to achieve varying degrees of isolation. Cork, like fiberglass, is not susceptible to solvents and can handle a wide temperature range. On the other hand, felt, which is made of organic material, can be very sensitive to solvents. Fiberglass isolators are not very useful at frequencies below 20 Hz, and cork materials do have a tendency to crumble with age.

Inertia blocks, such as isolated concrete slabs, are very useful in vibration isolation. They, along with steel structures of sufficient mass, limit motion by overcoming the inertial forces that are generated by the equipment mounted on them. They lower the center of gravity, thus adding stability, and increase mass, thus decreasing the vibration's amplitude along with minimizing any rocking motion. In addition, they act as noise barriers between the equipment mounted on them and the floor below them. They must be mounted on isolators, and they can be fairly cumbersome and costly to construct, especially after the studio has been completed, unless they are used in one of my favorites places—right outside the studio complex under the HVAC equipment.

How the Pros Do It

Professional vibration engineers begin dealing with a vibration problem by starting with a very simple model called "a single degree of freedom," which takes into consideration only the vertical displacement of the system. This very basic approach is used to point them in the right direction as far as seeking a cure to the problem. Then they also take into account: system mass, damping constant, spring constant, position, velocity, acceleration, inertial force, viscous damping force, linear elastic force, and external forces, including impulse and repeated (like sine wave) or random forces. If all this is much more than you want or need to get involved with, take heart, you've got a whole slew of allies at the ready to help you out. Start with the offending machinery. It's no easy task to weigh a several hundred pound AC unit or figure out its compressor-pump's rotation speed, and the same goes for a fan mounted in a wall just below the ceiling. Yet, copying down the manufacturer's name along with the unit's model number is a lot easier and don't be surprised to find that they have at the ready the exact weight and rotation speed information you seek along with isolators for that particular device in stock as part of their product line.

More Help

Even if the manufacturer has been out of business for several years or might as well have been because the help they are able to give you is nonexistent, you are still not alone, as there are all kinds of trade organizations just falling all over themselves trying to provide you with very helpful information. I kid you not! It is important to take advantage of this wealth of information because it's often based on many years of research plus a ton of practical experience, and it is all readily available. As an example, when dealing with air conditioning, vents, or any HVAC related noise, it's wise to start by contacting organizations, such as the Air Movement and Control Association International, Inc., 30 W. University Drive, Arlington Heights, IL 60004, (847) 394-0150, (847) 253-0088, amca@amca.org. Believe me, there is an almost unlimited number of trade organizations that make available information having to do with the control of noise. You can prove this to yourself simply by spending a little time in the local public library, especially if you get the help of the librarian on staff, and most of it is free for the asking! For a start, try the National Council of Acoustical Consultants (NCAC) at www.ncac. com, the Acoustical Society of America (ASA) at www.acoustics.org, the Institute of Noise Control Engineering (INCE) at www.allenpress.com/inceusa/, and the American Society of Mechanical Engineers (ASME) at www.asme.org.

The manufacturers of vibration isolators themselves are also a great source of help in the search for the right product to alleviate a particular problem, as they are quite frank and honest about their products' capabilities. Let's say for instance you need isolators suitable to use in isolating your large, in-wall, soffit-mounted speakers from the wall itself because otherwise you'd end up having vibration-induced low-frequency noise being picked up by microphones in the studio whenever you monitored at loud levels while recording (it *does* happen). A manufacturer's catalog, such as the one from Barry Controls, would steer you away from their series 1000 through 4000 Cupmount mounts. While offering good protection against shock and structure-borne vibrations, the manufacturer plainly states that these units provide good vibration isolation characteristics at frequencies above 40 Hz. This is not exactly your best choice if your speakers' low frequency response can handle 32 Hz or extend down to even below 20 Hz. Not convinced? Okay. How about a manufacturer steering a customer away from one product to another less expensive solution because the latter would handle the situation better? Well, it actually happens. I called the Airspring division of the Firestone Industrial Products Company to inquire about one of their Airmount vibration isolators, which reduce structurally transmitted noise very well. The plan was to float a mobile control room on top of computer-room type of flooring. The size of the floor was just under 150 square feet, almost 10 feet × 15 feet. The floor panels were 2 square feet each and totaled 36 in number. In order to place an isolator underneath each support, 48 mounts would be needed.

Next came a little time going through that manufacturer's product line catalog, which contains all the mathematical equations and data in the form of graphs, tables, and charts needed to make the proper selection. Later when I called their distributor to ask about pricing to get the complete picture, they asked for the technical requirements involved (vibration frequency, load, and size) and came back with another alternative. You guessed it, the mount they recommended I use did cost more. Almost twice as much in fact. However that wasn't their main point. What they stressed was that I go with a single common floor, either by supporting the computer-room flooring on a continuous platform or on something like steel I-beams. This would not only give the structure more lateral stability but also allow for the use of fewer larger-sized mounts. While the recommended mounts were almost twice as expensive, the alternative plan required only one-third the number of mounts, thus bringing the actual cost down by a third! The trade-offs? None. The original mounts would support 48,480 pounds inflated at 60 psi and would render room isolation down to 8.28 Hz. The recommended mounts would support the exact same 48,480-pound load filled at the same 60-psi pressure and would isolate down to 7.2 Hz. These folks ended up saving me money. I was skeptical so I double-checked all the above figures. When I recently called again to confirm those figures nothing, including the pricing, had changed.

Oh I should mention that in situations like this I never like to use the Internet, at least after the initial contact. A phone call may cost a little bit more but that one-to-one, easy give and take, means less of a problem explaining yourself, and it's so much better getting instant feedback from a real live human being. *That* cannot be duplicated via computer screen read-out. A hand-written note of thanks may not be quite as fast as E-mail, but which would you prefer receiving? All this free information and help that's available makes a huge difference for the musician or any non-acoustician who is trying to build a recording facility capable of delivering a professional quality product without breaking the bank. It's one thing to be aware that something is not sounding right, but it's a whole different thing knowing how to go about achieving the correction. It seems as if all the development that has taken place over the past 25 to 30 years in the audio industry has been leading up to a whole new non-secretive approach to acoustics. You'll never understand what a big deal this is unless you once tried, back in the late 1960s, to explain to a bona fide professional acoustician/noise-control engineer you hoped to glean some info from that: No, it would not be possible to "simply" decrease the electric guitarist's amplifier level or "for that matter remove" one of the two 4 × 12-inch speaker cabinets that comprised his stack, and "Yes, sir," the electric bass guitarist did feel it was appropriate to "thump" his fingers against the instrument's strings, thereby causing those low frequency shock waves of vibration.

CHAPTER 5

The Control Room Monitor System

The control room is a *very important component* of your monitor system! In order to consider a recording studio control room an accurate tool for judging the integrity of the music being reproduced as well as for dialog intelligibility, many requirements must be met. Among these are: having a short reverb time, high sound isolation from outside sources including other rooms in the facility, and a minimum amount of ambient noise emanating from the equipment within the control room itself. The height of recording studio splendor arguably occurred during the 1980s. Multi-room facilities were common with rooms interconnected both physically and electronically. No expense seemed excessive and many studios were built to the highest orchestral acoustic standards, meaning 16 foot ceiling height and 6,000 or more overall cubic feet of space. While those days are now long gone, it is a good idea to have a mental picture of what went into building those facilities, even if only to steal an idea or two. Way-back time again!

Ziiiiiiiiip .... Hey, what's behind that door in the rear wall of the control room? The door opens inward, so there must be some room back there. Check it out; this whole wall is one large diaphragmatic absorber. Its widely spaced mounting studs are sitting on rubber blocks and it consists of $\frac{1}{4}$ inch plywood with denim wallpapered to the side facing into the control room. I once saw the same kind of approach to eliminating bass buildup using a $\frac{1}{8}$ inch sheet of plywood that had been covered with thin industrial grade carpeting. I measured the room before and after installation and its absorption outside of a low-frequency peak came pretty close to being even with frequency. As with Helmholtz resonators, there are calculations that help predict the effect diaphragmatic absorbers will have on a rooms frequency response and reverberation decay time. These take into consideration not only the square footage of the panel, but also its total weight. They do not provide absolute predictions of the results, but can put you in the ballpark as far as the resonant frequency and bandwidth.

Hey, this rear bass trap must be five or six feet deep! That's a lot of space to give up, but it does free up a good deal of control room space by not having those two multitracks and two stereo tape recorders in there, not to mention the noise and heat they can add to a room. Speaking of heat, these folks had a great idea. They've used a length of flexible duct to tap into the suspended ceiling return on the far left and dropped it down to almost floor level. On the far right, they've added a large ceiling vent to complete the HVAC return routing. So for no cost, they have added slightly warmed-up air conditioning to their equipment room.

And just as I would expect, they have a video camera with a wide-angle lens aimed at the tape recorders. You can do everything "by the book": align your recorders by first calibrating each output, then adjust the meters to read 0 VU, and check that every console return has exactly that same reading. But the first time you hear something that does not add up level-wise, or if you're a little concerned about how an effects level is going down to multitrack tape, you'll just have to see those multitrack meters. A simple switch will give you a bird's eye view right there on the video monitor located between the two main monitor speakers. If you're smart, you'll wire into that video monitor's auxiliary audio input so that you can use its little speaker as a mono reference as to what the program material will sound like through television and transistor radio speakers.

From the looks of those two structures on the back wall, I'd say they were being used to box-out a couple of windows. So it seems this machine room is also used to provide additional sound isolation from the outside. This room's depth actually provides enough space to walk around the equipment racks a full 360 degrees. That's what I came in here to look for. I was curious as to why I did not see any ⅓-octave equalizers for the control room monitoring system, and here they are. This studio is using a Klark Teknik DN 30/30 dual 30-band graphic equalizer, but it could easily have been a White, dbx or these days a Fostex, Rane, MXR, Roland, or any one of a host of graphic EQs now offered. As long as it conforms to ISO center frequencies and ANSI bandwidths, while adding no noise or other anomalies to the program material, it doesn't matter. However, every graphic EQ used on a monitoring system must be set up using a spectrum analyzer. Do not trust this task to your own or anyone else's ears. Ziiiiiiip....

There are no quick fixes in acoustics and no two rooms are the same. Solving acoustical mistakes made after the room's design and construction stages can cost more to correct than the original construction (as now all previous work must be disassembled). The fact is, if you are thinking of spending thousands of dollars to construct a control room, you should start the project by getting the name of an accredited acoustician from either the National Council of Acoustical Consultants (a not-for-profit organization located in Springfield, NJ) or the Acoustical Society of America (asa@aip.org). Right here at the beginning is where the mathematical computation and the study of the room modes as well as the material selection according to absorption coefficients begins. Once the project is underway, repeated electronic tests and measurements will be used to confirm that the sound transmission and acoustical specifications are being met.

But if this were the case, (i.e., you have thousands of dollars to spend) you would more than likely not be reading this book. So in keeping with the "budget" theme, let's dive into the prospect of *correcting* the acoustics of inadequate studios and control rooms. Guess what? Right here at the beginning is where the mathematical computations begin with the study of the room modes and the beginning of material selection according to absorption coefficients. But before we move on to Part II, a few more points need to be made.

Setting the Monitor System Gain

I've seen folks set amplifier input levels 6 dB below unity gain to allow for enough extra headroom so that the 10 dB peak levels common with music programming, would not cause distortion. Some studios adjust their power amplifier's input attenuators to the full clock-wise position. This is not proper gain-stage adjustment because it often requires the console to be set at a lower than optimum level in order to avoid excessive levels and distortion. Additionally, these low console level settings will place more noise (S/N ratio) from the console output amplifiers into the monitoring system. Low headroom and high background noise can easily be avoided by proper gain-stage adjustment. Amplifier input attenuation should be set so that the volume level in the room is at a predetermined *normal* listening level when the console output meters are reading 0 VU. This, of course supposes that the mixer's output voltage has been calibrated to the correct level (meaning adjusted as per a voltmeter reading while connected to the amplifier). After that, the meter's driver amplifiers are adjusted to indicate a 0 VU reading at this level. What is the correct level? It is generally determined by the optimum input level of your recording equipment. But if your recorder has unbalanced inputs, meters that read 0 VU at −10 dB, and your board is calibrated to 0 VU = +4 dB, you are going to need, and more than likely already have, a device (active/electronic or passive/transformer) to interface the two. In this case, you will have a better signal-to-noise ratio by operating your room at +4 dB. If every piece of equipment is *talking* to each other at the same level and impedance, it becomes a matter of S/N ratio versus headroom optimization. Too low a setting increases the S/N ratio, while too high leaves you with less headroom and the possibility of distortion. This only has to be decided once, but requires regularly scheduled calibration checks to maintain. Here's the step by step process:

> Set the console output bus levels to unity gain while reading the console's output voltage on a voltmeter with the amplifier (the load) connected.

> Adjust or calibrate the consoles meters to read 0 dB at this level.

> Set the power amplifier's input attenuators for what is considered the studio's nominal listening level with the console level set at this 0 dB output level. Doing this gets you less noise, and more headroom. It is easier on the ears and makes for longer lasting equipment. Good payoff for a few minutes work, no?

Monitoring Systems Part II

Back to the control room. Before continuing this discussion of monitoring systems, it's important to get something straight; there is no such thing as a perfect transducer. While nobody would argue with that statement, some folks act as if some products are perfect, and a few end-users actually expect transducers to perform as if they were a short length of wire! If you actually believe that sound can be reproduced perfectly, try this little experiment. Take the world's best mega-dollar microphone of your choice. Hey, take a bunch for that matter! Grab an acoustic guitar by its neck right under the headstock, or finger a chord, as long as you can hold it out away from your body. Strum it and record the output. Then play it back. You can use any mic, any placement, any recording methodology, any amp, any cable, or any speaker system, in any room. It will not sound the same. Oh, it may be close, but it will not be exactly the same nor will it ever be. Once you accept this fact, you are free to deal with all the imperfections in audio, while totally disregarding the hype. I've been in this position for well over a quarter century; and believe me, it is not only a much nicer place to be, but my work is substantially better because of it.

Testing the Control Room/Monitor System

Audio Perception

For over one hundred years, folks have quested after perfect audio fidelity from their audio equipment, especially microphones and speakers. Forget about it! No microphone can pick up sound exactly the way our ears hear it. Even if it could, it would not be feeding the information to a brain for decoding and processing. Although they can do a decent job with trumpets and vocals, no speaker will ever be able to radiate sound exactly the way an acoustic guitar, piano, or any other sound source that propagates sound in multiple directions does. Our brain processes the audio information fed to it by our ears, and continuously updates this information, which further aids in our interpretation of the acoustic stimuli from the physical environment around us. In order for any audio reproduction system to perfectly match the original sound, not only must the system render an unaltered version of the original *direct* sound, but in addition, all the information pertaining to the acoustical environment must be presented to our ears as well. This vast amount of data *cannot* be conveyed through a pair of speakers.

Here's another experiment to try. Place a microphone right where you are standing while listening to a music instrument being played; then go into the control room and listen to what the mic picks up. It will not sound the same. Neither the instrument (the *direct* sound) nor the room acoustics will sound the same as what you heard, no matter what the audio system consists of, even if every piece of equipment in the chain has specs that are superior to those that would be necessary for reproducing that particular instrument. If you don't believe me, then try the same thing with speech from a little transistor radio placed out in the studio. You can get a good direct sound using mics positioned close in, but picking up the sound from far away is best attempted with shotgun, hyper-cardioid, or at least cardioid, mics because they help eliminate extraneous sounds. The human ear's pickup pattern is almost omnidirectional, but our brain processes the sound we hear so that we can discern one sound from another even from a far-off position. Our audio perception also includes the ability to learn about the acoustical environment and the location of the sound source, and adjust to it so that we are able to pick up only what we want.

Control Room Speakers

Flat Response?

The common goal is to have a system with a perfectly flat frequency response, and if this could be instituted and repeated in every control room it would make judging audio quality uniform. On the other hand, a perfectly flat response doesn't sound very good and has nothing to do with what people in the real world are listening to. Personal stereo headphones, automotive sound systems, club PA speakers, radios, televisions, and home systems all have frequency responses that are far from flat. People wouldn't buy them if they were. This means that the audio recordist must produce programming that can be adequately monitored on these less than perfect systems. This simply means that you have to understand the difference between the reference monitors that you are using and the playback characteristics of what "normal folks" are using. I've seen studios fall into the trap of overly augmenting their monitor system by boosting the low-frequency content while adding *air* or *crispness* to the highs. This may impress some clients, but a product judged to be acceptable using this kind of *biased* response may pale, and thus, embarrass the recordist when reproduced on normal systems.

If someone came up with a speaker system that produced full bandwidth audio with a single driver, referencing would be a lot simpler for the recordist and life would be a whole lot more luxurious for its exceedingly wealthy inventor. Don't hold your breath. It can't be done. The physical requirement of reproducing 20 Hz has very little connection to the physical demands placed on high-frequency drivers. Therefore, we must deal with the inadequacies that come with multi-driver speaker systems. These include uneven frequency response, phase smearing, crossover distortion, and power consumption that is less than 100 percent efficient.

Operationally, speakers are fairly simple transducers. The amplified audio signal is fed to a coil, which sits within a magnetic field. The audio being AC causes positive and negative changes in the coil's position within this field. The coil is attached to a cone, which is held in place by a *surround* and a *lower spider*. Both are pliable and attached to the driver's frame, or cage. The in and out movement of the coil is limited by the cone's maximum excursion. Cones are often made of paper, and the wire used for the coil is about one-hundredth of an inch (0.01) thick. At high frequencies, the coil can vibrate thousands of times a second, and at the low-end, excursions can reach as much as a full inch. Thermal failure is caused by high power heating the coil over long periods until the wire eventually breaks or simply melts. Mechanical failure can result from a single hit of excessive power. Here the coil can separate from the cone. The cone can also separate from the frame, or more commonly, the coil/cone connection will be deformed to some extent. This results in the coil rubbing against the side of the magnet's gap, which means distortion, friction, heat, and a very short life expectancy.

Serious power hits can rip cones, spiders, and surrounds. I've witnessed coils forced completely out of the magnet's gap. Then they returned off-center so that they ended up stuck outside the gap. Besides the fact that they can no longer produce sound, a coil in this position is exposed to air, has no resistance to power input, and receives no heat sinking from the magnet. Say your good-byes quickly. Metal dome tweeters can actually shatter by being flexed beyond their limit, due to excessive power inputs. So speaker failure can be caused by mechanical stress, long-term high-power operation, excessive peaks, such as powering up equipment on the input side of the amp after the amp itself has been powered up. With this in mind, the value of sequential power-up and delayed turn-on devices becomes apparent. Speaker specifications are all over the place. I've seen power rated according to peak and continuous operating levels, and according to *program* and *music power* levels. How about a specification like: "100 watts RMS continuous power at 1 kHz." If it is a three-driver speaker system, two of the drives may be able to handle only 60 watts. If so, operating at the rated level may not cause failure, but it will cause distortion.

Common Sense and Smaller Monitor Speakers

Everyone knows that having trusted reference monitors allows the engineer to accurately evaluate the integrity of the sound. But today small, close-range (near-field) monitors are more often than not the primary referencing source, and they may not prove adequate for hard core rock, symphonic, and film post-production work. The best positioning is to set the small monitors on stands away from the wall. Each should make up the base corners of an isosceles triangle with the opposite point located some distance behind the engineer's position. The *depth* of this point is determined experimentally and is chosen both to allow the engineer a good deal of lateral movement while still remaining in the "sweet spot," and to provide a position for the producer within this area. The height should be set for a head-tilt angle of about 15 degrees, as this has been found to be the most natural and comfortable for human hearing.

Damage to small monitors from excessive listening levels is common. Their diminutive size requires more power to produce the gratifying levels equated with some popular music. Many studios turn to fusing to reduce casualties, even though the distortion fuses produce is clearly evident on an oscilloscope's display. You may not notice it while monitoring thrashing heavy metal programming, but that violin solo that explodes in the middle of a ballad may make the hair on the back of your neck stand up. Other remedies include adding a capacitor. Normally, these are used to block low frequencies from getting into high-frequency drivers. But here values are chosen to stop low frequencies from getting into the bass speakers as well. Passing the full bandwidth through a phase-shifting, high-pass filter does not appeal to me at all. Common sense tells me that if high levels causes damage to the system, back off on the gain. But if high-listening levels are necessary, augment the system by doubling up on the number of small speakers. Placed side-by-side (as is common with PA systems), the sound quality will remain the same, but the monitoring levels can be increased without damage or distortion. But it's still not loud enough for the keyboardist's overdub position at the back of the control room. Add another set there as well. "Hey, now you're talking about a lot of money." Yeah, like driver reconing and speaker replacement is cheap. This multiple monitor remedy makes even more sense when you think about failures at inopportune times (such as in the middle of the night) and the need for quick recovery from downtime. That's what world-class studios used to do. They worked at minimizing downtime.

Even then, realistic comparisons often called for running out with the mix for a listen on an automobile system. Ever wonder why music sounds so good inside an automobile? You already know the answer. They are sealed acoustical enclosures so the low-frequency sound pressure is enhanced. The high frequencies benefit from the added dispersion from the many reflective surfaces, so the Fletcher-Munson research on human hearing sensitivity points to this listening environment as being a delightful audio experience. Science may have been proven right again, but it is impractical to bring an automobile interior into a recording studio facility. I know. I tried it once, with a '57 Chevy no less.

Control Room Acoustics and Monitoring

I believe all audio engineers should have a basic understanding of acoustic principles so that they have a better understanding of what they are up against (i.e., you walk into a control room and find that its shape is a perfect cube. Pull out them trusty headphones, partner). Even though most of the principles involved with room acoustics are based on the unchangeable laws of physics, other aspects peculiar to control room monitoring make things a little more ambivalent so that it is nearly impossible to put forth an exact formula. Here room size takes priority, then decisions on speaker placement. How reflective the surfaces around the monitors are definitely comes into play. There is also the possibility of interference from slapback delay from the back wall, or from reflections from the ceiling and side walls. And that's only half the story if you plan on recording acoustic instruments, vocals, or any other non-electronic source within that same room. But let's start with just the control room/objective monitoring side of the coin.

Some Facts on Control Room Acoustics

A spectrum analyze can give only hints as to phase response, crossover anomalies, speaker inadequacies, and room resonances. Most amps and speakers today provide high quality sound reproduction, but the room they are in can add up to 10 dB of frequency response variation due to mode resonances. Speaker manufacturers know all about anechoic and free-field testing, and as a result, speaker systems with near flat response have been available since the late 1970s. The main problem now is due to the boundaries and acoustics of the room they are placed in. It's a fact that reflections from video monitors and near-field speakers can affect the main monitor's output and cause comb filtering. Ambient noise in the control room is more apparent today because the electronic circuitry is quieter, microphones have greater sensitivity, and the dynamic range of recording equipment has expanded.

You know about diaphragmatic absorbers and how they resonate at low frequencies depending on their size, weight, thickness and the amount of air space behind them. There are equations that predict the outcome of constructing diaphragmatic absorbers as well as Helmholtz resonators (both are very effective at curing low-frequency problems), but they are most often built on an experimental basis or from past experience. The math isn't that difficult, and the improvement in the outcome is worth the effort.

Serious low-frequency build-up problems can be remedied with bass traps, but these often require a good deal of space. It can take up to three feet of air space to attenuate frequencies in the 100 Hz area. A problem down around 60 cycles may cost you as much as five feet. It's much easier on the wallet to get things right during the initial planning stages. This also points out the importance of avoiding *accidental* diaphragmatic wall, floor, and ceiling structures that were intended to be solid but were specified or constructed incorrectly. The erratic problems this causes not only wreaks havoc on a room's acoustic response, but can negate all the hard work and expense put into achieving a high level of noise transmission loss.

If noise transmission is not handled correctly at the start of the project, there will be no second chance at adjustment, as this area of studio/control room construction is dependant on the accuracy of design *and* the integrity of the implementation. Good control rooms are close to airtight, completely isolated from outside noise, and structurally solid. This is one of the reasons basement control rooms always seem to have great bass response. Ambient noise above 30 to 40 dB can be a problem. Once a quiet background is achieved, any change in the room's layout such as equipment additions can increase the background noise. The test procedure that initially qualified the room as acceptable should be repeated often as small changes in background noise can go unnoticed, especially by those who spend a great deal of time in the room.

Sound dispersion in a room can be checked by feeding pink noise into the system and using a ⅓-octave analyzer's display to search out anomalies. If plotted out, these will often point to the cause of the problem. As an example, I once tracked down the source of a 2 to 4 dB bump centered around 4 kHz to two structural ceiling beams. They were fairly close to each other (a couple of inches apart), and of course, parallel. They set up a standing wave that was actually noticeable enough to add to the high-frequency content of the signal in the room. Sound dispersion is normally better at low frequencies.

Due to their large wavelengths, they tend to just go around obstructions. Higher frequencies are more directional and reflected more easily. If you want to check how *even* a room's sound dispersion is, once again you use a spectrum analyzer, but now check the high-frequency content at multiple locations.

As already stated, room modes can add up to 10 dB to the level at resonant frequencies. Here the room's geometric shape can cause certain frequencies to be augmented or sustained. Sometimes you are better off dealing with the predictable problems of parallel walls than the often unpredictable results of splaying them. There's a lot more to it than lowest resonant frequency and ideal proportions, but people still want to have a go at it without hiring an accredited professional acoustician.

A material's sound transmission blocking capability is simply a function of its stiffness and weight, that is, until the acoustic seal is broken. A hole the size of a quarter in a 12-inch cinder block wall will pretty much negate the whole effort. Lead is heavy, and it is also a limp mass that easily conforms to the required shape to seal off openings. However lead sheet placed between two sheet rock panels in a wall will lose some of its sound blocking qualities when the wall moves as one unit. This is not exactly the case with impregnated vinyl materials because they add damping to the wall structure when placed in a staggered array between the two layers of sheet rock.

Things change fairly slowly in the field of acoustics, as truly new developments are rare. When something does change, it is often the result of building upon the previous work of many mathematicians, physicists, and acousticians. An example of this is the superior sound diffusers that are now available. Since the time of the early Greek outdoor theaters, acousticians have know that a concave surfaces focus sound to one particular point (just as parabolic reflectors do with microphones). Convex surfaces are better at dispersing sound, but the outcome is very unpredictable. Highly irregular walls like those made up of field stone masonry are very effective diffusers, but still unpredictable. This is due to the fact that diffusion is dependant on the size (both depth and width) of the irregularities in the surface as well as the wavelength and angle (incidence) of the sound impinging upon it.

The innovation in diffusion I am talking about began with the physics of optical diffraction no less. Next (during the 1800s) came Karl Gauss's sequential number theory. Later Mr. Manfred Schroeder linked light diffraction to sound diffusion. This particular instance of building on past innovations and accomplishments ended up as a new type of panel diffuser that spreads sound impinging upon it across a wide area. Its main achievement, however, was the fact that this scattering applied to a broad band of frequencies and most angles of incidence. I heard of this in the early 1980s. The folks at Syn-Aud-Con were all over this design, using their TEF systems to analyze its effects. I also vaguely remember an AES paper on the subject, but for me the real head turner was the BBC's use of this type of diffusion in their huge (8,000 cubic meter) Manchester music studio. This diffuser design covered all of the back wall and both side walls for the full depth of the orchestral seating area. The studio ended up requiring only a fraction of the normal amount of absorption to even out the frequency response, even though it had a reverberation time of a full 2.25 seconds. The broadcast quality improved, but most importantly every musician heard an even blend of everything. Today, sequentially spaced vertical indentations cover almost the full surface area of BBC Manchester's Studio 7.

These days you can order them from several manufacturers, and UPS will haul them in for you. I would not be surprised if complete modular control rooms were to become available. The prospective buyer could visit a sales rep, listen to some programming, select from a list of options, determine size, work out a payment plan, and make arrangements for delivery. Just like buying a car! This method has already been a common practice in the design of broadcast facilities for years.

Live End Dead End (LEDE)/Dead End Live End (DELE)

LEDE-designed control rooms are based on science. First they eliminate comb filter causing early reflections by the use of an absorptive monitor end that provides an anechoic-like path between the speakers and the mix position. Then the hard rear area diffuses sound reflections so as to blend them into the sound at the mix position at a time delay that will not interfere with the direct signal. Eliminating the perception of separate (not temporally fused) early reflections from the back wall is accomplished by paying strict attention to the Haas effect. When control rooms are large enough to be able to extend the time of the initial reflection beyond 10 to 30 milliseconds (meaning they would no longer be perceived as part of the signal coming from the monitors, but as separate sounds), reflectors are used to shorten the path, and thus, the propagation time back to the mix position.

The LEDE's rear diffusion adds to the control room's ambience and reverberation time so that there is a balance between absorption and diffusion. Purely dead studios and control rooms do not work. I know; I've built a few. A trick used by Chips Davis to determine the possibility of comb filtering caused by speaker reflection off of the console face is to place a mirror on the console's surface, sit in the mix position, and look in the mirror. If you can see the monitor, you have a reflection to deal with. He also placed the mirror on top and against the sides of the near-field monitors to see if the sound coming from the main monitors would be reflected into the mix position by those surfaces. As with any control room design, strict attention must be paid to having an inner surrounding structure that is symmetrical without parallel surfaces.

Another control room design approach could be termed DELE. Here the wall that the main speakers are soffited (flush mounted) into, as well as the ceiling and the walls immediately surrounding the monitors seem to have random diffusion, but they are designed to project the early reflections around the listening position to the back of the room. Here the late reflections are absorbed or dispersed so as not to cause any late sound arrival at the mix position. The idea is to break up the very early (less than 20 milliseconds arrival time) and late (more than 50 milliseconds arrival time) reflections, so that they do not interfere with the direct sound at the mix position. Having experienced some of this method's finer control rooms, I can describe the end result as a listening situation where nothing comes between the listener and the monitors. In either case (LEDE or DELE), the mix you "walk with" will be the same as you heard on the control room monitors, no matter which decent quality speakers it is played through. Both design methods work if executed properly, but until recently the LEDE approach had a more predictable outcome.

Finally, I'd like to straighten something out about rear wall *slapback* echo. The sound coming back into the listening position from the back walls must have a delay of at least 30 milliseconds ($\frac{30}{1000}$ of a second) for it to interfere with what you are monitoring by causing a distinct separate echo-like reflection. This means that the time it takes for the sound to travel from the listening position to the back wall and then back to the listening position must be at least 30 milliseconds. Sound travels at 1130 feet per second. Therefore your rear wall must be about 17 feet behind the listening position for this to be a problem. So rear-wall, slapback echo problems are *highly* unlikely in small control rooms.

The Studio Side of the Glass

The Cue System

Instead of room/speaker interface anomalies, studio headphone monitor systems present more down to earth problems. Variations in headphone impedances can cause one set of cans to be blaring loud while another's output is minuscule. Therefore, it's best when all the headphones used are identical. Then there's the imminent danger of the immediate cessation of audio reproduction from one side of the cans when someone mistakenly inserts a ¼-inch mono plug into the system. This power outage could be followed by possible amplifier destruction if the system has not been furnished with safety load resistors. These load or "build-out" resistors are placed in series with each of the amp's outputs (meaning four resistors for a stereo pair) to keep the amp from "seeing" a short circuit. The resistor's impedance rating should be the same as the headphones' and its wattage rating should be about one third of the amplifiers' rated output. Therefore, if the headphone Z or impedance is 8 ohms and the amplifier is rated at 225 watts, amp safety calls for an 8-ohm resistor rated for 75 watts. These guys can run hot. I've seen them burn the plywood they were mounted on, so attach them to some kind of metal heat sink, like a blank rack panel.

Dead Rooms

During the 1980s many studios were designed to be almost anechoic, or at least provide a large, dead area for the recording of multiple instruments. A non-reverberant atmosphere was sought for several reasons. The main benefit derived from multitrack recording is that it affords the musicians the ability to overdub or re-do parts they feel are not adequate. Not only can these parts be re-recorded at any time, but additional instruments, vocals, and effects tracks can also be added. Music production now becomes wide open. When the number of open tracks is insufficient for planned or unscheduled additions, *bouncing* (combining or mixing several tracks together to a single track) is used to open up new real estate. Easy as pie. However track bleed could be a major problem. If the rhythm guitar was picked up by the snare mic, the original performance by the guitarist would still be heard after the guitar track had been rerecorded.

Isolation could be furnished electronically with the use of gating and direct input interfaces, but the first line of defense is always at the source. Movable panels (gobos) were constructed to go in between instruments. Some were made tall to go around standing musicians (horns, vocalists, or whatever). Windows were included so that visual cueing would still be possible. The studio walls would be lined with absorptive materials. One of the favorites was velvet, theater curtains with deep folds (air space) over raw fiberglass. Itch city! Next came drum and other isolation booths to completely separate instruments. All this gave the engineer and producer a clean, separate, and dry signal to work with. Because of the newfound ability to add quality reverberation along with choices from a huge palette of additional effects at any time, things became exciting to say the least.

However, some things just weren't right. The sound of weighted drum skins and muted shells being played in a reverberant free environment not only lent itself to the addition of very spacious electronically generated reverb, but also became all the rage for certain forms of music like disco, reggae, dance hall, and rock ballads. But deadened drum skins do not provide adequate rebound in their response to stick hits. This not only slowed drummers down but also encouraged them to adopt a more heavy-handed playing style. The quick, light-touch dexterity of jazz drumming became noticeably absent.

Dead rooms do not work on many levels. During the early 1970s I built a rehearsal studio in a commercial neighborhood. Due to the closeness of the nearest neighbors, sound transmission was a major concern. The room's boundaries included three masonry walls, which was a big plus. The first step was to seal *every* crevice. This was followed by applying absorptive material. The owner's intention was to provide a space for rock musicians to rehearse "full throttle," which they were unable to do elsewhere. This equated to dual-Marshall stacks for guitars (one stack for the rhythm and one stack for the lead), two 18-inch driver bass cabinets (one front mounted and the other in a folded horn), a Hammond B-3 with dual Leslie speaker cabinets, and a half dozen smaller "combo " amplifiers for synthesizers and other miscellaneous instruments. Dual horn, and double-low-frequency drivers along with four separate stage monitors made up the PA system. The size of this room was only 25 × 30 feet!. With every instrument blaring, the PA levels would have to be set pretty high, so I figured mucho absorption was the way to go in order to avoid massive feedback levels. It worked! Everyone could set their amp levels all the way up and get the distortion and sustain they wanted, and you could still hear the vocals above the din.

Then the problems set in. To say the least, I learned a lot from this experience, the hard way. While all the amps, speakers and drums were on small stage-like risers that were set on vibration isolation pads, I hadn't correctly figured on the massive playing levels of these young power-crazed rockers. They were vibrating the building's masonry structure! This displeased our upstairs neighbor no end. Since this business was one of "ill-repute," the owners were not happy when their workers reported to them that the clientele was disturbed and distracted by the noise. This was explained to me in a very succinct manner.

The 1,500 BTU in-wall air conditioner had been outfitted with intake and exhaust ductwork to reduce sound leakage. During the summertime construction stage, it easily chilled the hard-working crew of which I was part. But sound absorption equals thermal insulation. Add in a completely sealed room and a whole bunch of tube amps glowing purple and red while a bunch of sweaty, excited young musicians jumped about practicing their energetic stage show and you end up with a whole lot of heat, humidity and electronic failure. Equipment was breaking down at an alarming rate! I ended up putting a governor on the PA, which induced the vocalists to plead for saner amp levels so that they could hear themselves. Luckily, another space became available that was larger and had more understanding neighbors. Central AC was the first investment.

Deadened recording studios don't exactly suffer the same plight, but some of the joy that comes from recording well-performed instrument playing is lost. Next time you're in a studio with well-balanced acoustics, have someone play the baby grand. Get a good earful and then wrap a towel around your head. Electronic reverb will help, but it will not give you back the lost high end. Things just sound better in a large studio whose overall sound is live.

Correct acoustic room response is essential for making quality recordings of orchestras and even small string sections. While woodwind and brass sections can be placed in smaller areas, these instruments especially suffer from dead-sounding acoustics, as they project their sounds multidirectionally. With a large live area, the engineer has the option of recording the drums set up in the middle of the room for a thunderous sound, or they can be relegated to a corner of the room or placed in a drum booth, depending on what the situation calls for as far as separation and the desired sound.

This corner area as well as the booth, can provide various acoustic settings with the addition of movable panels or curtains. This kind of acoustic variability also aids the sound of other acoustic instruments as well as amplified instruments. There's something special about being able to walk out into the studio and flip a hinge-mounted panel next to a Fender twin combo amp to change the guitar sound from a biting lead caused by its high frequencies being reflected from the panel's hard surface, to a mellower bottom-accented sound due to the added high-frequency absorption by the panels "soft" side. Acoustic panels can be made in an endless variety of shapes, sizes, and configurations to diffuse, reflect, or provide frequency selective absorption. There's tons of information out there that will tell you exactly how to construct your own along with countless manufacturers offering an almost endless variety of prefabricated panels. Variable acoustics can be critically important to what microphones pick up, and this should obviously be important to the professional audio recordist.

Want a Look into the Future?

Harman Professional's "Powered by Crown" comprehensive System Architect software runs on the current Apple operating system iOS, including iPad, iPhone, and even iPod Touch. It was designed for live audio venues. What's it do?

With Crown amplifiers and JBL speakers, it handles input selection and routing, monitor levels, and limiting; shows load status and incoming AC voltage; and provides input level adjustment (on smooth faders), switching inputs as available (analog, AES3, CobraNet) and the insertion of test signals, if the amplifier has a built-in signal generator. Used as a remote unit for troubleshooting, it promises to be real time saver. You can go anywhere in the venue and quickly boost or attenuate fills, add delay, or insert test tones to check drivers. Even when you are in the area where the amplifiers are located, this app will be handy when working behind the rack because you'll have the front panel's controls and display right there in front of you. The download costs $3.99. That's right, three dollars and 99 cents!

But it presently works only with Crown amps and JBL speakers.

While this is *mighty* impressive, it is a very short hop to self-amplified near-field monitors with a hand held i-whatever remote that allows you to change the level, the input, the attenuation, compression, and limiting; perform automatic fade-ins/outs; spectrum analyze; equalize; and generally completely optimize your monitor system.

Talk about Impressing Clients!

What's going to be the difference in getting this live venue development turned around to include the small studio owner? Numbers as in high demand. You got 'em so start making phone calls, sending e-mails and bugging Harman until they realize the amount of money they are *not* making by not offering this technology to the audio recording community! You have the numbers. You have the buying power. They *will* pay attention!

PART II

Brick and Mortar (Drywall and Stud) Studios

CHAPTER 6

My Studio—How Big and What Shape?

The radio broadcasting industry nourished the "talk booth" concept in the early years. The size of the speech studio only needed to be large enough to accommodate one person (or possibly two for an interview), a table, a few chairs, and a microphone. This sanctified telephone booth-sized studios with very serious built-in acoustical problems. To understand the reason for these problems, we must realize that a roomful of air is a very complex acoustical vibrating system. In fact, it is a series of many resonant systems superimposed upon each other forming a super complex problem.

Let us consider a rectangular studio 12 feet high, 16 feet wide, and 20 feet long (ratios 3:4:5) as sketched in Fig. 6.1. First we shall pay attention to the two opposite and parallel N-S walls, neglecting for the moment the effects of all the other surfaces. Even though acoustically treated to some extent, some sound is reflected from these surfaces. For sound to travel a distance of one round trip between the two walls, or 2L feet, takes a certain, finite length of time. This time is determined by the velocity of sound which is about 1130 feet per second (about 770 miles per hour). At a frequency of 1130/2L Hz, this pair of opposing, parallel walls L feet apart comes into a resonance condition and a standing wave is set up. For example, in the N-S pair of walls in Fig. 6.1, L 20 feet and the frequency of resonance is 1130/40 or approximately 28 Hz.

This resonance effect also appears at every multiple of 2L. Harmonics of 28 Hz appear at 56, 84, 112, 140 … etc. Hz. Although the term "harmonics" is not precisely accurate in this context, we'll use it for convenience. The single pair of walls then gives a fundamental frequency and a train of harmonics at which resonance effects also occur, called axial mode room resonances. The E-W walls (Fig. 6.1) give another fundamental frequency of 35 Hz and a train of harmonics. The floor-ceiling combination gives a fundamental frequency of about 47 Hz and a third series of harmonics. These modal frequencies determine the sound of a room. They yield bad effects only if they pile up at certain frequencies or are spaced too far apart, as we shall see later.

We have considered only axial modes of this room involving a single pair of surfaces. There are also tangential modes involving two pairs of surfaces and oblique modes involving three pairs of surfaces which have still different fundamental frequencies and harmonics (Fig. 6.2). Taken all together, these make the sound field of an enclosed space extremely complex. Fortunately, the effect of tangential and oblique modes is less than that of the axial modes. Basically, you can get by quite easily and reasonably accurately in designing a studio by considering only the axial modes, although you must depend

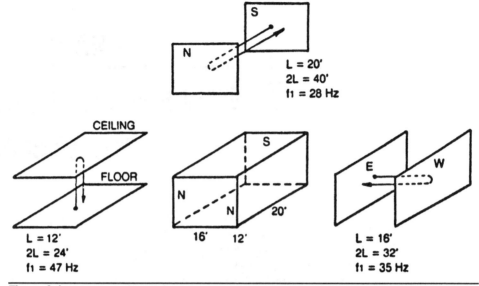

Figure 6.1   The six surfaces of a rectangular room are broken down into three pairs of opposite and parallel surfaces when considering axial mode room resonances. Each pair has its fundamental resonance frequency and train of harmonics.

on the resonance effects of tangential and oblique modes to do a certain amount of filling in between the axial modes.

Because of the standing wave effect the sound pressure at 28 Hz is far from uniform across the room, being very high near the surface of the N and S walls and zero at the center of the room. The situation is very much like the inside of an organ pipe closed at both ends. At the second harmonic near 56 Hz, however, the distribution of sound pressure in the standing wave is quite different, with two null points and a maximum in the center of the room as well as at each wall surface. When complex sounds of speech or music excite the fundamental and harmonics of a single series

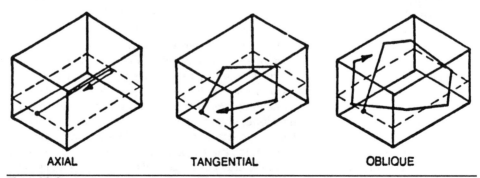

Figure 6.2   Axial modes involve two surfaces; tangential modes four surfaces; and oblique modes six surfaces. The axial modes commonly dominate small studio acoustics.

there is extreme complexity between a single pair of wall surfaces as the standing wave patterns shift. Adding the effects of the E-W and vertical modal series and adding the tangential and oblique modes results in a constantly shifting music or voice sound field, the complexity of which defies description.

How Big Should a Studio Be?

Each room resonance frequency has a certain bandwidth (or Q).[6,*] The ideal situation is to have adjacent resonances (fundamentals and harmonics) locked arm in arm with neighbors through these resonance skirts. This results in signal components of constantly fluctuating frequency being treated uniformly. If the spacing of these resonances is too great, some of the precious signal energy is boosted by resonances, and some which "falls in the cracks" is discriminated against. On the other hand, if three or four room resonances occur at the same frequency or are very close together, signal energy in this part of the spectrum receives an abnormal boost. Such pileups are inevitably accompanied by gaps elsewhere in the spectrum. Good studio sound requires careful attention to these resonance frequencies which are, in turn, controlled by room dimensions and proportions.

Room size determines how the low frequencies are treated. The larger the room, the lower the frequency components the room can support. Small rooms result in great spacing of room modes. A talk booth of 6 feet × 8 feet cannot support sound lower than about 70 Hz. Even though there is little energy in voice below 150 Hz, such a small room is unsuitable for recording because of excessive mode spacing. The BBC has concluded that any studio of less than 1500 cubic feet is not practical. Any saving in construction cost is outweighed by cost of correcting acoustical deficiencies—and usually successful correction of deficiencies is not feasible.

Distribution of Modes

We have considered the three axial frequencies and the three series of harmonics of the studio of Fig. 6.1. Now we must ask the question, "Are these frequencies properly distributed?" To answer this, each frequency must be computed and examined. This may be a bit tedious, but it involves only the simplest mathematics.

A convenient approach is to tabulate the series for each of the three room dimensions, such as in Table 6.1. The 141.3 Hz and 282.5 Hz occur in each column, and they bear a 2:1 relationship to each other. These coincident frequencies are called *degeneracies*. How well the other modal frequencies are distributed overall is difficult to see from columns of figures.

In Fig. 6.3 each modal frequency is plotted on a linear frequency scale. Each mode is represented by a vertical line, although actually each one has an average bandwidth of about 5 Hz (as shown for the lowest mode of Fig. 6.3). The triple coincident frequencies at 141.3 and 282.5 Hz are greatly spaced from their neighbors. The piling up at 141.3 and 282.5 Hz means that signal energy near these frequencies will be unnaturally boosted. Also, the great separation from neighboring modal frequencies guarantees

---

*All references are found at the conclusion of this book.

	N-S Walls (2L = 40 Ft.)	E-W Walls (2L = 32 Ft.)	Floor-Ceiling (2L = 24 Ft.)
f1 (fundamental)	28.3 Hz	35.3 Hz	47.1 Hz
f2 (2nd harmonic)	56.5	70.6	94.2
f3 (3rd harmonic, etc.)	84.8	105.9	(141.3)
f4	113.0	(141.3)	188.3
f5	(141.3)	176.6	235.4
f6	169.5	211.9	(282.5)
f7	197.8	247.2	329.6
f8	226.0	(282.5)	
f9	254.3	317.8	
f10	(282.5)		
f11	310.8		

TABLE 6.1   Axial Resonance Frequencies

that signal components in these gaps will be unnaturally depressed. This adds up to almost certain audible colorations at these two frequencies that are monotonous, and repetitive blasts of energy which distort music and are particularly obnoxious in speech.

Deciding on Best Studio Proportions

The 3:4:5 proportion of Figs. 6.1 and 6.3 were, at one time, highly recommended but they are ill-suited for studio construction because of poor modal distribution. What room proportions should be used? Numerous studies have been made on this subject. Three suggestions from each of two authors are presented in Table 6.2. In this table, studio dimensions following the suggested ratios are included based on a ceiling height of 10 feet. The studio volume, which varies from example to example, is included for each ratio.

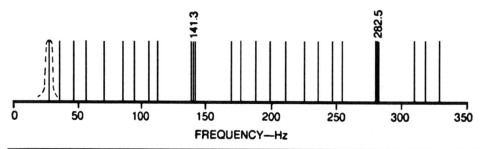

Figure 6.3   The axial mode frequencies of a rectangular room 12 × 16 × 20 feet are shown as lines, although each has a bandwidth of approximately 5 Hz as indicated on the extreme left. Three coincident frequencies at 141.3 and 282.5 Hz are separated slightly for clarity.

	Ratios	Ceiling Height, Ft.	Length, Ft.	Width, Ft.	Volume, Cubic Ft.
(A)[7]	1:1.14:1.39	10.0	13.9	11.4	1585
(B)[7]	1:1.28:1.54	10.0	15.4	12.8	1971
(C)[7]	1:1.60:2.33	10.0	23.3	16.0	3728
(D)[8]	1:1.90:1.40	10.0	19.0	14.0	2660
(E)[8]	1:1.90:1.30	10.0	19.0	13.0	2470
(F)[8]	1:1.50:2.10	10.0	21.0	15.0	3150

TABLE 6.2 Studio Proportions

For a visual comparison of these ratios, the fundamental and harmonic series is plotted for each in Fig. 6.4. Although there are scattered double coincidences, all six show better distribution than the unfortunate 3:4:5 choice of Fig. 6.1. Yet none shows the equally spaced modal frequencies desired in our ideal studio. Nor should this bother you too much. It is just as bad to place too much emphasis on room proportions as it is to neglect them completely.

The presence of people and furnishings in a room so affects the actual modal frequencies that it is futile to worry over minor deviations from some assumed optimum condition. A practical approach is to be alert to problems of coincidence and mode spacing; to do what can be done to optimize them and then relax. If an existing space is being considered as a studio or control room, check the modal frequencies after the fashion of Table 6.1 and plot them as in Fig. 6.4 to see if serious problems exist. For new construction, do the same for proposed dimensions of any sound-sensitive rooms. A liberal education in axial modes awaits the persevering student who varies one dimension of a room (on paper) while holding the others constant and noting how the modal distribution is affected. You'll soon conclude that eliminating coincidences is a major victory—beyond that, there is little to be gained.

In Table 6.2, the ceiling heights of the room are held constant and volume is allowed to vary. If the several proportions were adjusted so that the volume remained constant, the ratios would draw closer together. This is illustrated in Table 6.3 in which the ratios of Table 6.2 are adjusted so that all volumes are the same as that corresponding to the first ratio. In other words, when rooms of the same volume but of different proportions are considered, the six ratio examples we have been studying are not as different from each other as they first appeared to be. However, they are different, relatively speaking, and you cannot escape the fact that smaller dimensions yield higher fundamental frequencies.

The fundamental resonance frequencies and harmonics of the ratios of Table 6.3 adjusted for the same room volume are plotted in Fig. 6.5. Figures 6.4A and 6.5A are, of course, identical as they represent the common point of comparison in the two cases. Although frequencies are shifted, a family resemblance can be seen when comparing Figs. 6.4B and 6.5B, Figs. 6.4C and 6.5C, and other corresponding pairs. The coincident or nearly coincident pairs of Fig. 6.4 still exist in Fig. 6.5. In general, however, the larger rooms of Fig. 6.4 have the advantage of yielding closer average spacings.

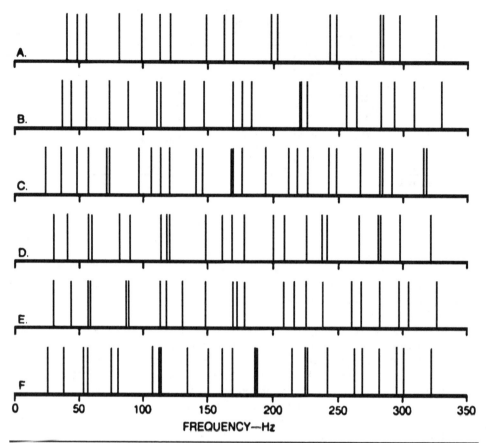

Figure 6.4 A plot of axial mode frequencies of six of the best studio dimension ratios listed in Table 6.2. Lines representing coincident frequencies are slightly displaced in several instances for clarity.

	Ratios	Ceiling Height, Ft.	Length Ft.	Width Ft.	Volume Cubic Ft.
(A)	1:1.14:1.39	10.0	13.9	11.4	1585
(B)	0.93:1.19:1.43	9.3	14.3	11.9	1585
(C)	0.75:1.20:1.75	7.5	17.5	12.0	1585
(D)	0.84:1.60:1.18	8.4	16.0	11.8	1585
(E)	0.86:1.64:1.12	8.6	16.4	11.2	1585
(F)	0.80:1.91:1.67	8.0	16.7	11.9	1585

TABLE 6.3 Varying Proportions with the Same Volume

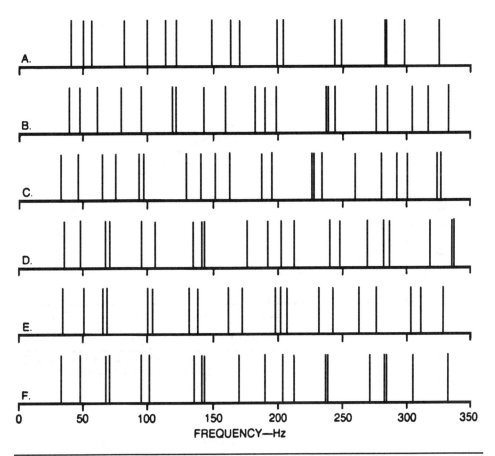

Figure 6.5 A plot of axial mode frequencies of the same six examples of dimensional ratios of Fig. 6.4 but adjusted for the same room volume of 1585 cubic feet. Some coincident lines are displaced slightly for clarity (refer to Table 6.3).

Studio Size and Low Frequency Response

An inspection of the left edge of the six plots of Fig. 6.4 reveals that the larger rooms have lower fundamental frequencies than the smaller ones. In Fig. 6.6 the fundamental room resonances corresponding to the longest dimension of the six cases are plotted against room volume. For the smallest studio having a volume of 1585 cubic feet (Fig. 6.4A), the lowest signal frequency which would have the advantage of resonance assistance is about 41 Hz. The studio of Fig. 6.4C, with its volume of 3728 cubic feet can handle signal components down to 24 Hz. For speech purposes, the smaller studio is quite adequate, because speech energy below 40 Hz is extremely low. The 3728 cubic foot studio adds almost another octave in the lows, which is quite advantageous for music recording.

This discussion has been based on the axial modes of a room. The tangential and oblique modes have somewhat longer paths and hence would tend to extend the low

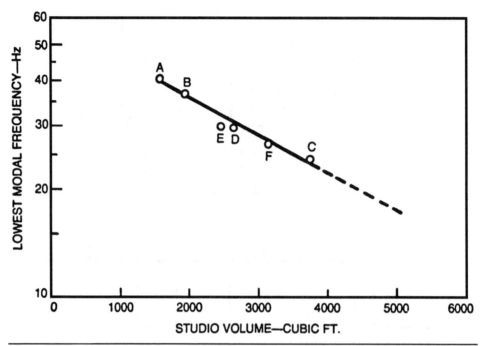

Figure 6.6   The six lowest modal frequencies of Fig. 6.4 as related to room volume. A 4000 cubic foot studio adds one octave to the room's low frequency response as compared to a studio of 1500 cubic foot volume.

frequency limit of a room. While axial modes have two reflections per round trip, tangential modes have four and oblique modes have six (refer again to Fig. 6.2). As energy is lost at each reflection, the reason for the lower amplitudes of tangential and oblique modal resonances is apparent.

However, axial reflections are perpendicular to the surfaces, the angle of incidence which gives the most efficient absorption. Tangential and oblique modal paths, while involving more reflections, are usually at smaller angles of incidence. This results in less absorption at each reflection. The limit is grazing incidence at which absorption is very small. Thus, the number of reflections for tangential and oblique is greater than for axial modes, but, acting in the other direction, there is the fact that the smaller angles of incidence result in less loss per reflection than the 90° incidence of axial modes. How do these opposing factors add up? Since the axial modes are more dominant than the tangential and oblique modes, the practical extension of a room's low frequency response due to tangential and oblique modes is limited.

Room Cutoff Frequency

Every studio has some frequency above which the modal frequencies are close enough together to merge into a statistical continuum. This is called the cutoff frequency of the room. At frequencies higher than the cutoff frequency, various components of the signal will be treated more or less uniformly and the room will act more like a large auditorium. At frequencies below cutoff, excessive spacing of modes exists with resulting uneven treatment of signal components.

The cutoff frequency of a room is a function only of its reverberation time and volume and may be computed approximately from the following statement:[9]

$$\text{Room cutoff frequency} = 20{,}000 \sqrt{\frac{T}{V}}$$

where

$$T = \text{reverberation time, seconds}$$

$$V = \text{volume of room, cubic feet}$$

Figure 6.7 is computed from this basic statement for volumes and reverberation times common in small studios. The larger and the more dead a studio, the lower the cutoff frequency and the less difficulty in handling the low end of the audible spectrum. In Figs. 6.3, 6.4, and 6.5 the plots of modal frequencies were terminated at about 300 Hz, which approximates the cutoff frequency of the average small studio above which modes are increasingly closer together.

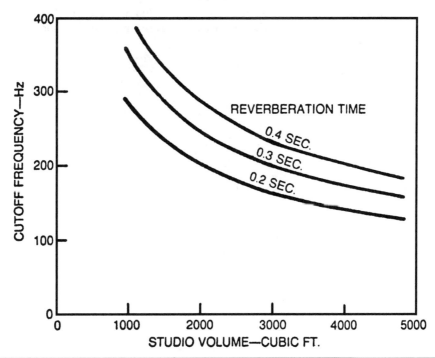

Figure 6.7 The cutoff frequency of a room is that frequency above which modes are numerous enough and close enough together to merge into a statistical continuum. Cutoff frequency is determined only by the volume and reverberation time of the room.

Summary of Room Mode Effects

- Rooms smaller than 1500 cubic feet are subject to insurmountable room mode problems and should be avoided for quality recording studios and control rooms. The larger the room, the closer the average spacing of modal frequencies and the more uniform the treatment of the various components of the signal.

- In choosing studio proportions, try to eliminate coincident frequencies below 300 Hz. Multiple coincidences in the region below 200 Hz are more apt to cause audible colorations than those between 200 and 300 Hz.

- Once multiple coincident frequencies are reduced or eliminated it is futile to carry modal analyses to extremes because occupants, furnishings, and other room irregularities result in great deviations from the idealized paper condition.

- The splaying of studio walls may be helpful in reducing flutter echo, but modal frequencies are only shifted to some unknown values. Splaying may also tend to break up degeneracies (coincident frequencies) in an otherwise symmetrical room, but keeping things under control by choosing proper proportions of a rectangular room is a satisfactory approach if flutter echo and diffusion are controlled by proper placement of absorbing material.

- What can be done to treat an unavoidable coloration in a studio? A tuned Helmholtz resonator can be introduced to tame the coincident frequency. The sharpness of tuning may have to be controlled to avoid the slow decay of sound in a high-Q structure. The construction and tuning of such absorbers is discussed in later sections.

CHAPTER 7

Elements Common to all Studios

E ach of the 12 studio plans studied in this book has certain elements in common with all the others. For example, all require protection from interfering noise, whether it originates outside or inside the studio. All studios require an observation window for visual contact between the control operator and those in the studio. It would be ridiculous to repeat descriptions of each of these common elements a dozen times, so this chapter discusses them all.

Sound Lock Acoustical Treatment

A typical sound lock corridor places two doors in series and two walls in series so that external noise must traverse both to penetrate to the quiet rooms. Many examples of sound locks sharing common principles and differing only in details are seen in the studio plans.

Functionally, the sound lock both isolates and absorbs. Wall construction (one way of achieving isolation), and the treatment of doors and door seals are considered later in this chapter.

As you enter a sound lock from the outside, the exterior noise momentarily floods the corridor. A sound level reading outside the open door is higher than one inside the sound lock, even with the door open, if the interior surfaces are highly absorbent. The sound transmitted through a sound lock corridor is significantly reduced if the corridor surfaces are properly treated.

Consider an untreated corridor (average absorption coefficient, say, 0.1) and the same corridor with surfaces treated (average absorption coefficient 0.9). By adding absorbing material the noise level in the sound lock is reduced as much as 9.5 dB from the bare condition. Even if both the exterior door and the studio door were open at the same time (an unusual occurrence), the exterior noise level in the studio would be reduced by about this amount due to the inner treatment of the sound lock alone. This is why it is so important to make sound lock surfaces as absorbent as possible.

Although sound locks come in all shapes and sizes, the treatment of Fig. 7.1 is representative. Heavy carpet and pad is the almost universal solution for the sound lock floor. Of the numerous approaches for the ceiling, the ubiquitous suspended ceiling supported by a T-bar grid offers many advantages. The better grade of lay-in panels offer absorption coefficients averaging 0.75 to 0.85 throughout the 125 Hz–4 kHz band. They give excellent low frequency absorption which is further increased by introducing a thick layer of household insulation into the cavity above the lay-in panels. Manufacturers specify coefficients for the standard Mounting 7 (or E-405 in the new system) in which

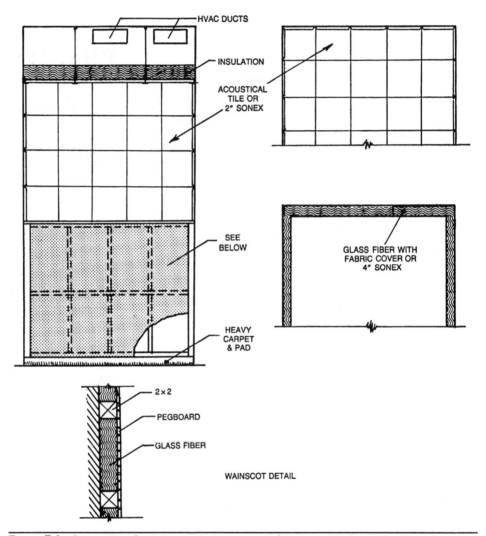

FIGURE 7.1 Suggestions for the acoustical treatment of the sound lock corridor. To minimize noise being transmitted through the sound lock it should be very absorbent, although the requirements otherwise are not critical.

the lay-in ceiling is 16 inches below the structural ceiling. Other distances may be used, of course, with some modest change in absorption. A lay-in ceiling of this type also provides an ideal hidden location for air conditioning ducts and electrical service runs.

Another approach is covering ceilings and upper walls with common acoustical tile or 2 to 4 inches of dense glass fiberboards. The latter require some sort of protective cover such as expanded metal, wire screen, or loosely-woven cloth. Cloth has the advantage of controlling the sloughing off of tiny irritating glass fibers.

Instead of old-fashioned acoustical tile or glass fiber with a protective cover, one of the new foam products such as Sonex (Illbruck), Sound-Sorber (Discrete Technology), or Acoustafoam (FM Tubecraft) may be used.* They are more expensive, but labor-

* See Appendix D at www.mhprofessional.com/shea4 for manufacturers' addresses.

saving. Their dramatic and colorful appearance may provide an intangible advantage in the reactions of clients or visitors.

The lower walls (wainscoting) must withstand considerable mechanical abrasion. There are dozens of proprietary panels which serve well in this location, but all are expensive. One straightforward and relatively inexpensive approach is to mount panels of Tectum (see Tectum and Gold Bond Products, Appendix D), a structural board of wood fiber with a cement binder. It stands up well under abuse and can be painted with a nonbridging paint with minimum effect on absorbing properties. A 1-inch layer of Tectum over a glass fiberfilled air space obtained by furring out on 2 × 2s gives good midband absorption as shown in Fig. 7.2. The panel acts as a diaphragmatic absorber.

Figure 7.1 shows another approach utilizing 2 × 2 furring and ordinary pegboard. The sound absorption of pegboard backed by glass fiber is shown in Fig. 7.2. The Tectum arrangement is less sharply tuned than the pegboard, which is really a Helmholtz absorber. Although the Tectum absorbs less at low frequencies, above 500 Hz it is reasonably

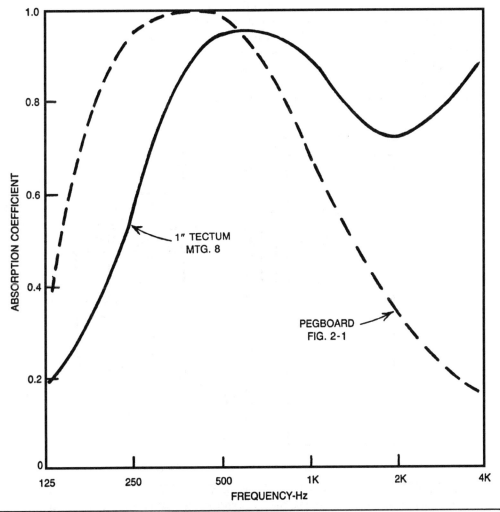

FIGURE 7.2 Approximate absorption characteristics of 1-inch Tectum and pegboard wainscot treatments.

uniform throughout the important audible frequencies. The absorption coefficients of Fig. 7.2 are only for the purposes of comparison. There is little justification for spending time on calculating the absorption of a sound lock corridor.

Acoustical Doors

The access to each studio-control room suite should be via a sound lock. All doors should open into the sound lock. The use of a sound lock relaxes door requirements since heavy loaded and laminated doors with awkward ice box type of clamping hardware are eliminated. However, at the opposite extreme, ordinary household doors are far too thin acoustically to be acceptable. A reasonably inexpensive intermediate solution is $1\frac{3}{4}$ inch solid core doors which are readily available. Such doors have solid wood or particleboard cores which provide the density necessary to impede the flow of sound through them. A typical particleboard core faced with $\frac{1}{8}$ inch hardboard has a density of 5.3 pounds per square foot of surface. This gives a transmission loss of 33 dB at 500 Hz on a mass law basis. In common with other barriers, its transmission loss is less than this amount below 500 Hz and greater above 500 Hz. This would be insufficient alone, but the sound lock principle places two such doors in series between a high noise area (exterior or control room monitor loudspeaker) and the studio.

A 10 inch × 10 inch window in each door (Fig. 7.3) reduces the possibility of injury by a door opened suddenly. It also allows visual appraisal of the situation in a room before entering. Such a window of $\frac{1}{4}$-inch plate glass, adequately sealed, does not seriously

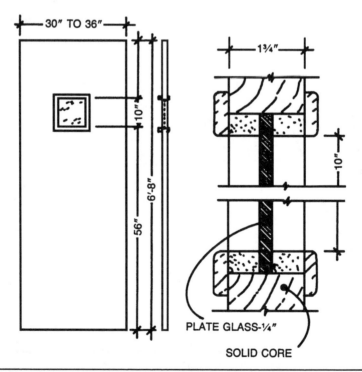

FIGURE 7.3   Plans for an acceptable solid-core door for studio, control room, and sound lock use. The sound lock places two such doors between high-level noise sources and sound sensitive areas. The window prevents collisions and allows appraisal of the situation in a room before entering.

deteriorate the acoustical quality of the door but adds materially to functional efficiency. Some people think that upholstering a door makes it "more acoustical." As far as transmission loss through the door is concerned, vinyl cloth covering over a sheet of foam rubber and studded with large-headed tacks is a waste of time and effort. However, such a covering does offer sufficient high-frequency absorption to reduce unwanted reflections or flutter echoes.

Weatherstripping Doors

A hermetically sealed door is the ideal type from an acoustical standpoint. Such a condition can be approached in sound lock doors by careful installation of common weatherstripping materials. These are available in a host of different configurations. Figure 7.4 illustrates representative types. Strips of foam or felt in the form of metal backed strips, rolled beading or gummed strips are shown in Fig. 7.4A. Strip magnets, enclosed in flexible plastic as shown in Fig. 7.4B, are attracted to mild steel strips imbedded in the door, effectively sealing the opening. These are commonly used in refrigerator doors and they can provide an excellent acoustical seal.

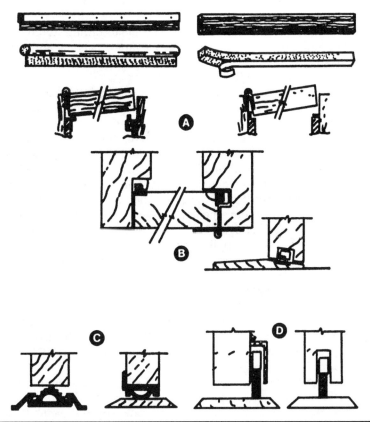

FIGURE 7.4   Some of the ways to weatherstrip doors to reduce sound leaks: (A) foam or felt strips, (B) magnetic seals such as used on refrigerator doors, (C) wiping rubber seal, and (D) automatic drop seal.

The sides and top of a door are easier to seal than the bottom. A wiping rubber seal is held in place by a shaped metal or plastic retainer in Fig. 7.4C. In Fig. 7.4D, a mechanical threshold closer automatically lifts as the door is opened and drops on closure. The types of weatherstripping in Fig. 7.4A are usually considered impractical for door bottoms because in such a position they take a beating from passing shoes. For this reason, one of the types shown in B, C, or D are far more effective at the threshold.

All of the door sealing methods pictured in Fig. 7.4 require periodic inspection to assure a continuing tight seal. Occasional adjustments are required by some and complete replacement by others as foam and felt deteriorate. Door seal maintenance should be added to the studio maintenance reminder list, which normally stops at such things as amplifier response, recorder head alignment and cleaning, etc.

HVAC Noise

It is a difficult word to pronounce, but HVAC stands for heating, ventilating, and air conditioning facilities. In this book the term *air conditioning* (A/C) includes heating and ventilating. With proper advance precautions, A/C equipment noise in the sound-sensitive studio and control room areas need not be a problem. In practice, it is very often a problem because the higher standards required in recording studios are infrequently encountered by architects, building contractors, and air conditioning equipment suppliers and installers.

What background noise level is acceptable in recording studios? Single figure noise levels, lumping the entire audible frequency range together, are of very limited value. For this reason noise criteria (NC) curves have been proposed for adoption as standards. Figure 7.5 illustrates a family of such curves.

The downward slope of these curves reflects both the ear's increasing sensitivity with increasing frequency and the spectral shape of common noises decreasing with increasing frequency.

An NC-15 contour is a reasonably stringent, though usually quite attainable, noise specification and goal for recording studios. An NC-20 contour is a more relaxed specification for less critical recording. The sound analysis contours of Fig. 7.5 are for octave bands of noise. Sometimes single frequency hums or whines associated with motors or fans stand out prominently from the general background of distributed noise. The peaks of such single frequency components determine the NC rating contour applicable. Thus, if the noise levels in all bands are below the NC-15 contour except one which reached the NC-30 contour, the NC-30 applies to that noise rating. The contours of Fig. 7.5 offer a means of specifying the maximum permissible noise from air conditioning facilities or from other sources.

The ducting arrangement of Fig. 7.6A illustrates a method of reducing crosstalk between adjacent rooms through the duct from grille to grille. By avoiding serving adjacent rooms directly from the same duct (supply or exhaust), the duct path is lengthened and attenuation via the duct path increased. The greater the length of lined duct and the more lined bends between one grille and the next, the greater the attenuation and the less the crosstalk.

Figure 7.6B isolates the effect of absorptive duct lining on the attenuation of sound through the duct. Note that high-frequency noise is absorbed better than low-frequency noise. The 2-inch thick liner attenuates sound down the duct better than 1-inch thickness.

Figure 7.6C pictures the effect of a lined duct bend. The attenuation is greater at high frequencies than low, unfortunately similar to the straight lined duct. With an arbitrary rule that a minimum of 15 feet of lined duct and two lined bends must be

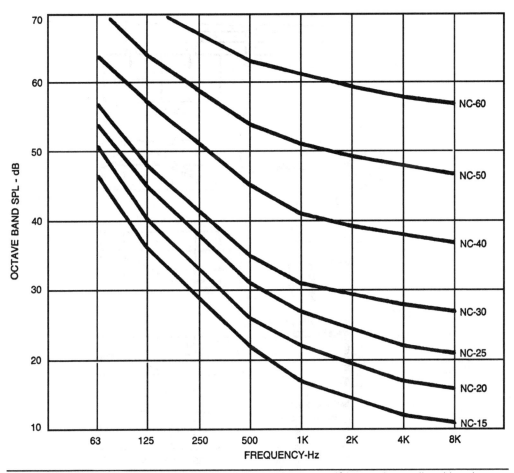

FIGURE 7.5   Noise criteria curves useful in assessing noise or specifying maximum allowable noise levels in studios and other sound-sensitive areas.

installed between any two adjacent rooms, what kind of attenuation of crosstalk from room to room can you expect? The curves help a rough estimate of the following:

	Attenuation		
	250 Hz	1 kHz	4 kHz
15' lined duct (1" lining)	15 dB	33 dB	27 dB
2 lined bends	4	10	20
Total	19 dB	43 dB	47 dB

This dramatically illustrates the fact that duct attenuation is easier to achieve at high frequencies than in the low-frequency region. Placing 30 feet of ducting between rooms increases the total 250 Hz attenuation to 34 dB. Hums and whines often occur in the 125 or 250 Hz bands. Another possibility not detailed here is to use a sharply tuned duct stub (acoustical band reject filter) to attenuate a troublesome single-frequency noise component.

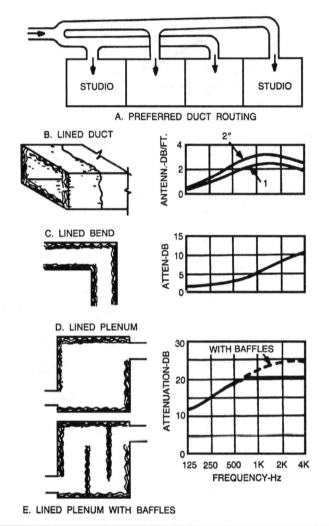

A. PREFERRED DUCT ROUTING

B. LINED DUCT

C. LINED BEND

D. LINED PLENUM

E. LINED PLENUM WITH BAFFLES

FIGURE 7.6 Means of reduction of air conditioner noise in studios and the reduction of crosstalk between adjacent rooms via a duct path.

It is usually easier to combat noises at their source than to introduce adequate attenuation downstream. The lined plenum of Fig. 7.6D, which offers excellent attenuation across the band, can be inserted in the duct between the A/C machinery and the sound sensitive areas. Baffles in the plenum increase the high-frequency attenuation, although as noted above, adequate high-frequency attenuation is usually available elsewhere. In some installations a large step-inside plenum or air chest is part of the basic equipment. If this plenum is lined with acoustically absorbent material, the noise traveling down the ducts is greatly reduced and the larger the plenum, the greater the low-frequency attenuation.

Air velocity is commonly higher in the budget A/C installations than in the more professional jobs. The air velocity should never exceed 500 feet per minute at the grilles to avoid generating excessive hissing noise due to air turbulence.

Wall Construction

The common 2 × 4 frame wall construction with a single layer of gypsum drywall on both sides is shown in Fig. 7.7A. It is included as a point of comparison only; its sound transmission class rating of STC 34-38 is too low for walls contingent with studios and control rooms. However, it could be used for sound lock walls not touching studios and control rooms. This wall can teach us several things. (Sound transmission loss of a wall varies with frequency. To achieve a convenient rating scheme a standard contour has been adopted which, when applied to the actual characteristic of a wall, results in a single number rating called the *Sound Transmission Class*, STC. Think of the STC rating number as a sort of average midband transmission loss in dB.)

For example, insulation in the cavity of this type of wall, with both faces closely coupled by being nailed to the same 2 × 4s, increases transmission loss only modestly. Another problem is that both faces vibrate as diaphragms. As they are identical they resonate at the same frequency and at this frequency there is a sort of acoustical hole in the wall. By making the faces of different thicknesses, of different densities, or supporting them in different ways, the two resonances are made to occur at different frequencies, improving the wall performance. The caulking of the perimeter, of at least one face but preferably both, seals the tiny cracks that are inevitable in normal construction.

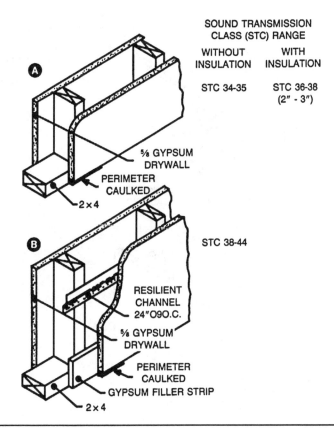

Figure 7.7   A comparison of wall constructions in regard to their ability to shield sound-sensitive areas against intruding noise: (A) common single-stud wall for comparison and (B) single-stud wall with resilient mounting of one face.

Resilient Mounting

There is a very substantial improvement in transmission loss if gypsum board on one face is screwed to a resilient channel as in Fig. 7.7B. This must be done carefully, with screws designed for the purpose, or else the resilient channel might be "shorted out."

For example, if a screw that is too long hits a stud, the resilience is lost. Acoustical elements, cabinets or shelves attached to such walls must be mounted carefully lest the extra expense and effort of using the resilient channel are destroyed by solid contact of the gypsum board with the studs.

If a double layer of gypsum board is required on a resiliently mounted wall, the base layer is attached vertically with screws and the face layer cemented in place following the manufacturer's recommendations.

Staggered Stud Construction

Figure 7.8A shows typical staggered 2 × 4 studs with 2 × 6 plate. This eliminates the solid connection of one wall diaphragm with the other except around the periphery.

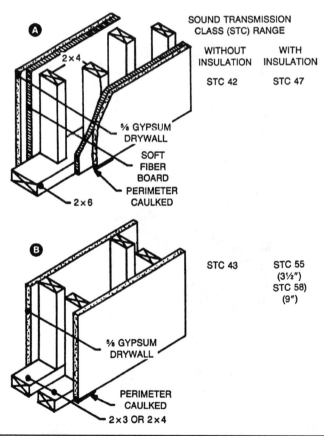

FIGURE 7.8   A comparison of wall constructions of (A) a staggered stud wall and (B) a double stud wall.

This does essentially the same thing as mounting one of the wall diaphragms resiliently as in Fig. 7.7B, and the STC results are somewhat comparable.

An additional feature of Fig. 7.8A is the use of soft fiberboards under each gypsum layer as sound deadeners. Because of their low density such soft boards contribute little to transmission loss directly, but they do serve as frictional elements in dampening vibrations of the gypsum diaphragms.

Nailing is the common method of supporting both the soft base layer and the gypsum face layer. In Fig. 7.8A both sides are identical which, as you have seen, is less desirable than making one side differ from the other. One satisfactory combination is a layer of $\frac{5}{8}$-inch gypsum board over a $\frac{1}{2}$-inch layer of soft sound deadening board on one side and a double layer of $\frac{5}{8}$-inch gypsum on the other side. With ample insulation fill, a staggered stud wall with such facings adequately caulked comes close to STC-50—a good value for normal studio walls. The effectiveness of the filler insulation depends on thickness, but is independent of density. Therefore, the cheaper household thermal type of insulation is quite adequate for filling acoustical walls.

Double Walls

Double wall frame construction is shown in Fig. 7.8B. There is only a minor difference between walls framed of double 2 × 3s and 2 × 4s. The two wall diaphragms are still connected at the periphery through a common foundation (concrete floor?) which is somewhat less coupling than that provided by a common plate in staggered stud construction. The double 2 × 3 wall, if carefully constructed and sealed, can reach STC-55 to 58 with proper insulation fill.

Concrete and Masonry Walls

In new construction and in some cases of renovation, concrete or masonry walls are a viable choice. Ratings that apply to several common walls are found in Table 7.1. Practical

Wall	Sound Transmission Class
Concrete—4 inches	STC-48
Concrete—8 inches	STC-52
Concrete blocks—4 inches	STC-40
Painted both sides	STC-44
Plastered both sides	STC-44
Concrete block—8 inches	STC-45
Painted both sides	STC-46-48
Plastered one side	STC-52
Concrete block—8 inches	
Voids filled with well-rodded concrete and plastered both sides	STC-56

TABLE 7.1 Transmission Loss In Concrete And Masonry Walls

concrete and masonry walls are quite comparable to framed walls in their STC ratings. The hard walls are somewhat inferior in another way—that of efficiently conducting structure-borne impulse noises from afar and reradiating them into sound-sensitive areas.

Floor-Ceiling Construction

If the space above a studio is occupied and people are stalking around with hard-heeled shoes, the situation calls for careful attention. Impulsive sounds of this type penetrate to an extent that noises of other types seem tame by comparison. Floor-ceiling construction becomes very important in such cases. The construction in Fig. 7.9A is

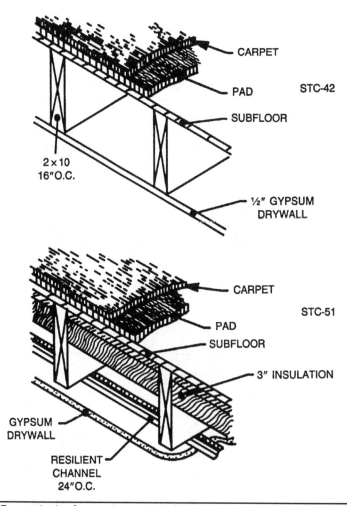

CARPET

PAD   STC-42

SUBFLOOR

2 x 10
16"O.C.

½" GYPSUM
DRYWALL

CARPET

STC-51

PAD

SUBFLOOR

3" INSULATION

GYPSUM
DRYWALL

RESILIENT
CHANNEL
24"O.C.

FIGURE 7.9   Two methods of protecting a studio from noise from the floor above with frame construction: (A) with normal gypsum board ceiling and (B) with resiliently mounted ceiling and insulation in the air space.

very common and the carpet helps to reduce footfall noise. The STC-42 rating, however, is marginal for most budget studios. The construction in Fig. 7.9B yields STC-51 by adding a resilient ceiling below and some insulation in the cavity between the floor joists.

Electrical Wiring

Building a 50 dB wall and then loosely mounting electrical boxes back to back is an exercise in futility. A surprising amount of sound can leak through a very tiny opening and through small areas of thin spots in a wall. Electrical boxes are necessary, however, and Fig. 7.10A suggests staggering them and using copious quantities of acoustical sealant to seal openings and beef up the boxes. Surface boxes for microphone connectors reduce compromising the wall and they may be handled as shown in Fig. 7.10B. In addition to sealant at the ends of the metal conduit, it is a good idea to also pack glass fiber tightly around the audio pairs to avoid sound traveling through the conduit itself.

Lighting

If fluorescent lighting is considered, the ballast reactors should be removed from the fixtures and mounted in a metal box in the sound lock or completely outside the suite.

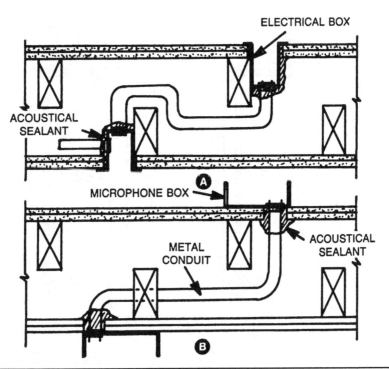

FIGURE 7.10   Treatment of wiring boxes with acoustical sealant to reduce sound leaks: (A) recessed electrical boxes and (B) surface mounted microphone boxes.

Although this takes more wire, it removes the electrical and acoustical buzzes these reactors are famous for generating outside the recording and sound evaluating areas.

Track lighting fixtures have the advantage of flexibility in concentrating the light where it is needed and hiding the light source from the eyes of those in the other room. This is the proper way to eliminate troublesome reflections in the observation window glass.

Light dimmers of the selenium controlled rectifier type create electrical noises which are likely to give trouble in the low-level microphone circuits.

Observation Window

An observation window plan for staggered stud and double wall construction is shown in Fig. 7.11A. The window frame is in two parts, one nailed to the studs on the control room side and the other to the studs on the studio side. In this way the glass on each side is an extension of its own wall and has no solid connection to the other side. Positioning felt or sponge strips in this gap between the two frames prevents accidental solid contact between them. A comparable wall for single-leaf construction is shown in Fig. 7.11B. Beads of nonhardening acoustical sealant seal off tiny cracks between the window frame and the walls in both types.

Each glass plate is a resonant system as is the air cavity between. By using glass plates of different thicknesses the effects of plate resonance are minimized by preventing

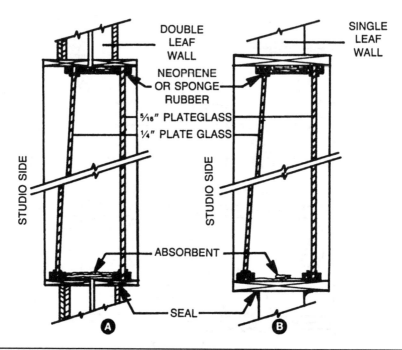

FIGURE 7.11 Plans for observation window having a reasonably high transmission loss: (A) for a double-leaf wall (staggered stud, double-stud, or double-masonry walls), and (B) for single-leaf wall.

them from occurring at the same frequency. The cavity resonance is controlled by utilizing an absorbent reveal periphery. This can be acoustical fiberboard or glass fiber of the 703 type (Owens-Corning Fiberglas Corporation Type 703 industrial glass fiber semirigid boards of 3 pounds per cubic foot density, used widely in the studio designs to be described) with a cloth cover, or even strips of heavy carpet.

In any event, the periphery between the glass plates should be black or of a dark color to avoid attracting undue attention to it. Each glass plate is isolated from its retaining stops and the frame by strips of neoprene or sponge rubber. The strips bearing the weight of the glass plates should not compress more than about 20 percent under load, but the side strips can be much more pliant. Fasten the stops on one side at least with screws so that the inside surfaces of the glass plates can be reached for cleaning if necessary.

Absorber Mountings

All of the absorption coefficients listed in the appendix have been measured in reverberation rooms following standard procedures (ASTM:C-423-84a or the equivalent). A 72 square foot patch of the material to be measured is arranged on the floor of the reverberation room. The specimen is mounted to reflect the requirements of the test. For example, the absorbence of the material mounted on a hard surface may be desired, or a 16-inch air space may be needed in the case of material for suspended ceilings.

Mounting designations have been recently changed (ASTM: E-795-83) and the industry is now in a transition between the old and the new. Fortunately, the old and the new are quite comparable; only the designations have been changed significantly. A comparison of old and new mountings is given in Table 7.2. Drawings of the six most commonly

New Mounting Designation*		Old Mounting Designation**
A	Material directly on hard surface	#4
B	Material cemented to plasterboard	#1
C-20	Material with perforated, expanded, or other open facing furred out 20 mm ($\frac{3}{4}$")	#5
C-40	Ditto, furred out 40 mm ($1\frac{1}{2}$")	#8
D-20	Material furred out 20 mm ($\frac{3}{4}$")	#2
E-405	Material spaced 405 mm (16") from hard surface	#7

*ASTM Designation: E-795-83
**Mountings formerly listed by Acoustical and Board Products Manufacturers Association, ABPMA, (formerly the Acoustical and Insulating Materials Association, AIMA).

TABLE 7.2   Relationship between Old and New Mounting Designations

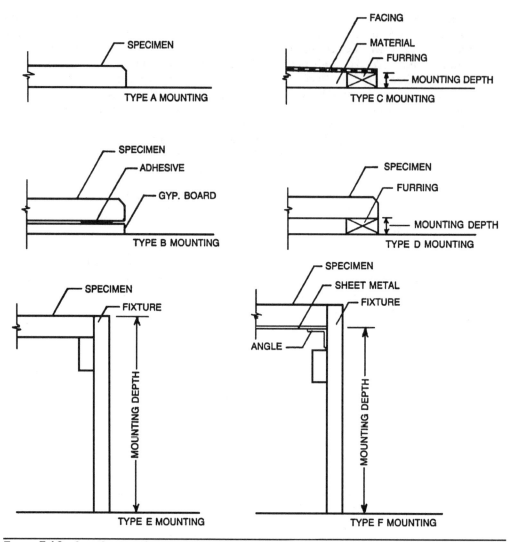

FIGURE 7.12   Specifications for mounting absorbing material for reverberation room measurements (see Table 7.2).

used mountings are shown in Fig. 7.12. With the aid of this table and this figure, absorption coefficients from recent or older sources can be compared and understood.

Reverberation Time

Applying the concept of reverberation time to small rooms, such as the 12 designs to follow, has come under considerable scrutiny since the first edition of this book appeared. The argument runs like this: The reverberation time equations have been derived on the assumption that a random sound field exists. Since a true reverberant field seldom exists in a small room, the concept of reverberation time should not be applied.

It is a valid criticism, but practical considerations dictate their continued use, while being aware of their limitations. The normal modes of a small room decay, even though they are too few and too far between. The decay of these modes, excited by voice or music signals, constitutes the sound of the room.

Calculating the reverberation time of a small room gives a basis of apportioning the areas of various absorbing materials to give uniform overall absorption throughout the audible band. This is the procedure followed in the upcoming chapters.

Sabine's early work resulted in the reverberation time equation bearing his name. Eyring and numerous others published different versions of Sabine's equation, some of which are supposed to give more accurate results in small rooms. Young[43] has pointed out that absorption coefficients, such as those in the appendix, are Sabine coefficients which can only be directly applied in the Sabine equation. For this reason, only the simple Sabine equation is used in the chapters to follow.

Foams

Numerous attractive and effective foam panels of various dramatic surface textures are now available under such names as Sonex (manufactured by Illbruck Company), Acoustafoam (manufactured by FM Tubecraft Support Systems, Inc.), and Sound-Sorber (manufactured by Discrete Technology). These companies may be located through the listing in Appendix D.

None of the designs in this book specify any of these foam panels, although they may be used. They are more expensive than glass fiberboards for the same absorption (sabins per dollar). However, they are easier to apply, and are more pleasing to the eye. Compare sound absorption coefficients for yourself and feel perfectly free to make substitutions if the expense seems to be justified.

Corner Absorbers

In recent years several products have become available which have special significance for small room acoustics. These are portable absorbers intended to stand (or be mounted) in the corners of the room. All room modes terminate in the corners of the room. Absorbers placed in the corners therefore act on all modes, not just some of them.

Among these corner absorbers are Tube Traps® (manufactured by Acoustic Sciences Corporation) and Korner Killers® (manufactured by RPG Diffusing Systems). Both companies can be located through the listing in Appendix D.

Although acoustically quite applicable to most of the designs to follow, permanent installations have been favored over the portable units. The latter, admittedly, allow you to "trim" a room by ear until satisfactory conditions prevail, but not everyone is qualified to make such judgments. Try incorporating such interesting new products in an experimental program.

Construction Permit

It is imperative that a construction permit be obtained before work is started. This requires plans and specifications. During construction, inspections can be expected covering structural, electrical, and plumbing installations. Obtaining the permit gives evidence that zoning restrictions are met and gives assurance that fire and other insurance will not be invalidated at a later time.

CHAPTER 8

Audiovisual Budget Recording Studio

Here is the problem presented by the client: to build a small, repeat small, studio and control room suitable for producing sound tracks for audio-visual presentations such as filmstrips, slide sets, and 16 mm motion picture film shorts. It is to be placed inside a large prefabricated building with ample headroom, but floor space is at a premium. On top of this, the cost must be held to an absolute minimum. Quality performance; bottom dollar. This message is familiar enough and occurs often enough to suggest a detailed treatment of the solution.

Studio

Speech is the predominant sound to be recorded, to which is added canned music and sound effects from subscription disc or tape libraries in the editing process. This means that the 1500 cubic foot minimum room volume discussed earlier is acceptable for the studio. The floor plan of Fig. 8.1 provides a studio with a floor area of 158.5 square feet and, with a 10 foot ceiling, a volume of 1585 cubic feet. The dimensional ratio of 1:1.14:1.39 distributes axial modes quite well as shown in Fig. 6.4A and Table 6.2A. The two closest modes near 283 Hz are high enough in frequency to be unlikely to cause voice colorations. A studio of these dimensions has a response down to 40 Hz, which is more than adequate for voice.

Control Room

The high ceiling of the prefabricated building allows a control room ceiling of any reasonable height to be specified. This suggested the possibility of standing the control room on end, so to speak, to minimize floor space. The dimensional ratio of Fig. 6.4B and Table 6.2B of 1:1.28:1.54 was selected over the dimensions in Table 6.2A because a longer (or, in this case, higher) room results.

The elevation sketch in Fig. 8.1 shows the ceiling of the control room at 13 feet 9 inches. The 9.0 feet × 11.4 feet × 13.75 feet dimensions of this control room are slightly different from the case of Table 6.2B (10.0 feet × 12.8 feet × 15.4 feet), but are close enough for us to use Fig. 6.4B to get a qualitative view of mode spacings.

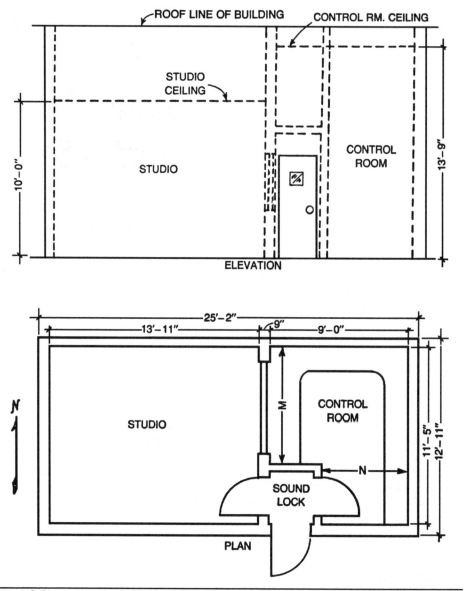

FIGURE 8.1   Plan and elevation of a budget studio suite for the production of audiovisual materials. Because of limited floor space the control room "stands on end" to obtain the requisite room volume and room proportions.

The two modes near 124 Hz, only 2.3 Hz apart, alert us to possible voice colorations at that frequency. The three just below 250 Hz are probably high enough in frequency to be less troublesome. Remember, however, that voice recording is done in the studio; this considers only factors which might affect listening conditions in the control room.

Sound Lock

The sketch in Fig. 8.1 includes a tiny sound lock corridor to control the effects of both entering from the noisy exterior, and traffic between the studio and control room. This places two doors in series for any given path which eliminates the need for expensive, special acoustical doors with their awkward clamping hardware and seals. Also, it allows studio access and egress during recording without a blast of noise from monitor loudspeakers or outside. Sound locks should always be a part of any studio suite intended for professional or quasiprofessional use.

Stealing space for the sound lock from the control room affects the modal situation in the control room. The pristine rectangular room with the usual three axial modes now has two other modes added, those associated with the M and N dimensions of Fig. 8.1. These dimensions of about 8.0 and 5.5 feet reduce average mode spacing, which is good, but increase the possibility of coincidences, which is bad. Because these small dimensions resonate at higher frequencies we can expect their effect to be less noticeable. A detailed examination in such cases always alerts you to a potential coloration problem.

Work Table

A control room dedicated to audiovisual activity needs worktable space. For this reason a built-in workbench with some drawer and cabinet space below is suggested. The mixing console for such an activity is normally one of the simple desktop models and the recorders are of the advanced audiophile type which require table space rather than floor space.

Studio Contracarpet

When the client says, "Carpet on the studio floor," the acoustical consultant gulps and bravely says, "Can do!" Carpet plays a dominant role in the acoustics of the studio because the floor area is a substantial part of the total surface area of the room; the problem is that carpet absorbs well at higher frequencies and very poorly at lower frequencies. Carpet on the floor dictates compensating absorbers peaking at low frequencies and this often means tuned Helmholtz units.

In the elevation of Fig. 8.1 the studio and control room walls run all the way to the roof. This is necessary to prevent flanking sound traveling from one room to the other via the "attic." Establishing an acoustical ceiling in the studio can be easily done by suspending it with common metal angles and Tees and wires. However, instead of the usual 24 inch × 48 inch soft fiber lay-in panels, the special *contracarpet* panels detailed in Fig. 8.2 are used.

BBC engineers have used the term *anticarpet* to denote an absorber to compensate for the carpet deficiencies. However the term *contracarpet* seems to be somewhat more descriptive, at least in this case where the contracarpet units are opposite the carpet. The contracarpet units are Helmholtz resonators about 2 inches thick fabricated in the familiar 24 inch × 48 inch size. The side facing the studio is $\frac{3}{16}$-inch plywood or Masonite perforated so that about 0.1 percent of its area is holes. Holes of $\frac{3}{16}$-inch diameter spaced 6 inches on centers give a perforation percentage of about the proper magnitude. The

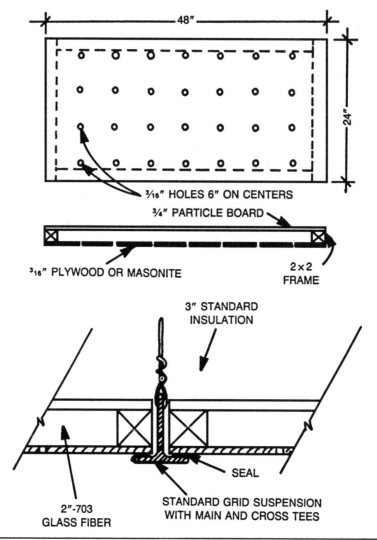

FIGURE 8.2   Contracarpet panels are used in suspended ceiling of studio to compensate for unbalanced sound absorption characteristics of the carpet. They operate on the Helmholtz resonator principle.

back (top) of each unit is of ¾-inch particleboard (chipboard) which is somewhat denser than plywood. This particleboard constitutes the acoustical ceiling which should be established 10 feet from the floor.

Sections not designated for contracarpet panels (Fig. 8.3) are filled with panels of ¾-inch particleboard. Thus the acoustical ceiling height varies by 2 inches from place to place. Both types of lay-in panels are set on a continuous bead of nonhardening acoustical sealant on the suspended Tee frames. This makes a virtual hermetic seal between the studio and the "attic" space and reduces the possibility of rattles.

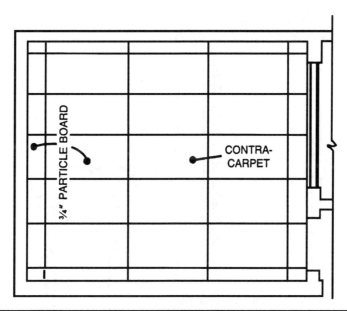

FIGURE 8.3   Projected ceiling plan of studio showing location of the seven contracarpet panels. Frame sections not holding contracarpet units contain blank panels ¾-inch particleboard.

A 2-inch thickness of 3 pounds per cubic foot density glass fiber is jammed into the approximately 1¾-inch space determined by the 2 × 2s between the particleboard and the perforated facing. (Throughout the book glass fiber of this density is repeatedly specified. Owens-Corning Type 703 is admirably suited. It is available in thicknesses of 1, 1½, and 2 inches, but building up a thickness of, say, two 2-inch thicknesses to obtain a 4-inch thickness is acoustically equivalent. Johns-Manville has a product, Series 1000 Spinglass of 3 pounds per cubic foot density which is also acceptable.) This glass fiber broadens the low frequency absorption peak. On top of the contracarpet and blank panels a blanket of common house insulation material of approximately 3-inch thickness is laid. If paper is attached, it should be placed downward. The purpose of the insulation layer is not so much to make the ceiling more impervious to sound as to deaden the space resonances in the "attic."

Studio Wideband Wall Units

The third acoustical element in the studio (in addition to the carpet and contracarpet) is a series of identical wall units constructed as shown in Fig. 8.4. Each of these is nothing more or less than patches of 4-inch thick glass fiber of 3 pounds per cubic foot density, each having an acoustically effective surface of 12 square feet. These give essentially perfect sound absorption at 125 Hz and above.

The frame is of ordinary 1-inch lumber. The backing board of ³⁄₁₆-inch or ¼-inch plywood or Masonite is only to strengthen the frame and to make each unit a self-contained entity which can easily be mounted or removed. The cloth cover serves both

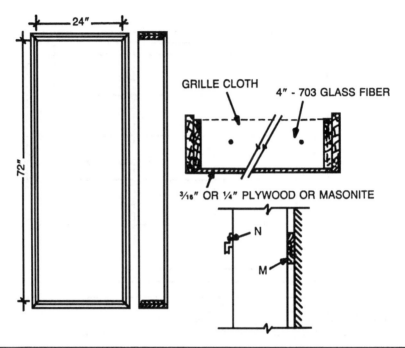

FIGURE 8.4   Construction details of wall modules having wideband absorption characteristics. Used in both studio and control room.

as a cosmetic function and as an aid to contain the irritating glass fibers. This cloth should be treated with fire retardant chemicals for safety. Loudspeaker grille cloth is ideal, although relatively expensive. Ordinary burlap or other open-weave cloth of light weight can be employed. This fabric cover presents an excellent opportunity for color emphasis in the decor of the studio (pink noise penetrates even a purple grille cloth)!

The method of mounting the wideband units to the wall is left to the ingenuity of the builder, although a simple suggestion is shown at M in Fig. 8.4. If molding M is a strip running the length of a wall, the units may easily be positioned laterally anywhere on the wall. Further, a metal hook N on each front edge of the frame allows reversal of the unit.

In this way complete flexibility is realized: mounting or removing, positioning and reversing. With the soft side out, reverberation time is decreased. With the hard back exposed, it is increased, yet retains the advantage of a rectangular protuberance for diffusion of sound and for a measure of control of flutter echoes. There are limits to such adjustments of the acoustical properties of the room, but this degree of flexibility comes with negligible cost.

The suggested locations of the wideband modules on the walls of the studio are shown in Fig. 8.5. Three modules on the west wall oppose the window and the door on the east wall. The pair of modules on the north wall opposes bare areas on the south wall, and vice versa. At first glance, Fig. 8.5 seems to show module opposite module, but remember that these wall elevations are the view one has facing the wall from inside the room and one must "do a 180°" between looking at the north wall and looking at the south wall.

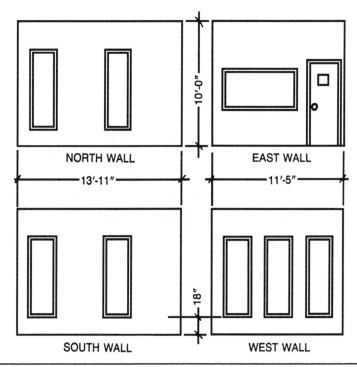

FIGURE 8.5   Studio wall elevations showing location of wideband modules.

Studio Drywall

If a structural element contributes significantly to sound absorption in the studio, it must be considered as part of the acoustical treatment. The type of wall construction utilized in this studio is illustrated in Fig. 8.10. A layer of gypsum drywall panels is applied to one face of the wall and a double layer to the other face. As far as noise isolation is concerned, either face can be on the studio side. In the ensuing calculations it is assumed that the single layer of drywall is toward the studio and the control room although there would be only a minor difference in absorbing effect if it were the other way around. The gypsum panels on both sides of the wall vibrate as diaphragms on the cushion of air contained between them. The sound absorbed is greatest near the resonance frequency of the panel which, in turn, is a function of the depth of air space and mass per unit area of panel.

Absorption coefficients are available for $\frac{1}{2}$-inch gypsum board on 2 × 4 framing which resonates at about 61 Hz. Using $\frac{5}{8}$-inch instead of $\frac{1}{2}$-inch and a nominal air space of 8 inches instead of 4 inches shifts the resonance frequency down to about 38 Hz. This reduces the absorption coefficients in the 125 Hz to 4 kHz range somewhat.

In Table 8.1, however, the available published values for the $\frac{1}{2}$-inch thickness and 4-inch airspace are used to avoid complicating the procedure. Both the contracarpet and blank panels of the ceiling contribute slightly to low-frequency absorption as diaphragms, over and above the contracarpet Helmholtz resonator effect. This compensates somewhat for the fact that the wall used differs from the one to which the coefficients strictly apply.

SIZE .... 11'5" × 13'11" × 10'0" Ceiling
FLOOR .... Heavy carpet and pad
CEILING .... 7 – 2' × 4' Contracarpet modules (see Fig. 8.3)
WALLS .... 7 – 2' × 6' Wideband modules (see Figs. 8.4 & 8.5)
SURFACE AREA .... 823 square feet
VOLUME .... 1,585 cubic feet

Material	S Area Sq. Ft.	125 Hz a	125 Hz Sa	250 Hz a	250 Hz Sa	500 Hz a	500 Hz Sa	1 kHz a	1 kHz Sa	2 kHz a	2 kHz Sa	4 kHz a	4 kHz Sa
Carpet	159.	0.08	12.7	0.24	38.2	0.57	90.6	0.69	109.7	0.71	112.9	0.73	116.1
Drywall	665.	0.10	66.5	0.08	53.2	0.05	33.3	0.03	20.0	0.03	20.0	0.03	20.0
Contracarpet	56.	0.90	50.4	0.54	30.2	0.30	16.8	0.16	9.0	0.12	6.7	0.10	5.6
Wideband Modules	84.	0.99	83.2	0.99	83.2	0.99	83.2	0.99	83.2	0.99	83.2	0.99	83.2
Total Sabins, Sa			212.8		204.8		223.9		221.9		222.8		224.9
Reverberation Time, seconds (Sabine)			0.36		0.38		0.35		0.35		0.35		0.35

TABLE 8.1 Studio Calculations

Studio Computations

A bit of figuring gives us the required data for the studio: surface area = 823 square feet, volume = 1585 cubic feet. With this we can enter the *sanctum sanctorum* of the *Sabine equation* to determine the absorption required to realize our desired reverberation time, which is 0.35 second. The *Sabine equation* is:

$$T_{60} = \frac{0.049 \ V}{(S) \ (a)}$$

where

T_{60} = reverberation time, seconds
V = volume of studio, cubic feet
S = surface area of studio, square feet
a = average absorption coefficient

The next step is to determine the number of absorption units in the room required to give us the desired T_{60}, 0.35 second.

$$\text{Total absorption units} = (s) \ (a) = \frac{(0.049) \ (1585)}{0.35}$$

$$= 222 \text{ sabins}$$

Now, what does this mean? Simply that 222 square feet of perfect absorber ($a = 1.00$) in the room yields the desired 0.35 second reverberation time. The original perfect absorber conceived by pioneer acoustician Wallace Sabine was an open window.

All the sound falling on an open window is surely absorbed as far as the room is concerned, but the practical absorbing materials we have to work with are something less than perfect, especially if you add the requirement, "throughout the range of audible frequencies."

The room computation process requires some of what is euphemistically called *engineering estimating*. This is nothing more than guessing, but engineers become better and better guessers as their years of experience pile up. The guessing comes in deciding how much of what land of absorbing materials will give the 222 sabins for each frequency point throughout the band.

Carpet is specified, so there is no guessing about that. The carpet area is entered in Table 8.1. The absorption coefficients for the carpet selected are entered for each frequency. By multiplying the carpet area by each coefficient, the absorption in sabins is found for each frequency and entered in Table 8.1.

By plotting the carpet absorption points in Fig. 8.6, graph A is obtained. As drywall is the other fixed element, its absorption is calculated for each frequency and entered in Table 8.1. By adding carpet and drywall absorption and plotting the resulting sums on Fig. 8.6, graph B is obtained.

The drywall partially makes up for lack of carpet absorption in the low frequencies, but not enough. A few trial calculations show that seven contracarpet ceiling units give us graph C which is reasonably horizontal at roughly 140 sabins. This must be raised to the vicinity of 225 sabins and it is the function of the seven wideband wall modules (essentially perfect absorbers in the 125 Hz to 4 kHz range according to the manufac-

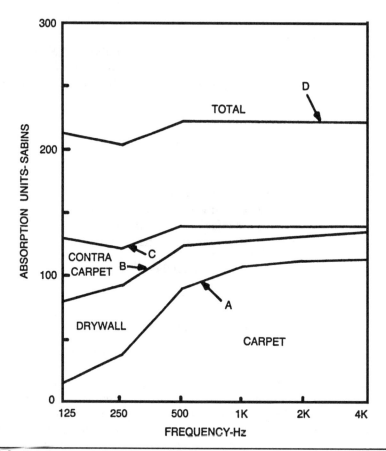

FIGURE 8.6   Relative sound absorption contributions of various elements used in treating the studio. Note that the contracarpet panels and drywall construction compensate for the low-frequency deficiency of the carpet.

turer's measurements) to do this (graph D of Fig. 8.6). The wiggles of graph D have only a minor effect on reverberation time.

Table 8.1 shows that reverberation time varies only from 0.35 to 0.38 second as a result of the fluctuations of graph D, Fig. 8.6. Our precision is not good enough to justify pursuing such calculations further. The calculated studio reverberation time variations with frequency are shown graphically in Fig. 8.7.

Control Room Treatment

The control room is generally admitted to be a work room, especially in audiovisual work. For this reason there is usually a minimum amount of opposition to vinyl tile floors which are especially practical for rolling equipment around the room.

Reverberation time for a control room should be somewhat shorter than that of the studio being monitored. The reverberation associated with studio sounds reproduced on the monitoring loudspeaker are then heard without being masked by control room

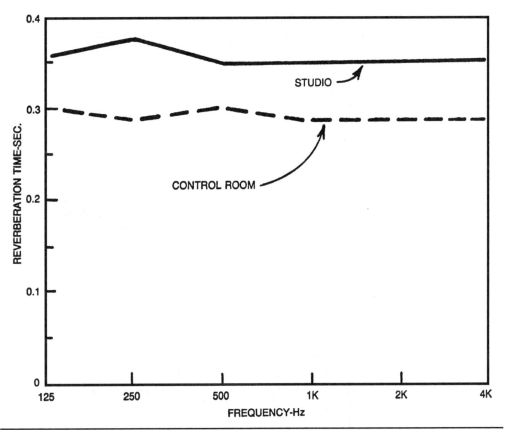

FIGURE 8.7   Computed reverberation time characteristics of the studio and control room. The evaluation of studio sounds monitored in the control room is aided by having lower reverberation in the control room.

reverberation. Listening rooms require relatively uniform reverberation time with frequency as do studios.

Control Room Ceiling Treatment

A standard suspended ceiling is specified to oppose the reflective vinyl floor. This ceiling is dropped 16 inches below the acoustical ceiling of drywall which is at a 13 foot 9 inch height (Fig. 8.8).

The distinction must always be made between the visual and the acoustical dimensions of a room. The space between the suspended lay-in ceiling and the solid ceiling at a 13 foot 9 inch height is acoustically active.

For example, there are specific resonance effects between the lay-in panels and the air space above them which result in good low-frequency absorption. This makes the overall absorption relatively uniform with frequency (Fig. 8.9E).

By using the usual soft fiber lay-in panels with an NRC (*noise reduction coefficient*) rating between 0.75 and 0.85, the suspended ceiling absorption coefficient varies between 0.65 and 0.92 (Table 8.2). By using the 16-inch drop instead of some other distance, the

FIGURE 8.8 Wall elevations of control room showing placement of standard suspended ceiling, 2-foot × 6-foot wideband wall modules and acoustical tile.

coefficients supplied by the manufacturer can be used with some confidence. Deviations from this standard 16-inch drop results in no known coefficients to depend upon.

Control Room Drywall

The diaphragmatic absorption of gypsum board surfaces must also be figured into the control room. The floor is vinyl tile covering concrete, so that is not included. Because of the complicating effect of the suspended ceiling we shall neglect the drywall ceiling above it. The walls alone (including doors and window which also act as diaphragms) total 561 square feet. The sum of drywall and suspended ceiling absorption from Table 8.2 is plotted as graph F in Fig. 8.9. Slight overcompensation now prevails in the low-frequencies.

SIZE .... 9'0" × 11'5" × 13'9" Ceiling
FLOOR .... Vinyl tile
CEILING ...... Standard suspended ceiling, panels NRC 0.75–0.85
WALLS .... 42 Acoustical tiles 12" × 12" × 3/4" (Fig. 8.8)
              7 Wideband modules 2' × 6' (Fig. 8.8)
SURFACE AREA . 742 sq. ft.
VOLUME .... 1,243 cu. ft.

Material	S Area Sq. Ft.	125 Hz a	125 Hz Sa	250 Hz a	250 Hz Sa	500 Hz a	500 Hz Sa	1 kHz a	1 kHz Sa	2 kHz a	2 kHz Sa	4 kHz a	4 kHz Sa
Suspended Ceiling	90.4	0.65	58.8	0.78	70.5	0.67	60.6	0.82	74.1	0.89	80.5	0.92	83.2
Drywall	561.	0.10	56.1	0.08	44.9	0.05	28.1	0.03	16.8	0.03	16.8	0.03	16.8
Acoustical Tile	42.	0.09	3.8	0.27	11.3	0.78	32.8	0.84	35.3	0.72	30.2	0.64	26.9
Wideband Modules	84.	0.99	83.2	0.99	83.2	0.99	83.2	0.99	83.2	0.99	83.2	0.99	83.2
Total Sabins, Sa			201.9		209.9		204.7		209.4		210.7		210.1
Reverberation Time, seconds (Eyring) (Sabine)			0.30		0.29		0.30		0.29		0.29		0.29

TABLE 8.2 Control Room Calculations

133

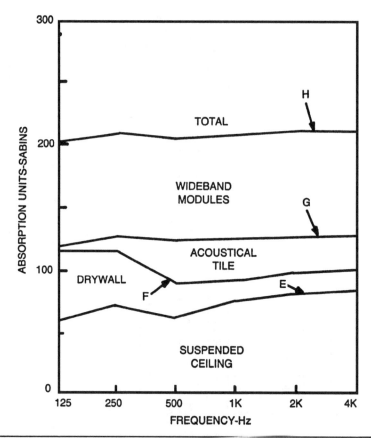

FIGURE 8.9   Relative sound absorption contributions of the various elements used in treating the control room. The unavoidable absorption of the drywall construction used results in too much low-frequency absorption which is corrected by the acoustical tile.

Control Room Acoustical Tile

Common acoustical tile is probably the most abused and misused product in the annals of sound treatment because people expect too much of it. However, it is inexpensive, easy to apply, and an excellent product if properly used. It is characterized, as is carpet, by good high-frequency absorption, but little at low frequencies.

This is exactly what is needed in the present case. The addition of 42 acoustical tiles 12 inches × 12 inches × ¾ inch to the control room brings the absorption to an approximately uniform 125 sabin level throughout the frequency range of interest as shown by graph G of Fig. 8.9. It is desirable not to have all of a given type of absorber in a room active in only one mode. Figure 8.8 shows the acoustical tile acting on both the N-S and E-W modes of the room.

Control Room Wideband Modules

The 125 sabin level of graph G in Fig. 8.9 is about 83 sabins below the 208 sabins required for a reverberation time of 0.25 second. Seven 2 foot × 6 foot wideband modules provide

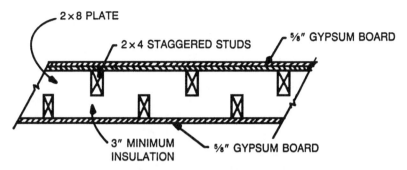

FIGURE 8.10   Construction details of studio and control room walls to protect against external noise and to provide adequate isolation between control room and studio.

this 83 sabins, bringing the total absorption up to graph H of Fig. 8.9 which hovers close to the 208 sabin goal. The control room reverberation tine of Table 8.2 is compared graphically to that of the studio it serves in Fig. 8.7.

A grand total of 14 of the 2 foot × 6 foot wideband modules detailed in Fig. 8.4 are now required, seven for the studio and seven for the control room. Some economy in effort and expense should result from mass production.

Noise Factors

The level of noise outside the studio suite and background noise standards set for inside the studio determine the type of wall construction. A nearby printing press, buzz saw, or another such noisemaker may require exceptional measures.

However, if this is to be a budget studio, wall construction costs must be kept in line. Most small organizations attracted by the budget approach are willing to do some horse-trading. If the printing press operates only part time, the audiovisual people can schedule their recording time accordingly. Perhaps a flashing red light at the buzz saw during a take would suggest to the carpenter that this is a good time for a cup of coffee.

Having to repeat three takes a year because of low-flying helicopters is far cheaper than building a building within a building to get 80 dB transmission loss.

For this budget studio complex, an economical wall that offers good protection (about STC 50) against outside noise is illustrated in Fig. 8.10. The staggered stud principle gives two independent walls attached only at the periphery to the 2 × 4 studs 16 inches on centers on a 2 × 8 plate. A single layer of ⅝-inch gypsum board is nailed on one side and a double layer of ⅝-inch gypsum board on the other. Thermal type insulation of at least a 3-inch thickness minimizes resonances in the space between the walls. The effectiveness of even this staggered construction depends upon tight sealing. To assure this, a bead of nonhardening acoustical sealant should be run around all intersections of walls with ceiling, floor, and other walls. All exterior walls and the wall separating the control room from the studio should be of the construction shown in Fig. 8.10. The north and east walls of the sound lock and the sound lock ceiling may be of normal single stud construction with single layers of gypsum board. The control room ceiling at a 13 foot 9 inch height is standard ⅝-inch gypsum board on frame construction with the suspended ceiling dropped 16 inches below this.

CHAPTER 9

Studio Built in a Residence

$\mathbf{B}$uilding a studio and control room in the average modern single family dwelling presents major problems: thin walls, low ceilings, and limited floor area. In this case, however, the residence is not average. It has concrete floors, stone and brick walls, and ample headroom. Needless to say, it is located outside the United States. There are a number of lessons to be learned from this case, however, and the solutions to specific problems to be considered are quite applicable to other situations.

Floor Plan

The *as found* floor plan is shown in Fig. 9.1. Walls are either 17-inch thick stone or 10-inch thick brick or glass. The first two warm the acoustician's heart but the last one is mentally placed high on the *get rid of it* list. The dining and living rooms are separated by a 4-foot barrier. The floor of the dining room is 8 inches lower than that of the living room. This is the house that is available. Can it be converted into an effective studio and control room without major alterations?

Figure 9.2 illustrates the changes made. Basically, the living room was visualized as the studio and the dining room as the control room. This required pouring enough concrete into the dining room to bring its floor up flush with the living room floor. The north wall of glass was eliminated and a 10-inch thick brick wall established at the outer center column, enlarging the width of the studio more than 6 feet. A wall of brick to hold the observation window was placed at a 45 degree angle to give the control operator a good view into the studio and to provide certain acoustical advantages in both rooms. This angled wall makes the control room unsymmetrical which would be considered a disadvantage in a professional recording studio. However, in the present case the advantages outweigh the disadvantages.

A sound lock with brick walls was located as shown in Fig. 9.2 and the inner glass wall of the entrance hall was eliminated in the process. The external glass wall of the entrance hall survived as the entrance hall function remained unchanged. Door A in the west wall of what is now the control room was bricked up. This routed all traffic between the studio and other parts of the house either through door B in the west wall or by outdoor paths. After objections were voiced by the consultant at having a second door into the control room, it was left to the client to either establish a second sound lock in the west hallway or route all traffic outdoors and through the entrance hall. As this residence is in a tropical country, this latter should create no major problems.

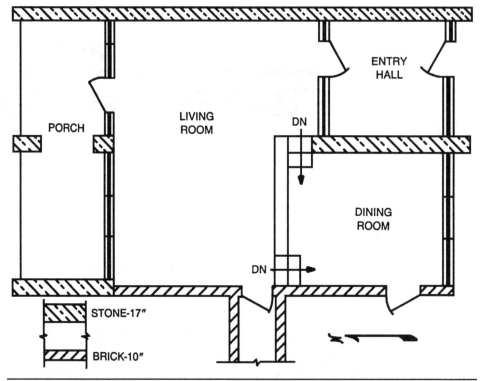

FIGURE 9.1 Floor plan of residence before conversion into studio complex.

The space in the sharp angled northwest corner of the control room was made into a closet for the storage of tape stock, recorded tape library, etc. Note also that the operator's position is set back at least 5 feet from the glass surface. This gives the operator an acceptable angle with the monitoring loudspeaker(s) and an acoustically better position for critical listening. The pillar in the studio area was slated for removal if not loadbearing. However, it could remain, causing some inconvenience but little adverse acoustical effect. The added studio volume obtained by pushing out the studio north wall is very important acoustically.

Studio Treatment

Stepping into the untreated studio of Fig. 9.2, one is met by a great expanse of brick and stone walls, concrete floor, glass observation window and the wooden underside of the roof. At this time you cannot escape the thought that it would make an excellent reverberation chamber. The first step in making a studio out of it is to determine the correct number of absorption units (sabins) required. Measurements reveal a volume of 4,614 cubic feet and an inside surface area of 1,597 square feet.

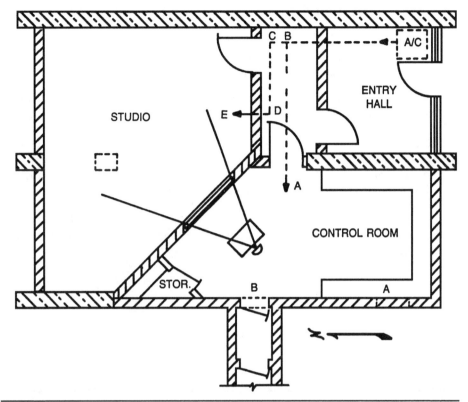

FIGURE 9.2   Floor plan of residence converted into studio and control room with sound lock.

What reverberation time should be adopted? As both music and speech are to be recorded in this room, a compromise value of about 0.48 second seemed in order.[2] Cranking these values of volume, and reverberation time into the Sabine equation it is computed that 470 absorption units, or sabins, are required. This is a point of departure and a start in building Table 9.1.

Ceiling

At the high frequencies the carpet is a dominant factor, supplying almost half the required absorption while contributing practically nothing at the low frequencies (Fig. 9.6 on page 144). This poses the classical problem of introducing other absorbing elements which have the opposite effect. This time semicylindrical panel units are chosen over the Helmholtz resonator approach. Such cylindrical units on the studio ceiling between the concrete roof beams, as shown in Figs. 9.3 and 9.4, will contribute in the following ways:

- They augment a thin roof in protecting against outside noise.
- They act as excellent diffusers of sound in the studio.
- They absorb sound in the studio in a way which tends to compensate for the carpet deficiencies at low frequencies.

These cylindrical elements are basically a thin skin of $\frac{3}{16}$-inch plywood or masonite stretched over bulkheads cut as segments of a circle. The radius and chord of this segment are carefully adjusted so that the arc is 48 inches—the standard width of plywood and masonite. The skin of these cylindrical elements, which vibrates vigorously in response to sound in the room, must not rattle. As protection against rattles a bead of nonhardening acoustical sealant, or better yet, a thin strip of felt, is applied to the edge of each bulkhead before the skin is bent over and nailed in place. The functioning and construction of such cylindrical elements (often called *polycylindrical diffusors* or *polys*) as well as absorption coefficients for units of different sizes are detailed in a companion volume.[3] The space within the semicylindrical units can be stuffed with common thermal type mineral wool or glass fiber.

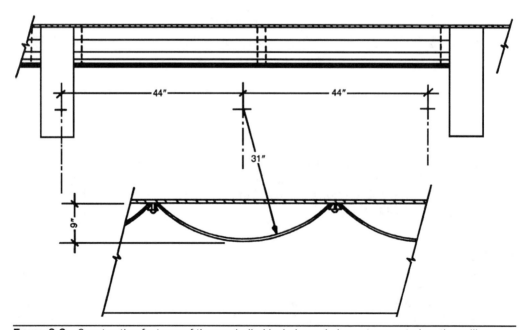

FIGURE 9.3 Construction features of the semicylindrical plywood elements mounted on the ceiling between the concrete roof beams. The radius and chord are chosen so that the arc is 48 inches—the plywood width.

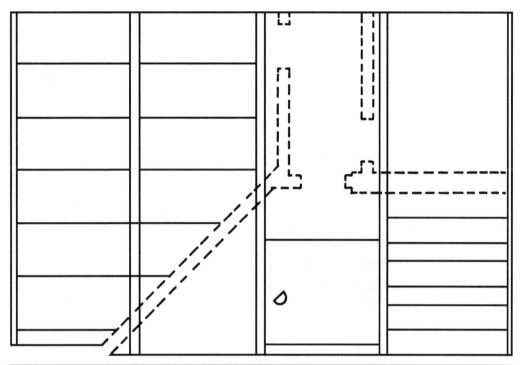

Figure 9.4 Projected ceiling plan for both studio and control room showing placement of semicylindrical plywood elements in the studio and patches of 4-inch glass fiber in the control room.

Walls

The compensation for carpet deficiencies by the cylindrical elements, as shown in Fig. 9.6, is quite good except for 125 Hz and below. The effects of this on reverberation time are considered more fully later. To approach the required 470 sabins, about 256 square feet of wideband absorber are required. This absorption can be supplied by 4 inches of 703 glass fiber or its equivalent. The suggested wall modules have several advantages. They:

- Contribute to the diffusion of sound in the room, thus making microphone placement less critical.

- Allow for greater trimming of acoustics if measurements indicate the necessity for this.

- Result in economy by being used in both studio and control room.

A total of sixteen 2 foot × 8 foot units provide the required 256 square feet. They are positioned on the walls as shown in Fig. 9.5.

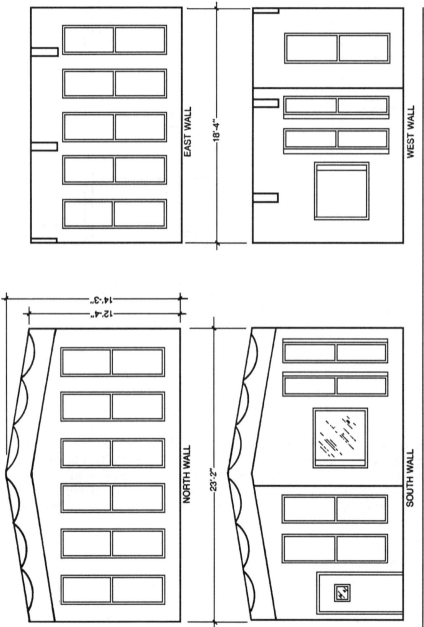

FIGURE 9.5 The four wall elevations of the studio showing placement of the wideband absorbing modules.

		SIZE .... 18'4" × 23'1" (corner cut) × 13'3" ave. ceiling ht. FLOOR .... Carpet CEILING .... Semicylindrical plywood elements WALLS .... 16'2" × 8' Wideband modules VOLUME .... 4,614 cu. ft.											
	S Area Sq. Ft.	125 Hz		250 Hz		500 Hz		1 kHz		2 kHz		4 kHz	
Material		a	Sa	a	Sa	a	Sa	a	Sa	a	Sa	a	Sa
Carpet	349	0.05	17.5	0.15	52.4	0.30	104.7	0.40	139.6	0.50	174.5	0.60	209.4
Semicylindrical Plywood	328	0.45*	147.6	0.57	187.0	0.40	131.2	0.25	82.0	0.20	65.6	0.20	65.6
Wideband Modules	256	0.99	253.4	0.99	253.4	0.99	253.4	0.99	253.4	0.99	253.4	0.99	253.4
Total Sabins, Sa			418.5		492.8		489.3		475.0		493.5		528.4
Reverberation Time, seconds (Sabine)			0.54		0.46		0.46		0.48		0.46		0.43

*Mankovsky, Ref. 40

TABLE 9.1   Studio Calculations

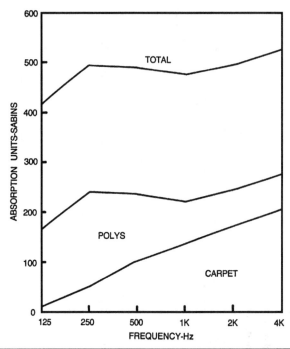

FIGURE 9.6   Absorption contributions of the three materials used in the studio. The semicylindrical plywood ceiling elements compensate for the carpet deficiencies quite well, except at 125 Hz.

Studio Reverberation Time

Table 9.1 brings together the specific absorption contribution of the carpet, the cylindrical ceiling elements, and the 16 wall modules at each frequency. The resulting reverberation time is plotted in Fig. 9.7. There are some deviations from the goal of 0.48 second. The slight drooping at high frequencies is no problem. It is actually preferred by many.

The bass rise shall be examined a bit more closely. BBC engineers have looked into this with characteristic thoroughness.[10] They found that the degree of impairment of speech quality by such bass rise was affected by the voice of the person speaking, the type of microphone used, and the distance between the speaker and the microphone (which determines the relative effect of room reverberation). As a tentative conclusion for the average situation, they suggested that a rise of reverberation time at 125 Hz over the 500 Hz value of no more than roughly 20 percent should be allowed for voice work. The rise in Fig. 9.7 is very close to this amount. They suggested that no more than about 90 percent rise of the 63 Hz reverberation time over that at 500 Hz be allowed. Table 9.1 stops at 125 Hz but we know that absorption of both ceiling elements and wall modules falls off below 125 Hz. It seems unlikely that the reverberation time at 63 Hz, however, would be greater than the allowed 0.9 second.

Control Room Treatment

Deciding in favor of a hard floor (concrete, vinyl tile, parquet wood, etc.) reduces the treatment of the control room to a single type of absorber, 4 inches of 703. Table 9.2

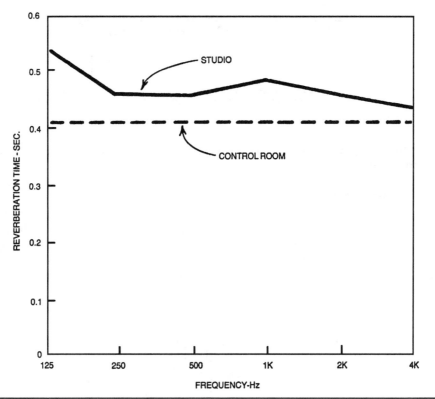

FIGURE 9.7   Reverberation time characteristics of studio and control room. The bass rise in computed studio reverberation is within BBC limits.

combines the 124 square feet of 703 assigned to the ceiling, the 184 square feet in wall modules on the east and west walls, and 48 square feet in wall modules on the south wall. It treats the 356 square feet total together. This total absorption, of course, was calculated from the Sabine equation by inserting the desired reverberation time of 0.41 second, and the volume of 2,964 cubic feet, turning the crank and coming out with a total of 352 sabins.

Ceiling Treatment

First, it is important that the ceiling over the operator be absorbent. The operator's position is indicated in Fig. 9.4 by the same symbol used in Fig. 9.2. Remember that this ceiling slopes as shown in Fig. 9.5 which is favorable for the floor-ceiling mode. The bare ceiling area must be distributed for the best diffusion effect and the pattern of Fig. 9.4 was selected.

The 4 inches of 703 is held to the ceiling in the manner detailed in Fig. 9.8. There is really considerable freedom in how this is done. If 703 of 4-inch thickness (instead of two 2-inch layers) is employed, the glass fiberboard is rigid enough to allow straight-forward cementing to the ceiling. This method does not provide for the cosmetic grille cloth cover nor does it offer protection from small glass fibers falling, but acoustically it is excellent. It may be desirable to stretch zig-zag wires between the edges of the

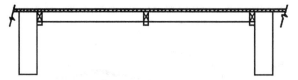

FIGURE 9.8   Method of mounting the 4-inch glass fiber on the ceiling of the control room.

frame of Fig. 9.8 under the cloth cover to hold the 703 in place and to keep the cloth cover from sagging.

Wall Treatment

The 703 glass fiber is applied to control room walls in the same modular form as shown in Fig. 8.4. In the present case both 8-foot and 6-foot modules are used, both of similar construction. The 8-foot module really needs a divider in the middle for strength and to help the unit keep its shape. Unlike the 6-foot module of Fig. 8.4, a similar central divider is suggested in this case for a more uniform appearance. The wall modules are positioned as shown in Fig. 9.9.

FIGURE 9.9   The four wall elevations of the control room showing placement of wideband wall modules.

SIZE ....11'0" × 16'2" × 13'4" ave. ceiling ht.
FLOOR ....Vinyl tile or concrete or parquet
CEILING ....Patches of 4" 703 glass fiber
WALLS ....10'2" × 8' and 6'2" × 6' Wideband modules
VOLUME ....2,964 cu. ft.

Material	S Area Sq. Ft.	125 Hz		250 Hz		500 Hz		1 kHz		2 kHz		4 kHz	
		a	Sa	a	Sa	a	Sa	a	Sa	a	Sa	a	Sa
Ceiling: 703 Glass Fiber	124												
E and W Walls Wideband Mods	184												
S Wall Wideband Modules	48												
Total Area	356	0.99	352.4	0.99	352.4	0.99	352.4	0.99	352.4	0.99	352.4	0.99	528.4
Total Sabins, Sa			352.4		352.4		352.4		352.4		352.4		352.4
Reverberation Time, seconds (Sabine)			0.41		0.41		0.41		0.41		0.41		0.41

TABLE 9.2 Control Room Calculations

Reverberation Time

With only 4 inches of 703 entering into our calculations for the control room, only a uniform reverberation time between 125 Hz and 4 kHz can be expected. In Fig. 9.7 this uniform reverberation time for the control room is shown by the broken line. It is somewhat lower than that of the studio it serves, which is as it should be.

Observation Window

The construction of this most important part of the studio complex must be carried out carefully, following the general plan of Fig. 7.11 except that with 10-inch thick brick walls, the frame need not be divided as with the staggered stud wall. The frame should preferably be made of sturdy 2-inch thick lumber and mounted as the brick wall is being laid. The frame should be supported in the center by bracing during bricklaying so that the weight of bricks and mortar does not distort the window frame. Beads of acoustical sealant must be run between frame and mortar and plaster on both sides as hairline cracks commonly develop as the mortar and plaster dry.

Air Conditioning

Figure 9.2 shows a possible location for the air conditioning equipment behind the entry door. A suspended ceiling in the entry hall would hide the ducts in this room. Metal ducts for both supply and exhaust lined with acoustically absorbent board are suggested. The paths of the supply ducts are shown by broken lines in Fig. 9.2 as an example. This supply duct for the studio follows the path b-c-d-e. The supply duct to the control room follows the path b-a. These duct routings would accomplish the following:

- Give one 90 degree bend at b between the control room grille and the HVAC unit plus about 20 feet of lined duct length.
- Give two 90 degree bends at c and d between the studio and the HVAC unit plus about 20 feet of lined duct length.
- Give 90 degree bends at b, c, and d between the control room and studio plus about 20 feet of lined ducting.

This ducting plan should reduce HVAC machinery and fan noise to an acceptable level and minimize the *speaking tube* effect between the control room and studio. The air velocity at the grilles should be kept below 500 feet per minute, a limit easily met. The exhaust duct routing should follow a similar plan to that of the supply ducts.

CHAPTER 10

A Small Studio for Instruction and Campus Radio

M any institutions of higher education have communication departments and most of these departments teach courses in electronic media (radio and television). Students in radio broadcasting need hands-on experience and this requires at least a small recording facility to serve as a practical laboratory. It is common for such students to produce programs to be broadcast over the campus system. The studio and control room described in this chapter are for precisely this purpose.

Studio Plan

A classroom was made available for space for the studio and control room. Dimensions of 18 feet 10 inches × 21 feet 8 inches with a 10 foot 3 inch ceiling height are not what you would call munificent, but they represent a volume of almost 4200 cubic feet, which is fairly generous. The ceiling height of 10 feet 3 inches gives relief from 8-foot heights often encountered in budget renovation jobs. The floor plan of Fig. 10.1 was settled upon after a bit of horse-trading. Many students were to be accommodated in the studio at one time as observers and performers, fewer in the control room. A studio volume of 2525 cubic feet, a control room volume of 1138 cubic feet, and a sound lock were carved out of the classroom. This means that the control room volume was below the recommended minimum of 1500 cubic feet. This sacrifice made it possible to have a larger studio.

The angled wall separating the control room and the studio reduces the chances for flutter echo in both rooms, tends toward spreading out of model resonances, and gracefully provides for reduction of the volume of the sound lock.

As for penetration of outside noise, concrete walls, ceiling, and floor were comforting, but almost the entire west wall was taken up by four windows overlooking a very busy thoroughfare with many trucks growling up a steep hill. These windows were plugged by four thicknesses of ¾-inch chipboard (particleboard) as shown in Fig. 10.2. This ¾-inch chipboard has a surface density of about

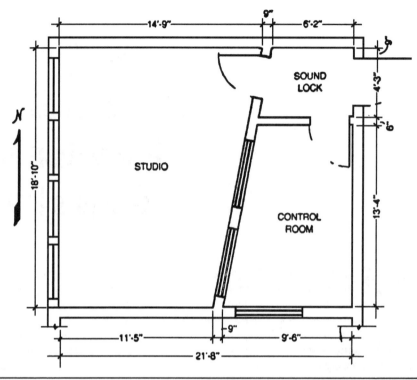

FIGURE 10.1   Floor plan of a studio suite designed for student instruction in radio broadcasting in a college communication department. A former classroom was converted for this purpose.

3 pounds per square foot. Four thicknesses bring the surface density to about 12 pounds per square foot. Considering only the mass and neglecting for the moment any resonance effort, such a well-sealed window plug should give a transmission loss of about 38 dB at 500 Hz, less for lower and more for higher frequencies. The frame holding the chipboard panels was sealed tightly to the concrete window opening. The four chipboard panels were then sealed by a soft rubber strip as the panels were pressed into place. The optional carriage bolt arrangement makes possible the nondestructive removal of the panels if necessary.

Studio Ceiling Treatment

Common suspended ceilings are rarely seen in studios. Before the end of this chapter is reached, some of the reasons for this state of affairs may be apparent. In the present case, such a ceiling was selected because it was attractive, cheap, and promised some easily obtained low-frequency absorption. Because of eye appeal, Johns-Manville Acousti-Shell Textured Vault 3-dimensional 24 inch × 24 inch ceiling panels were selected. Laying out the

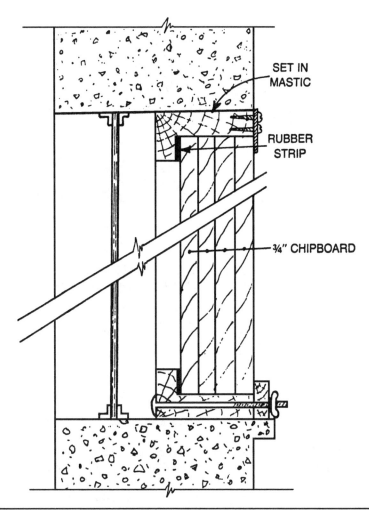

SET IN
MASTIC

RUBBER
STRIP

¾" CHIPBOARD

FIGURE 10.2   Nondestructive and inexpensive window plug designed to protect the studio from exterior noise.

suspension grid on paper for such an odd shape emphasized the need for flat panels around the edges as shown, in Fig. 10.3. The coordinated Acousti-Shell Textured Flats logically fill this role.

Semicylindrical Unit

For the south wall a large diffuser/absorber of semicylindrical shape was selected. The construction of this unit is detailed in Fig. 10.4. The skin is of ¼-inch plywood over which a very thin veneer was cemented for appearance. The frame is of 2 inch × 2 inch lumber and the space behind the skin is divided into nine sections by dividers of 1-inch

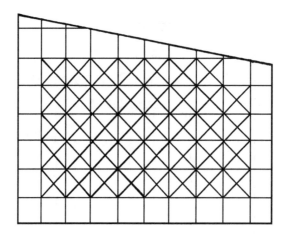

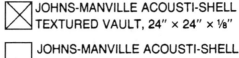

FIGURE 10.3   Layout plan for suspended ceiling in studio.

lumber and curved bulkheads of 2-inch lumber over which the skin is stretched. This also gives the cylindrical segment its shape.

Bulkheads and dividers are sealed where they meet the plastered wall so that each of the nine segments is essentially airtight. A self-adhering foam rubber strip was applied to the edges of the 2-inch thick shaped bulkheads before the plywood skin was mounted. This made for a tight seal and made the structure rattle-free.

Reversible Wall Panels

The west wall required some absorbing and diffusing elements to go over the plugged windows. Reversible panels of nominal 4 foot × 6 foot size (Fig. 10.5) were selected. The inside dimensions of the frame were adjusted to accept 24 Johns-Manville ¾-inch Temper-Tone acoustical tiles of 12-inch × 12-inch size without cutting. These comprise the acoustically *soft* or absorbent side. The *hard*, reflective side is made of ¾-inch plywood. The center of this reflecting surface is higher than the edges, to minimize flat reflections back to nearby microphones. The slope of this reflective panel determines, of course, that such a diffusing effect is primarily at the higher audible frequencies.

Although these are called *reversible*, it was recognized in advance that once hung they would probably never be changed. However, the dual-sided approach kept open some options during the testing phase. Either simple angle brackets can be used in mounting such wall elements or, if reversals are probable, two pins in the bottom of

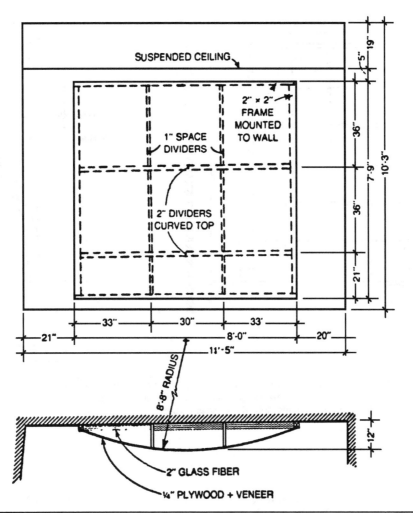

FIGURE 10.4 Constructional details of a large semicylindrical diffuser/absorber mounted on the south wall of the studio.

each unit can fit into holes in metal brackets affixed to the wall to carry the weight. A simple latch arrangement can also be installed to hold the top of the unit against the wall.

Figure 10.6 shows all four studio wall elevations and the treatment for each. The north wall has 36 Johns-Manville Temper Tone 360 acoustical tiles of ¾-inch thickness mounted in two patches. The south wall is dominated by the 8-foot semicylindrical element previously described. The east wall is almost completely taken up by two observation windows and the door. The west wall has three of the reversible wall elements of Fig. 10.5. The location of the plugged windows is indicated by broken lines.

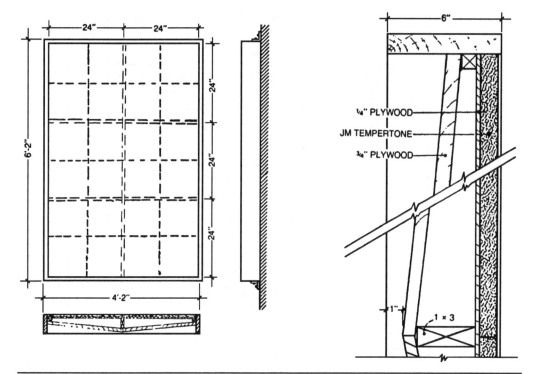

FIGURE 10.5 Construction details of reversible units on west all over the plugged windows. The reversible feature is intended more for allowing some final adjustments in room acoustics during measurements than for changing between program types.

The suspended ceiling line is indicated on each wall elevation 19 inches below the structural ceiling. This distance allows the accommodation of air conditioning ducting. This means that some uncertainty is introduced in the absorption coefficients which are given for a standard Mounting #7 (new designation, E-405) distance of 16 inches.

Studio Calculations

For those who wish to follow through on the calculations of studio reverberation time, Table 10.1 is included. It is really "gilding the lily" to separate out absorption of door and observation window glass in view of the uncertainties in all coefficients of absorption, but no harm is done by including them as long as limitations of our overall precision is kept in mind. Computations are included for the three units on the west wall with both the reflective (hard) and absorptive (soft) sides facing the room. Calculated reverberation times for these two conditions, taken from Table 10.1, are shown graphically in Fig. 10.7. By turning the units from soft to hard, an increase in reverberation time of about 30 percent is obtained

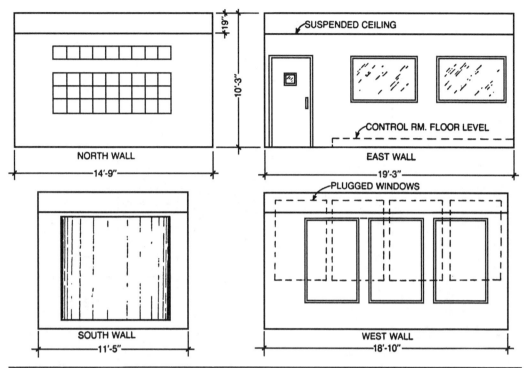

FIGURE 10.6   Placement plan for acoustical elements on four walls of studio.

for frequencies above 500 Hz. Looking at things the other way around, by changing from hard to soft a reduction in reverberation time of about 23 percent results for the higher frequencies.

Control Room

The control room, as mentioned previously, is substandard in volume, a compromise designed to accommodate a greater number of students in the studio. Figure 10.8 illustrates the very practical equipment arrangement utilized. Room for two standard racks for 19-inch panels of ancillary equipment was allowed at the north wall. Along the south and west walls are work surfaces with built-in drawers and storage cabinets below. The window in the south wall looks into the television studio. Although the television studio has its own control room, this window, along with interconnecting tie lines, makes it possible to use both studios for special productions. Normally this south window is covered by drawn drapes.

The floor of the control room was raised about 12 inches and a 4-inch × 4-inch trough runs around the east, south, and west walls for interconnecting cables. In spots not covered by cabinets, access to the trough is by hinged lids flush with the floor.

SIZE .... 18' 10" × 14' 9" × 11' 5" (one wall splayed)
CEILING .... Johns-Manville Acousti-Shell TV&TF
.... 24" × 24" × 1/8" Suspended 19"

FLOOR .... Vinyl tile
.... plastered concrete
WALLS .... Treated as below
VOLUME .... 2525 cu. ft.

Material	S Area Sq. Ft.	125 Hz a	125 Hz Sa	250 Hz a	250 Hz Sa	500 Hz a	500 Hz Sa	1 kHz a	1 kHz Sa	2 kHz a	2 kHz Sa	4 kHz a	4 kHz Sa
J-M Acousti-Shell Textbred Vault	156	0.64	99.8	0.66	103.0	0.67	104.5	0.75	117.0	0.72	112.3	0.70	109.2
J-M Acousti-Shell Textbred Flat	90	0.70	63.0	0.69	62.1	0.66	59.4	0.80	72.0	0.84	75.6	0.83	74.7
Floor, Vinyl Tile	246	0.02	4.9	0.03	7.4	0.03	7.4	0.03	7.4	0.03	7.4	0.02	4.9
Glass	30	0.05	1.5	0.03	0.9	0.02	0.6	0.02	0.6	0.03	0.9	0.02	0.6
Plaster	458	0.02	9.2	0.03	13.7	0.04	18.3	0.05	22.9	0.04	18.3	0.03	13.7
Door	20	0.24	4.8	0.19	3.8	0.14	2.8	0.08	1.6	0.13	2.6	0.10	2.0
Cylindrical Element, South Wall	62	0.50	31.0	0.35	21.7	0.22	13.6	0.14	8.7	0.11	6.8	0.10	6.2
J-M Temper Tone Tile, North Wall	36	0.09	3.2	0.25	9.0	0.70	25.2	0.85	30.6	0.83	29.9	0.89	32.0
			217.4		221.6		231.8		260.8		253.8		243.3
West Wall Panels, Hard	72	0.28	20.2	0.19	13.7	0.14	10.1	0.11	7.9	0.08	5.8	0.05	3.6
Total Sabins, Hard, Sa			237.6		235.3		241.9		268.7		259.6		246.9
Reverb. Time, sec.			0.52		0.53		0.51		0.46		0.48		0.50
West Wall Panels, Soft	72	0.2	14.4	0.5	36.0	1.0	72.0	1.0	72.0	1.0	72.0	1.0	72.0
Total Sabins, Soft, Sa			231.8		257.6		303.8		332.2		325.8		315.0
Reverb. Time, sec.			0.53		0.48		0.41		0.37		0.38		0.39

TABLE 10.1 Studio Calculations

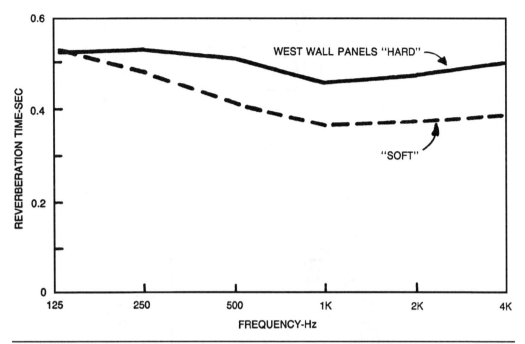

FIGURE 10.7   Effect on calculated studio reverberation time of reversing the three panels on west wall.

Measurements

Listening tests in the studio revealed that with the hard surfaces of the three west wall panels exposed, the sound was somewhat too bright; that is, the high frequency components of music and speech were too prominent. For a studio of 2525 cubic feet the optimum reverberation time for music is about 0.5 second and for voice about 0.3 second. This studio, to be used for both, should have a compromise reverberation time somewhere between the two. Flipping the west wall units first one way for a music program and then the other for speech is just too much trouble; many programs involve both. The object, then, is to determine by actual measurements which way these panels give the best compromise effect and then leave them that way.

Reverberation measurements were performed using interrupted octave bands of pink noise to evaluate the accuracy of the computations and to determine the proper exposure of the west wall panels. In Fig. 10.9 the measured and calculated values of reverberation time are compared for the west wall panels hard condition. Between 500 and 1,000 Hz the agreement is perfect, but calculated values are too high at low frequencies and too low at high frequencies. This means that there is greater absorption at low frequencies and less absorption at high frequencies than the coefficients of Table 10.1 indicate. These measured results surely explain why the sound was too bright with hard panels!

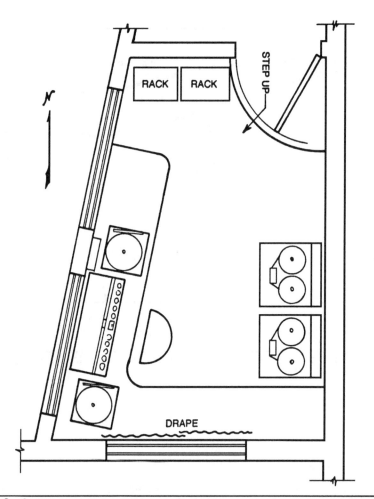

FIGURE 10.8 Control room equipment layout. Control room space is considerably less than ideal in order that many students can be accommodated in the studio at one time.

With the west wall panels exposing the Johns-Manville Temper Tone acoustical tile (the soft condition, see Fig. 10.5), the calculated values are again too high at low frequencies and too low at high frequencies as shown in Fig. 10.10. In the mid-frequency region there is excellent agreement at 500 Hz and 1 kHz.

Figure 10.11 compares measured reverberation time between the hard and soft conditions of the west wall panels. The measured comparison of Fig. 10.11 shows the same type of separation for the higher frequencies as the calculated comparison of Fig. 10.7. Otherwise the agreement is not too good.

Glancing down the various materials of Table 10.1 and the absorption coefficients used for each, you may ask, "Which ones are in error?" One relatively unfamiliar component was the suspended ceiling. In Table 10.1 good absorption is attributed to it both at low and high frequencies. Is this substantial portion of the total absorption actually

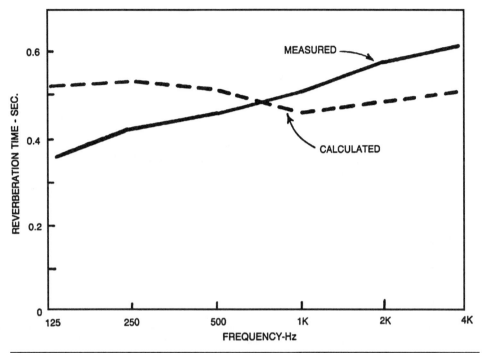

FIGURE 10.9 Comparison of measured and estimated reverberation time of studio in exercise to illustrate basic problems involving sound absorption coefficients in computations.

realized in practice across the band? Measurements were made both with the suspended Acousti-Shell panels in place and with them removed from the room with the results shown in Fig. 10.12. Aha! There is the expected absorption in the low frequencies, but essentially none at 2 and 4 kHz!

Is the measured low-frequency absorption really the expected amount? A simple computation can settle that question. The measured reverberation time at 125 Hz with the suspended ceiling in place was 0.33 second. With the ceiling removed, and no other change in the room, it was 0.65 second. By feeding these values back into Sabine's equation, corresponding values of 375 and 190 sabins of absorption for the two conditions are obtained. The difference of 185 sabins can only be attributed to the suspended ceiling. In our calculations of Table 10.1, 99.8 + 63.0 = 162.8 sabins were assigned to the ceiling. This means the suspended ceiling yields a modest 14 percent more absorption at 125 Hz than the manufacturer's coefficients would indicate.

The Acousti-Shell ceiling is actually 19 inches from the structural ceiling rather than the standard 16 inches for which the coefficients were obtained. This may account for the difference. Yet in Figs. 10.9 and 10.10 the calculated low-frequency reverberation times are too high, not too low. Where does all this absorption at 125 Hz come from? The semicylindrical element on the south wall is suspect. The assumed absorption coefficient of 0.5 may be too low. If this unit were resonant at 125 Hz, the highest possible coefficient would be 1.0. If this were the case, the calculated reverberation time for the

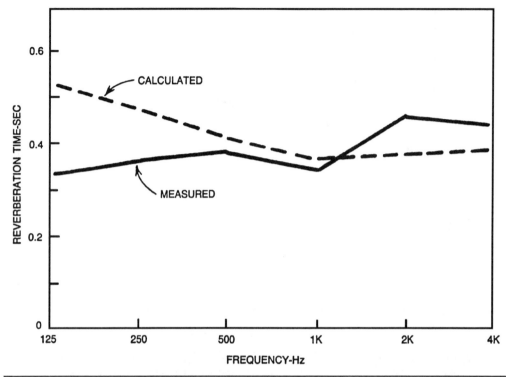

FIGURE 10.10 Comparison of measured and calculated reverberation time of studio with west wall panels soft side out. Calculated values of reverberation time are the same as those in Fig. 10.7.

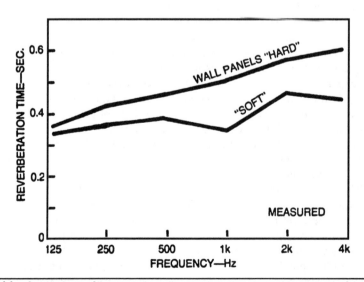

FIGURE 10.11 Comparison of measured studio reverberation time with reversal of west wall panels. The soft graph is the same as in Fig. 10.10.

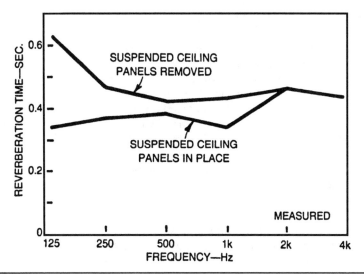

FIGURE 10.12   Measurements of studio reverberation time with and without ceiling panels in place.

soft west wall panel condition would be brought down to about 0.47 second. This is in the right direction but not enough to account for the difference.

Where else can you look for some unexpected low-frequency absorption? The graphs of Figs. 10.9, 10.10, and 10.12 are for the west wall panels in the soft position with the Johns-Manville Temper Tone acoustical tile facing the room. The coefficients used in Table 10.1 for the Temper Tone tile on the north wall are those provided by the manufacturer for the tile cemented to a solid wall. In Fig. 10.5 there is a substantial air space behind the tile on the panels. This would surely enhance both low- and high-frequency absorption of the west wall units with the Temper Tone tile exposed.

The coefficients used in Table 10.1 for the west wall units, soft side out, are wild estimates designed to allow for the effect of the air space backing as well as the diaphragm effect of the ¼-inch plywood backing. The absorption coefficient assumed for these tiles, soft side out, at 125 Hz, is 0.2. If it could be stretched to the limit of 1.0, most of the remainder of the lost absorption at low frequencies would be accounted for. Perhaps 100 percent absorption at 125 Hz is a rather optimistic assumption for both the semicylindrical unit on the south wall and the three panels on the west wall, but something close to 100 percent for these and some slack in other 125 Hz coefficients helps correlate measurements and calculations. It also reveals the flimsy nature of some of these coefficients. So much for disparity between low-frequency measurements and calculations.

In both Figs. 10.9 and 10.10 the measurements reveal that less absorption is realized in the high frequencies than the coefficients of Table 10.1 give. The measurement of Fig. 10.12, with and without the suspended ceiling, reveal that for some reason the Acousti-Shell elements seem to give no appreciable absorption at 2 kHz and 4 kHz. In Table 10.1, 109.2 + 74.7 = 183.9 sabins are attributed to the suspended ceiling at 4 kHz. If these sabins are eliminated, only 131 sabins remain for the soft wall panel condition which gives a reverberation time of 0.89 second, far higher than the 0.44 second measured.

Total absorption of 249 sabins is required to account for the 0.44 second measured reverberation time. It would appear that something like 66 sabins are being obtained either from the ceiling or some of the other sound absorbing elements listed in Table 10.1.

This has been a tedious excursion into the field of practical absorption coefficients. As James Moir, prominent British acoustician, has said, "Anything that is obvious in acoustics is nearly always wrong."[11] Perhaps the tedium is worthwhile if we learn only one thing: that absorption coefficients are the insecure basis of reverberation time calculations and that the computed results are no more dependable than the coefficients used.

Absorption coefficients supplied by manufacturers for their proprietary materials may or may not be realized in practice, depending on how closely the practical mounting and surroundings approximate those of the measuring conditions.

For the nonproprietary acoustical elements, such as the semicylindrical south wall unit, finding even approximate coefficients is a great problem and we are even more vulnerable to error. Computations, carefully done, serve only as a rough guide. Measurements and subsequent tuning adjustments are essential to accurate acoustical treatment of studios, control rooms, and other critical listening spaces.

CHAPTER 11

Small Ad Agency Studio for AVs and Radio Jingles

An advertising agency had a small, makeshift recording facility which was both cramped and poorly laid out. Expansion of agency business required moving up one floor in their large commercial building to provide the space needed. In the process the recording facilities were also to be enlarged and restructured. As is so often the case, those doing the actual recording work found themselves on the wrong end of the totem pole with their space shrinking daily as front office ideas grew during the planning stage. A space at one end of the floor between two concrete walls was eventually designated for studio use. The walls effectively blocked expansion north and south.

The primary use of this facility is in the production of radio advertising announcements. A secondary (but growing) activity is production of audiovisuals, principally slide sets and filmstrips. Two rooms were envisioned, one to be devoted principally to the recording and audiovisual functions, the other to be a combination control room for recording and a general work room in which mixing, editing, and dubbing would also be done. Several people, each working on a different project or at least different aspects of the same project, were bound to get in each other's way from time to time but the "gigantic step forward for mankind" which the new facility offered over the old one made such conflicts seem trivial.

Floor Plan

The floor plan emerging from the smoke and fire of space allocation, incorporating the best functional ideas of production personnel and the acoustical consultant, is shown in Fig. 11.1. It incorporates some basic problems, such as volumes below the 1500 cubic foot minimum (but not much below). Because sound lock space had to be taken from the studio, a cavity is created in the recording studio near the observation window. The temptation to sit in this cozy indentation at the built-in table near the window is great, especially for those nurtured on the radio tradition of the announce booth. The better position for the narrator during recording is back in the main part of the room and not in this indentation. The indentation would not have existed if sound lock space could have been allotted outside the studio area.

The control work room has a built-in work surface along the south and west sides of the room. This bench carries a mixing console as well as numerous advanced audiophile-type magnetic recorders.

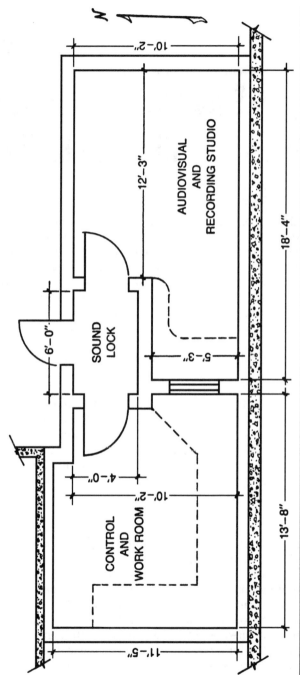

Figure 11.1 Floor plan of work room and recording studio wedged in between two existing concrete walls. Due to space limitation it was necessary to take the sound lock space from the studio.

Room Proportions

As discussed in Chapter 1, axial mode distribution is something of a problem even when we are free to specify the three dimensions of a room. In this case, two of the three dimensions of both rooms were fixed by circumstances and tight constraints were placed on the third. All that can be done in such circumstances is to study the distribution of fundamental resonance frequencies and harmonics of the space in an attempt to evaluate the threat of colorations on paper before construction is started.

Figure 11.2A is a plot of modal frequencies for the audiovisual/recording studio. The solid lines are associated with the basic 18-foot 4-inch length, 10-foot 2-inch width, and 8-foot 11-inch ceiling height (the length of the lines of Fig. 11.2 holds no significance). The broken lines are associated with the 5-foot 8-inch alcove in the N-S mode and the 12-foot 3-inch step in the E-W mode.

Some of these secondary dimensions within the room (broken lines) occur in rather large gaps between major resonances (solid lines), which is favorable, while others are almost coincident with major dimension modal frequencies, which can be unfavorable. The triple coincidence at about 277 Hz is probably no threat because few colorations are found to be problems above 200 Hz. The three or four double pileups or near pileups below 200 Hz may or may not be troublesome. These will require the application of a keen ear for evaluation.

The solid lines of Fig. 11.2B represent the modal resonance picture for the control work room major dimensions of a 13-foot 8-inch length, 11-foot 5-inch width and 8-foot 11-inch ceiling height. The broken lines are associated with the 10-foot 2-inch N-S secondary step in the room width. With the exception of one at 167 Hz, all the secondary resonances land nicely between the major dimension resonances. In our

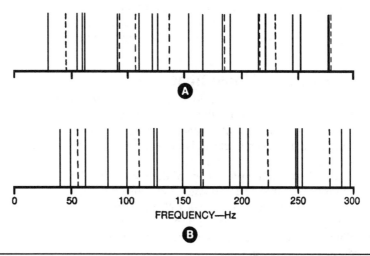

FIGURE 11.2 Distribution of axial modal frequencies: (A) for the audiovisual recording studio and (B) for the control work room. The solid lines are associated with the basic dimensions of the rooms, the broken lines with secondary dimensions of alcove and step.

keen ear analysis of this room, particular attention should be given to the possibility of colorations due to the double coincidences near 125 Hz and 166 Hz.

Wall Construction

It was immediately recognized that other diverse and noisy activities in the building could easily be carried to and radiated in the sound sensitive spaces by the concrete structure of the building. For example, elevator equipment mounted securely to the structure sends impulses into the reinforced concrete walls and pillars and these can be radiated into the studios by concrete surfaces acting as diaphragms. Isolating against such noises took on a high priority. Floating concrete floors were ruled out by budget limitations, but something could be done about walls and ceilings.

Walls paralleling existing concrete surfaces are set back, creating a 3-inch air space filled with glass fiber insulation. Metal studs 2½ inches thick form the framework and a double layer of ⅝-inch gypsum board make up the mass of the wall surface facing the studio. Other walls, including the one in which the observation window is set, are constructed as double metal stud walls separated by 3 inches. The space is filled with glass fiber insulation. Double gypsum board on both faces yield an overall wall 10½ inches thick with a rating in the vicinity of STC-50. The ceiling is suspended from the structural ceiling with a vibration isolation hanger on each wire. A black iron angle frame holds the double ⅝-inch gypsum board ceiling. All gypsum board edges are staggered and all joints caulked with nonhardening acoustical sealant.

This plan provides a reasonable degree of isolation from building sounds on all surfaces but the floor. It was decided that if the floor did become a problem, a wooden floating floor could be added at a later time at minimum cost.

Such structure-borne sounds can be a serious problem. For example, I visited a fabulous new government broadcasting house in a certain foreign land. The architect had claimed 90 dB transmission loss as protection against nearby jet landing pattern noise by the studio-within-a-studio technique. Stepping into one of the beautifully treated and decorated studios, however, a distinct hammering noise from another part of the building was clearly heard to the embarrassment of the engineer-host.

Audiovisual Recording Studio Treatment

The acoustical treatment of the audiovisual recording studio involves four basic elements: the carpet, wideband wall units (2 feet × 4 feet), midband wall units (2 feet × 2 feet), and the low frequency ceiling units (4 feet × 5 feet). The absorption of the gypsum board walls and ceiling has been neglected in the discussion to follow, but will be treated later in the chapter.

Figure 11.3 shows the placement of the wall units, Fig. 11.4A the placement of the ceiling units and Table 11.1 tabulates the computations for this room. Entering into Sabine's formula the room volume of 1390 cubic feet and the reverberation time goal of 0.3 second gives a required absorption of 227 sabins. The problem now becomes one of juggling the areas of the four types of absorbers to give close to 0.3 second reverberation time across the band.

To obtain a uniform reverberation time throughout the audible spectrum requires a constant number of absorption units (sabins) with frequency. In Table 11.1 the total

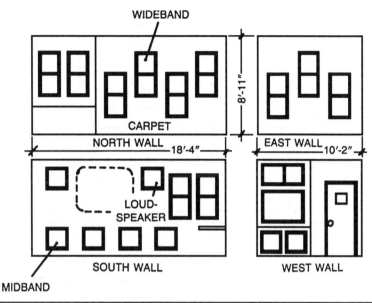

FIGURE 11.3 Wall elevations of audiovisual recording studio showing placement of wideband and midband absorbers.

sabins at each frequency is fairly close to the calculated 227 required, varying from 195 to 211.3. To see more clearly how absorption of each type of material varies with frequency, the data of Table 11.1 is graphically presented in Fig. 11.5. The greatly unbalanced carpet absorption (Fig. 11.5A) is quite well compensated by the equally, but opposite, absorption of the low frequency ceiling units. However, there is a sag at midrange frequencies of the low frequency plus carpet curve (Fig. 11.5B) and the midrange units, tuned to the 500 Hz-1 kHz region, are designed to straighten out the carpet + LF + midrange curve (Fig. 11.5C). Once this is done, enough wideband absorber is introduced to raise the total to approximately the 227 sabin level (Fig. 11.5D). This gives close to 0.3 second reverberation time across the audible range shown in Fig. 11.6.

Low Frequency Units

The low frequency units are most properly placed opposite the carpet for which they compensate, in the usual contracarpet position. The area required for these LF units means that 100 square feet, or 64 percent of the total ceiling area be covered with these 4 foot × 5 foot boxes, but leaving enough space for illumination fixtures. The facings of the low frequency units are quite reflective at the higher frequencies, but vertical flutter echoes are controlled by the carpet absorption.

The construction of the ceiling low frequency units is detailed in Fig. 11.7. The frame and center divider are made of 1 × 8 lumber strengthened by a backing of ½-inch

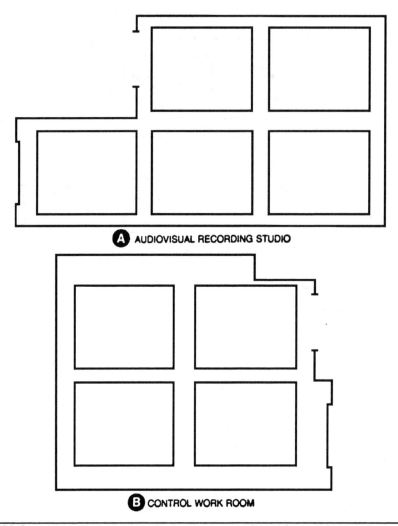

FIGURE 11.4 Projected ceiling plans showing placement of the 4-foot × 5-foot low frequency units on the ceiling: (A) audiovisual recording studio and (B) control work room.

chipboard on plywood. A facing of $\frac{3}{16}$-inch tempered hardboard perforated with $\frac{3}{16}$-inch holes spaced $2\frac{9}{16}$ inches on centers covers the entire frame. In intimate contact with the perforated cover inside the box is a 4-inch thick layer of Owens-Corning Type 703 Fiberglas of 3 pounds per cubic foot density.

If the glass fiber material is loosely fitted, something is needed to hold it against the perforated facing. The 1 × 4 spacers with fine wire tacked to the edges in a zigzag form will do this in a very positive way, but if gravity and friction can be depended upon to hold the glass fiber snugly against the back of the perforated cover, so much

	SIZE .... 10'2" × 18'4" × 8'11" Ceiling ht.
	FLOOR .... Carpet
	CEILING .... 5 Low frequency absorbers
	WALLS .... 12 Wideband, 6 midband absorbers
	VOLUME .... 1390 cubic feet

Material	S Area Sq. Ft.	125 Hz		250 Hz		500 Hz		1 kHz		2 kHz		4 kHz	
		a	Sa	a	Sa	a	Sa	a	Sa	a	Sa	a	Sa
Carpet	157	0.05	7.9	0.15	23.6	0.30	47.1	0.40	62.8	0.50	78.5	0.60	94.2
Low Freq. Absorbers 5 – 4' × 5'	100	1.0	100.0	0.68	68.0	0.39	39.0	0.17	17.0	0.13	13.0	0.10	10.0
Midband Absorbers 6 – 2' × 2'	24	0.35	8.4	0.63	15.1	0.88	21.1	0.84	20.2	0.66	15.8	0.35	8.4
Wideband Absorbers 12 – 2' × 4'	96	0.99	95.0	0.99	95.0	0.99	95.0	0.99	95.0	0.99	95.0	0.99	95.0
Total Sabins, Sa			211.3		201.7		202.2		195.0		202.3		207.6
Reverb. Time, seconds		0.32		0.34		0.34		0.35		0.34		0.33	

TABLE 11.1 Audiovisual Recording Studio Calculations

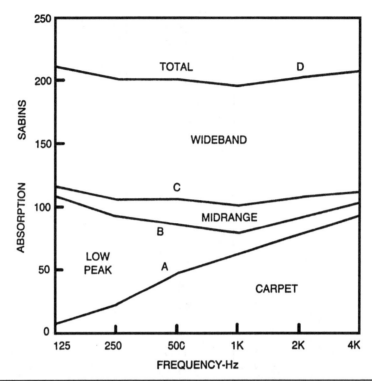

FIGURE 11.5 Distribution of room absorption between the four principal types of absorbers used in treating the audiovisual recording studio.

the easier and cheaper. The air space plays an active part in the performance of this absorber. The boxes can be mounted to the ceiling in any convenient way. Painting these units and the exposed parts of the ceiling flat black will render them visually unobtrusive, especially if the illumination fixtures direct the light downward. Track lights are ideal for this.

Hand drilling almost 500 holes in each of the nine ceiling box covers can be a staggering job. The obvious way to minimize this is to stack all covers, drilling all with one set of holes. Each cover can be split at the center divider if desired so that 18 pieces of 30 inch × 48 inch hardboard could be stacked for drilling.

Midband Units

The midband units have a relatively minor, but important, role to play in the overall treatment of the audiovisual recording studio as illustrated in Fig. 11.5. They are mounted on the wall under the window table and along the lower edge of the south wall in Fig. 11.3. Their simple construction is detailed in Fig. 11.8.

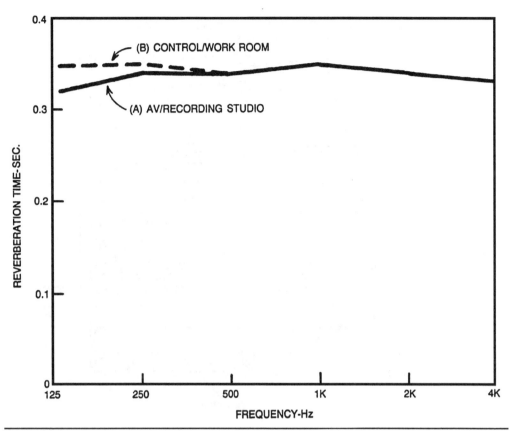

FIGURE 11.6   Reverberation time of the two rooms as a function of frequency.

The covers are Johns-Manville Transite panels which come perforated with 550 $3/16$-inch holes per square foot. These are autoclaved asbestos cement boards $3/16$-inch thick and their 24 inch × 24 inch size determines the size of the supporting frame. This frame is made of 1 × 3 lumber with 1 × 2 spacers inside. The 2-inch dimension should be met to accommodate the 2-inch thickness of the Owens-Corning Type 703 Fiberglas.

Wideband Units

Acoustically speaking, the 2 foot × 4 foot wideband units are nothing more or less than 4 inches of 703 Fiberglas of 3 pounds per cubic foot density. The rest is mechanical mounting and cosmetic cover. Figure 11.9 shows how the 1 × 6 frame, 1 × 4 divider and spacers, and the $1/2$-inch chipboard backboard fit together.

The manufacturer of the glass fiber stops short of attributing 100 percent absorption to 4 inches of 703, listing 0.99 as the coefficient from 125 Hz to 4 kHz.

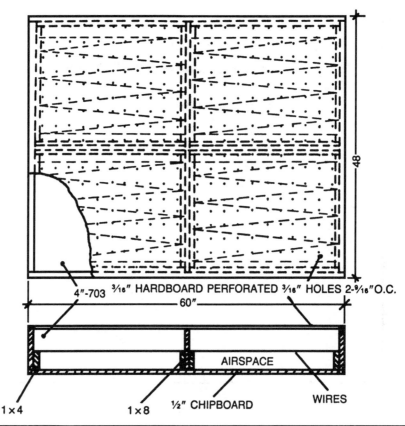

4"-703   ³⁄₁₆" HARDBOARD PERFORATED ³⁄₁₆" HOLES 2-⁹⁄₁₆"O.C.

60"

48

AIRSPACE

1×4   1×8   ½" CHIPBOARD   WIRES

FIGURE 11.7   Constructional details of the Helmholtz low-frequency absorbers to be mounted on the ceiling.

Below 125 Hz the absorption does fall off, of course, but in this low frequency region the diaphragmatic absorption of the five gypsum board surfaces tends to compensate. The double ⅝-inch gypsum board surfaces resonate well below 125 Hz. Calculations indicate that the walls paralleling the existing concrete walls on the north and south sides with 5-inch air space have an absorption peak at about 32 Hz. The other walls with their 8-inch air space peak near 26 Hz. These cavities are filled with insulation which increases the breadth of the absorption region markedly, but the resonance frequency little. No values of measured absorption coefficients are available for double ⅝-inch gypsum board walls with these cavity depths. However, by taking the values available for ½-inch gypsum board on 2 × 4s 16 inches on centers and shifting them to take into account the different resonance frequencies involved, an absorption coefficient of 0.05 to 0.08 is estimated for 125 Hz.

With a gypsum board area of about 580 square feet involved in the audiovisual recording studio, something like 30 to 40 additional sabins may be at work at 125 Hz because of wall absorption. This would reduce the reverberation time at 125 Hz to

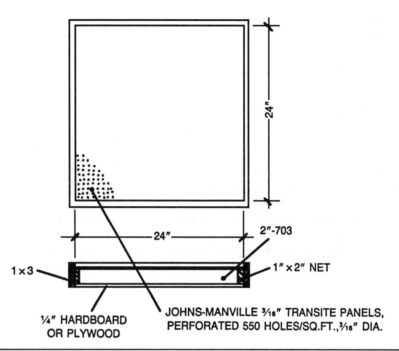

FIGURE 11.8   Constructional details of the midband absorbers having peak absorption in the 500 Hz-1 kHz region.

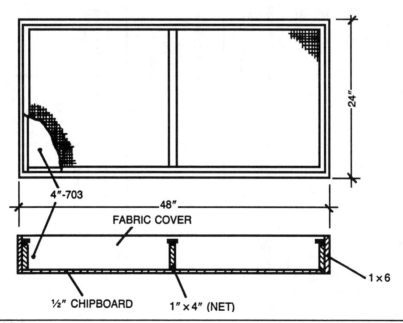

FIGURE 11.9   Construction details of the 2-foot × 4-foot wideband modules. These are basically 4 inches of 703 glass fiber with mechanical protection and cosmetic cover.

something like 0.23 second and be less effective than this at 250 Hz and above. This possible 23 percent reduction of reverberation time at 125 Hz requires measurements to tie it down specifically, but this discussion points out that such wallboard absorption might reduce the required area of ceiling low frequency absorbers.

Control Work Room Treatment

The control work room is treated with the same four elements as the audiovisual recording studio: carpet, wideband and midband wall units, and ceiling mounted low frequency absorbers. Figure 11.10 shows the suggested placement of each unit on the walls and Fig. 11.4B the placement of the four low frequency units on the ceiling. Table 11.2 lists vital statistics of this room as well as the absorption units (sabins) expected of each of the four elements at each of the six frequencies.

There is little need to plot the contribution of each of the four elements. Although the values differ somewhat, the general apportionment principle revealed in Fig. 11.5 for the audiovisual recording studio applies to this room as well. The calculated reverberation times of Table 11.2 are plotted in Fig. 6.6 for ready comparison with those for the other room.

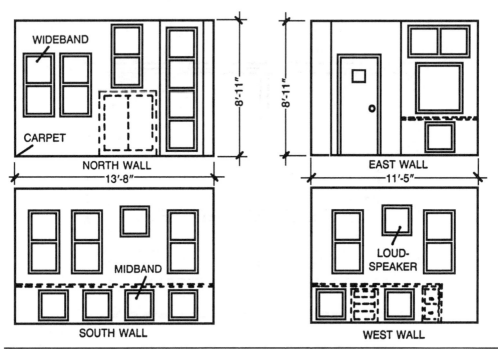

FIGURE 11.10 Wall elevations of control work room showing placement of wideband and midband absorbing units.

SIZE .... 11'5" × 13'8" × 8'11" Ceiling ht.
FLOOR .... Carpet
CEILING .... 4 Low frequency absorbers
WALLS .... 11 Wideband, 7 midband absorbers
VOLUME .... 1304 cubic feet

Material	S Area Sq. Ft.	125 Hz		250 Hz		500 Hz		1 kHz		2 kHz		4 kHz	
		a	Sa	a	Sa	a	Sa	a	Sa	a	Sa	a	Sa
Carpet	147	0.05	7.4	0.15	22.1	0.30	44.1	0.40	58.8	0.50	73.5	0.60	88.2
Low Freq. Absorbers 4 – 4' × 5'	80	1.0	80.0	0.68	54.4	0.39	31.2	0.17	13.6	0.13	10.4	0.10	8.0
Midband Absorbers 7 – 2' × 2'	28	0.35	9.8	0.63	17.6	0.88	24.6	0.84	23.5	0.66	18.5	0.35	9.8
Wideband Absorbers 11 – 2' × 4'	88	0.99	87.1	0.99	87.1	0.99	87.1	0.99	87.1	0.99	87.1	0.99	87.1
Total Sabins, Sa			184.3		181.2		187.0		183.0		189.5		193.1
Reverb. Time, seconds			0.35		0.35		0.34		0.35		0.34		0.33

TABLE 11.2   Control Work Room Calculations

CHAPTER 12

Multitrack in a Two-Car Garage

There has been a proliferation of small studios in basements, barns, garages, and other locations in and around private residences. Some of these are built by the members of new bands who figure that such a facility will help them develop their musical techniques and enable them to record demonstrations and other records at leisure without high studio charges. Some are built by advanced audiophiles who have become jaded to further improvements in the living room hi-fi, but are challenged by recording techniques. Such a private studio is a logical next step after experimenting with multichannel recording with a four-track consumer type tape recorder. This type of experimentation soon runs headlong into the frustrations of household noise, limitations in the number of tracks, and problems created by haphazard, temporary lash-ups.

Some may look to a private recording studio as a stepping stone to getting into professional recording. As far as gaining experience is concerned, excellent, but if renting the studio to musical groups is contemplated, beware. Most communities look with disfavor on commercial activities in residential areas. Construction of the studio to be described in this chapter definitely requires a building permit and the usage planned for the facility is sure to come up. Now that this point has been made, the studio has limited commercial possibilities. The greater the experience one has, the greater the emphasis on that word limited.

Floor Plan

The two-car garage to be converted into a multitrack studio is described in Fig. 12.1. It is almost square and is covered with a simple A-roof. The open ventilation louvers in the north wall, a 15-foot 3-inch overhead door opening in the east wall, and a small rear door in the south wall emphasize that the garage is just a wide open shell with unfinished walls inside.

The first step is to make the garage into a tight structure and to make provisions for a monitor control room. Figure 12.2 shows one way of distributing the 464 square feet of total area between the studio and control room. This gives a studio floor area of 352 square feet and a control room area of only 85 square feet. The studio size is over twice the minimum prescribed volume of 1500 cubic feet but the control room is only about half the minimum.

This may be sufficient incentive to drive the recording technician to using high quality headphones instead of monitor loudspeakers. The fact is that with a square

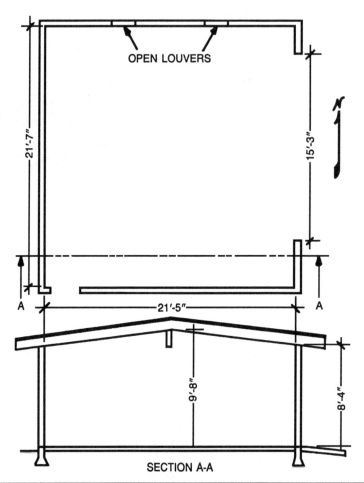

FIGURE 12.1   Plan and sectional views of garage before its conversion to a multitrack recording studio.

garage, any location for the control room other than a corner results in serious degrading of studio space. The floor plan of Fig. 12.2 favors the studio. Perhaps that demo record may have greater impact with a reasonable studio and operator wearing headphones than a tiny studio with its poor separation and more accurate monitoring room.

The use of high quality headphones,[12] if not a first choice, is at least a viable alternative for listening critically in acoustically difficult situations. They are being improved much faster than monitoring loudspeakers.

The louver ventilators in the north wall are abandoned and the opening framed in and covered to conform to the other external walls. The overhead door probably should be retained for external appearance, although the bulky hardware may be removed and stored for possible future use. A new frame closes off the 15-foot 3-inch door opening. A door (3 feet wide to accommodate instruments) is cut in the east wall for access to the studio. This should be a 1¾-inch solid core door and well weatherstripped. The sound

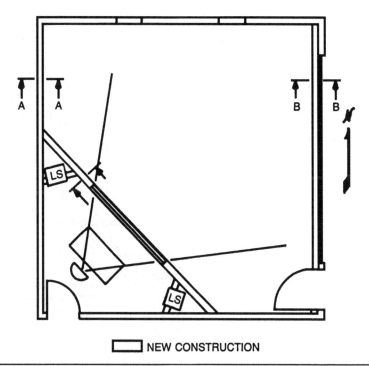

NEW CONSTRUCTION

FIGURE 12.2   Plan for conversion of garage of Fig. 12.1 to a multitrack recording studio.

lock corridor in this case is the great outdoors. The existing doorway in the south wall serves the control room, but the hollow core door is replaced by a 1¾-inch solid core door and also weatherstripped.

Wall and Ceiling Construction

The internal wall and ceiling surfaces are covered with ⅝-inch gypsum board as shown in Fig. 12.3. Great care should be exercised to assure tightness as this layer is the chief assurance against complaints from the neighbors. This requires filling of all cracks and taping of all joints as well as a liberal use of nonhardening acoustical sealant at all intersections of surfaces. This drywall layer goes in the control room as well as the studio. The diagonal wall between the studio and the control room has a ⅝-inch gypsum board on each side as shown in Fig. 12.3, Section C-C.

The conversion of this garage to a studio is fraught with compromises. The result will be midway between the living room and a first class studio. The wall between the studio and control room is such that studio sounds will sometimes be heard in the control room without benefit of amplifiers and loudspeakers. Neighbors can probably hear a lively number being played in the studio, but hopefully at a low enough level that they will not call the police.

Unless the two single doors are made impervious to sound, there is little advantage in strengthening the walls. A good, tight wall such as shown in Section C-C of Fig. 12.3 has an STC rating of 30 to 35 and the stucco plaster of Section A-A is only slightly better.

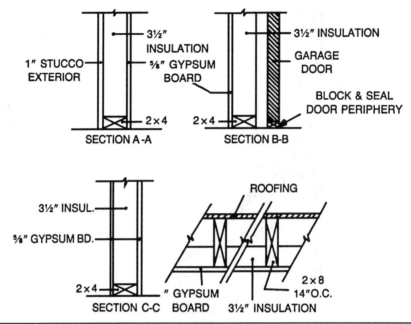

FIGURE 12.3 Wall construction details in converted garage studio. Sections refer to Fig. 12.2.

Great care must be exercised in weatherstripping doors to get STC 30. The 3½ inches of insulation fill indicated in all sections of Fig. 12.3 contributes very little (about 2 dB) to the transmission loss. It does help some in discouraging cavity resonances. Because of its minor effect, it could be omitted if the budget is very tight.

Studio Treatment

Multitrack recording requires acoustical separation of the sounds of instruments or groups in instruments which are recorded on separate tracks. One way of achieving such separation is to physically separate the sources and place a microphone close to each source. Space is limited in this studio, but this logical and desired approach is more effective in an acoustically dead space than in a live one. There are other ways separation can be achieved, such as the use of microphone directivity, or baffles.

Acoustical Goals

Musician reaction places a limit on the deadness of such a studio because they must hear themselves and other musicians to play effectively. The studio of Fig. 12.4 has been made as dead as practical to allow the achievement of reasonable track separation, even though the space is small for this type of recording.

Floors and Ceiling

Heavy carpet and pad are applied to the entire concrete floor except the drum booth area. This opposes the sloping, reflective ceiling surfaces which are bare gypsum board.

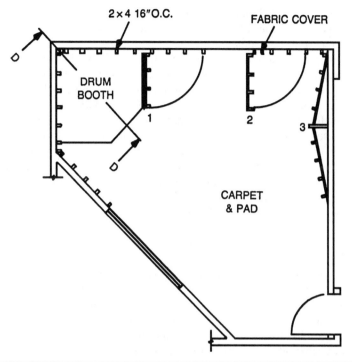

FIGURE 12.4 Studio details in garage converted to multitrack studio. Swinging doors 1 and 2 serve as baffles between instruments. Element 3 on east wall helps to absorb very low frequencies.

Reflections from this ceiling could contribute to leakage between tracks. If this proves to be a limiting factor, absorbent material could be applied to critical areas of the ceiling.

Walls

Much of the wall is faced with 4 inches of 703 semirigid glass fiber boards. These glass fiber panels are inserted between vertical 2 × 4s which are mounted against the gypsum board wall covering and run from floor to ceiling. Figure 12.4 shows the 2 × 4s spaced 16 inches center to center, lined up with the studs of the exterior wall to which they are nailed. After the glass fiber is installed between these inner studs, a fabric cover is stretched over all, tacked in place, and finished strips nailed on the edge of each 2 × 4 to complete the floor to ceiling job. The south half of the east wall and the south wall are left reflective to provide an area in the southeast corner of the studio, near the door, which would have somewhat brighter acoustics than the other areas near absorptive walls. Such localized acoustics can be of great help in instrument placement.

On the north wall are two swinging panels 4 feet wide running from an inch above the carpet to a height of 6 feet to 8 feet. These panels are framed of 2 × 4s with ¾-inch particleboard or plywood backs for strength holding 4 inches of 703 glass fiber covered with fabric like the walls. The back of panel 1 in Fig. 12.4 in open position presents the drummer with his only reflective surface apart from the floor. The space between swinging panels 1 and 2 can be occupied by one instrumentalist (or group) while the space between panel 2 and the east wall can be occupied by a second. Others will have to be positioned in the open area.

On the east wall the same 2 × 4 framing filled with 4 inches of 703 is followed except that here a space behind is provided to augment absorption in the very low frequencies (element 3). At these frequencies the sound penetrates the 4 inches of glass fiber, causing the ¾-inch plywood or particleboard to vibrate as a diaphragm, absorbing sound in the process. It may be found that the instrumentalist in space 1 is too close to the high level drum sound. It would be quite acceptable to move the low frequency element to the north wall and the instrument alcoves 1 and 2 to the east wall if this would meet separation needs better. Such a move would increase the distance between instrumentalists (and their microphones) and the drum kit. Barriers 1 and 2 would provide separation only between instruments, not between instruments and drums unless more complicated double-hinged panels were installed.

Drum Booth

The corner area for the drum booth is indicated in Fig. 12.4. The concrete floor of the booth is left bare under the drums to give the desired effect. Another reflective surface for the benefit of the drummer is the surface of swinging panel 1. Apart from these, all surfaces around the booth are highly absorptive to contain the drum sounds and thus improve separation from the sounds of other instruments.

Without adequate separation the advantages of multitrack recording disappear. There is general agreement on two points:

- That sounds from the drum kit are hard to contain.
- A good drum sound is basic to any group.

This is justification for doing something extra for the drummer, even in a budget facility such as this. It would be nice if we could do the same for the vocalist.

Leaving the bare ceiling over the drum booth would defeat the whole purpose of the booth. Drum sounds must be absorbed rather than allowed to float over the entire area. A canopy of 2 × 4 framing is dropped down from the ceiling to a point 6 feet above the concrete floor as shown in Fig. 12.5. The shape of this canopy follows the general drum booth front edge shown in Fig. 12.4.

The face of the canopy toward the studio is covered with a double layer of ⅝-inch gypsum board, but the inside of the entire 2 × 4 framing on 16-inch centers is left open to receive 4 inches of 703 glass fiberboard.

A fabric cover is then applied and held in place with finished strips nailed to the 2 × 4s. This 4 inches of 703 treatment faces the drummer on all sides including walls, underside of canopy, and inside of canopy lip. The construction leaves an *attic* between the canopy ceiling of 4 inches of 703 and the 4 inches of 703 affixed to the uppermost ceiling gypsum board. Because drums have a hefty low frequency content, the attic can be made into an effective absorber for very low frequencies, frequencies lower than the 4 inches of 703 can handle. The term *basstrap* has been applied to such absorbers. Such a catchy term, which seems only to confuse the populace, is not used in this book, at least until *trebletrap* is accepted to describe the acoustical effect of carpet. Low frequency absorbers, however, do come in varying degrees of *lowness*. The 4 inches of 703

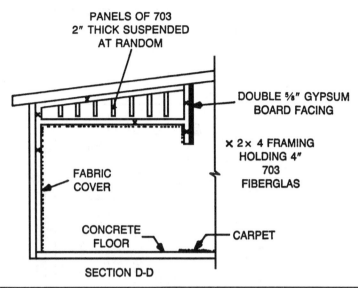

FIGURE 12.5 Section of Fig. 12.4 through the drum booth. All drum booth surfaces except the concrete floor are highly absorbent to contain drum sounds.

is essentially a perfect absorber down to 125 Hz and its absorbing effectiveness decreases as frequency is lowered. For this drum booth attic, a trick of the builders of early anechoic, *free-field chambers* or *dead rooms* is used. This is the hanging of spaced absorbing panels to extend the useful low frequency range of the room. The spacings of the panels of 2-inch thickness 703 may be random rather than following meticulous rules used in those early days.

Computations

Reverberation time, of itself, does not play too important a role in multitrack recording because our primary goal is adequate track separation. Distinctive sound, not naturalness, is the goal of rock recording. Following through on the computations of Table 12.1, however, gives a good feel for comparing treatment of this type of studio with the more traditional speech and music studios. The drywall, carpet, and 4 inches of 703 areas are estimated. The extra bass absorption effect of the drum booth attic and the structure on the east wall are neglected as their effect over and above that of the 4 inches of 703 is largely below 125 Hz, the lowest frequency of Table 12.1.

The calculated reverberation times of the studio range from 0.30 to 0.22 second as compared to 0.35–0.60 second if the same studio were treated for recording speech and traditional music (Fig. 12.6). This comparison emphasizes the general deadness of multitrack studios in the quest for track separation.

Studio deadness, of course, is not the only step toward adequate track separation. The use of baffles, microphone directivity, close placement of microphones, and other factors have their important effects.

SIZE .... 21'5" × 21'7" with corner cut
FLOOR .... Carpet and pad (except for drum booth)
CEILING .... ⅝" Gypsum board
WALLS .... ⅝" Gypsum board, partially covered with 4" 703
VOLUME .... 3,170 cubic feet

Material	S Area Sq. Ft.	125 Hz		250 Hz		500 Hz		1 kHz		2 kHz		4 kHz	
		a	Sa	a	Sa	a	Sa	a	Sa	a	Sa	a	Sa
Drywall	1,000.	0.1	100.	0.08	80.	0.05	50.	0.03	30.	0.03	30.	0.03	30.
Carpet	310.	0.05	15.5	0.15	46.5	0.30	93.	0.40	124.	0.50	155.	0.60	186.
4" 703	500.	0.99	495.	0.99	495.	0.99	495.	0.99	495.	0.99	495.	0.99	495.
Total Sabins, Sa			510.5		621.5		638.0		649.0		680.0		721.0
Reverberation Time, sec.			0.30		0.25		0.24		0.24		0.23		0.22

TABLE **12.1**   Studio Calculations

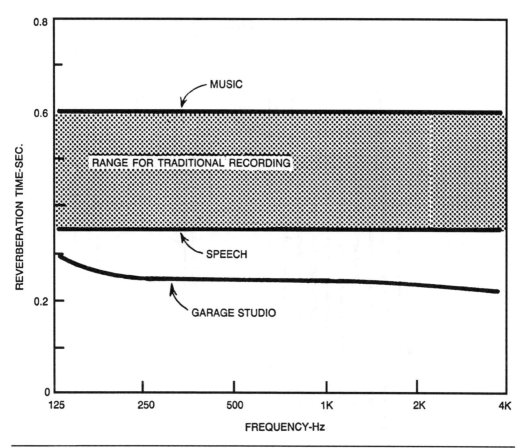

FIGURE 12.6   The reverberation time of the garage multitrack studio is far shorter than that for more traditional recording.

Control Room Treatment

Figuring the axial mode frequencies of a triangular room is far beyond the scope of this book, but the average ceiling height of about 9 feet would yield a fundamental of around 60 Hz; the others would not be too far from this. Cutting off the sharp corners near the window provides a shelf for the loudspeakers. To remove the bad effects of the cavity in which the loudspeakers sit, a heavy plywood baffle should be fitted around the face of the loudspeaker (Fig. 12.7).

The triangular space below the loudspeaker may be utilized as a low frequency absorber to help counteract the carpet effect. To coin a euphonious phrase, this is a slit resonator utilizing the slots between the slats and the cavity behind. It is made simply of 1×4s spaced about $\frac{1}{4}$ inch with 2 inches of 703 pressed against the rear of the slats. The cavity itself serves only to contain the springy air.

On the walls behind the operator about thirty 12 inch × 12 inch × $\frac{3}{4}$ inch acoustical tiles are cemented to each wall in a 6 × 5 array. This totals 60 square feet. Carpet covers the floor and the ceiling is drywall. The two resonators plus some 350 square feet of

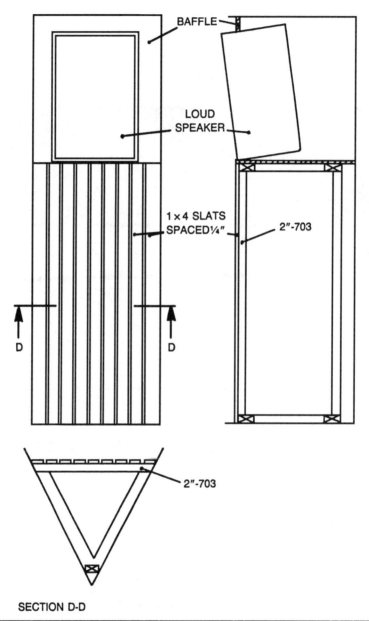

BAFFLE

LOUD
SPEAKER

1 × 4 SLATS
SPACED¼"

2"-703

D   D

2"-703

SECTION D-D

FIGURE 12.7   Corner treatment in control room of garage multitrack studio. Helmholtz resonators having a peak of absorption in the low frequencies are built into the corners. These are slat type resonators.

gypsum board compensate for the low frequency deficiencies of the tile and carpet, but not completely. This brings the reverberation time of the control room to about 0.34 second, rising to about 0.46 second below 250 Hz. With all the compromises involved there seems to be little justification for further acoustical adjustment.

Outside Noise to Inside

No matter how high the average and peak levels of music are in the studio, there are those soft, sweet, and sentimental passages for contrast. The barking of a neighbor's dog heard on the vocal track during such a passage is guaranteed to raise the emotional level even higher than the vocalist could hope for. Therefore, concrete walls 6 feet thick are longed for at such times, but they are just too expensive.

The simple walls of Fig. 12.3 are all this budget could stand. Statistics are on the side of the amateur or low budget recording job. How often do soft passages occur? How often do screaming motorcycles or other interfering noises occur? The permutations and combinations are such that redoing a very occasional ruined take is usually the answer for this type of studio. However, it is quite a different story if it costs $5,000 in studio time and for talent to redo the passage.

Inside Noise to Outside

Very square neighbors have been known to identify that beautiful stuff being laid down in multitrack within the studio as noise. The difference in point of view can bring the police. In some communities the noise ordinance is as broad and general as this one:

It shall be unlawful for any person to make, cause or permit to be made, any loud or unusual noise which directly causes an unreasonable interference with the use, enjoyment and/or possession of any real property owned or occupied by any other person.

In an increasing number of communities a certain maximum allowable noise level is set for the boundary of the property on which the studio rests. In brief, the sounds escaping from the studio may give far more trouble than exterior sounds spoiling takes. Walls offer the same transmission loss both ways, and the walls of Fig. 12.3 certainly offer minimum transmission loss with respect to noise going either way.

Multitrack Recording

Multitrack recording techniques require studios quite different from the traditional kind. Numerous books have been written on the multitrack subject[13-15] as well as many articles in the technical press.[16-26] I refer you to them for information on the host of pertinent points which cannot be covered in this book.

CHAPTER 13

Building a Studio from Scratch for Radio Program Production

Many small studio projects must be warped around to fit into space in an existing building. This means compromises, compromises, and more compromises. When this particular client said that the studio building was to be built from the ground up, the news was received with delight. That was the good news. Then came the bad news: The new building was to be jammed between existing buildings on two sides and a brick wall on the third side. But even this means a certain amount of shielding from exterior noise by these masonry structures.

The space available was 18 × 6 meters (the site is in a foreign country) or about 59 feet 0 inches × 19 feet 8 inches. This was taken as the outside dimensions of the building. Besides a studio and control room, space was required for tape and film storage, toilet facilities, and an office.

The floor plan of Fig. 13.1 was developed with some give and take. Walls of spaced double brick tiers were specified to surround the studio and control room as protection against outside noise. The studio size and shape was the first task. What maximum studio length could best fit a maximum inside width of 18 feet with a practical, economical ceiling height? Several factors were considered: one variable (length); one constant (width); and one variable with definite constraints (ceiling height).

When the opportunity to splay two of the four walls presents itself, as it did in the present case, it is wise to do so. This eliminates (or at least reduces) the chance for flutter echoes between parallel surfaces.

Placement of acoustical materials can also reduce flutter echoes, but if it can be done with geometry independent of the acoustical treatment, a greater degree of flexibility accrues in the placement of such materials.

The east wall containing the control room window was inclined at about 5 degrees from its rectangular position as shown in Fig. 13.2. The south wall was given two triangular protuberances with sides inclined 8.6 degrees. Any angle between 5 degrees and 10 degrees is usually sufficient for control of flutter echo. This inclining of wall surfaces takes care of the N-S and E-W flutter modes. The

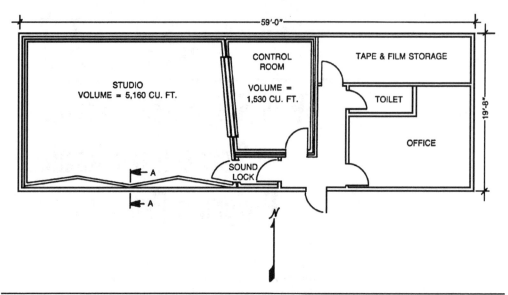

FIGURE 13.1   Floor plan of studio complex which includes service areas as well. Splayed east and south walls help control flutter echo.

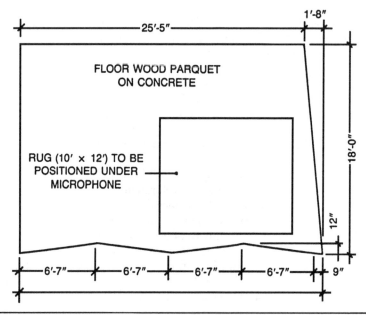

FIGURE 13.2   Splaying plan of studio. The splaying angles are about 5 degrees (east wall) and 8.6 degrees, for the triangular south wall.

vertical mode flutter could be cared for by constructing wrinkles of some sort in the ceiling (or the floor!) but other methods will be used.

Distribution of Modal Resonances

Again, considering only the axial modes of the studio and disregarding the less influential tangential and oblique modes, let us select studio length and ceiling height which will give reasonable distribution of room resonances. With a splayed east wall and a south wall broken up by four splayed surfaces, the average length is taken to be 26 feet 3 inches and the average width 17 feet 6 inches. With a height of 11 feet 3 inches, the distribution of modal resonances is as illustrated in Fig. 13.3A.

One very great advantage of a larger studio is that the average spacing of room resonances is reduced. The average spacing of resonances in this studio (5160 cubic feet) is 11 Hz. The average spacing of resonances in the 1530 cubic feet control room (Fig. 13.3B) is about 16 Hz. This closer spacing, if not too close or coincident, tends toward fewer colorations, thus better quality.

The small numeral 2 above certain room resonance lines in Fig. 13.3 indicates a coincidence of two modes at that frequency. You can disregard the two coincidences in the studio at higher frequencies and the one at 65 Hz because of the BBC experience that audible colorations are rare in those frequency regions.[6] This leaves the coincidence at 129 Hz and the close pair near 150 Hz as possible threats.

In the control room (Fig. 13.4) coincidences at 249 Hz are high enough to not cause concern. The close pair at 166/169 Hz is aggravated by wide spacings on each side. As sound decays in the room, these two could beat together at a 3 Hz rate.

Other factors must be favorable before coincidences result in observable colorations, hence they may or may not be troublesome. It is impractical to push further an analysis such as this; suffice it to say that there is the advantage of a warning of potential problems at 129 Hz and 150 Hz in the the studio and 166/169 Hz in the control room.

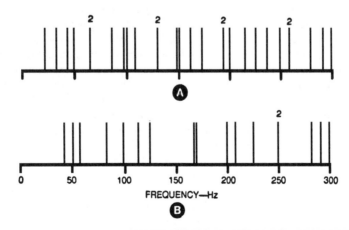

FIGURE 13.3   Distribution of axial room resonance frequencies for (A) studio and (B) control room. Note that for the larger studio the average modal spacing is about 11 Hz; for the control room about 16 Hz. The small 2 indicates two resonances which are coincident.

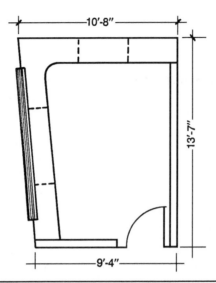

FIGURE 13.4 Floor plan of control room. Built-in work surface extends the length of two walls.

Noise Considerations

How heavy should the walls be made? In general, the heavier the walls the better protection offered against external noise. On the other hand, the heavier the walls the greater the cost. In the present case, brick, a favored and economical local building material, seemed both logical and economical for walls. The question is, single tier wall or double tier?

The answer is to be found in two categories: (a) the level of environmental noise at the studio site, and (b) the level of noise allowable inside the studio. The walls (and other parts of the enclosure, such as doors) are called upon to reduce (a) to (b).

The hoped for lowness of studio noise may be expressed by selection of a standardized noise contour. In this example the NC-15 contour was selected which is the lower curve in Fig. 13.5. This is a reasonably stringent requirement. Studio noise levels somewhat above this (e.g., NC-20) would not seriously impair most types of recording.

A lower background noise level is required for mono recording than for stereo. As this studio is engaged only in mono recording at present, with plans to convert to stereo in the future, it seemed wise to select the more conservative NC-15 criterion. Because of the increase in sensitivity of the human ear as frequency is increased, the NC-15 contour has its characteristic downward slope.

Evaluating the environmental noise at the studio site in any specific, direct way is a rather complicated procedure. This noise is usually anything but steady, having peaks and low points between peaks. There is usually a variation with time.

This studio is located behind an office building, about 100 feet from a city street carrying typically heavy downtown traffic. Because actual around-the-clock measurements were not feasible, similar measurements made by others and reported in the literature were used as the basis of estimation.[27] This procedure gives the upper curve in Fig. 13.5. The shaded area between the two curves represents the transmission loss which the walls must give to reduce the exterior noise to the NC-15 level in the studio.

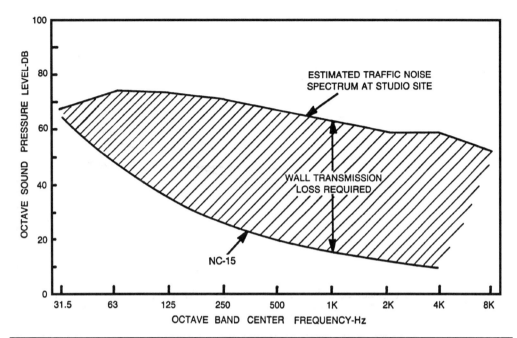

FIGURE 13.5   Studio wall requirements are based on the environmental noise level at the site and the noise contour selected for inside the studio. The shaded area between these two curves represents transmission loss which the walls must provide.

The differences between the two curves of Fig. 13.5 at selected frequencies gives the transmission loss required which is replotted as the heavy line in Fig. 13.6. Adequate data on the transmission loss of brick walls is difficult to find, but a paper published in Europe gives measurements on single and double tiered brick walls which are adapted in Fig. 13.6.

The double tiered brick wall has one face cement plastered, the single tiered wall has no plaster. If the transmission loss offered by these brick walls is greater than the transmission loss required, well and good. If below, the walls are falling short of the requirement. In Fig. 13.6 the single brick wall falls far short of the required loss from about 100 to 1,000 Hz. The double brick wall is considerably better, falling short a maximum of only 6 dB between 100 and 400 Hz. The site noise level and the selection of the NC-15 contour are not precise enough to make much of a fuss about this 6 dB.

Further, plastering both the exterior and interior faces of the double brick wall increases its transmission loss to the point where the 6 dB difference is partially made up. The conclusion, then, is that a double brick wall, plastered on exterior face and studio face, should bring the studio background noise level to about the NC-15 contour, a satisfactory level for the recording work contemplated.

Wall Construction

Figure 13.7 is Section A-A of Fig. 13.1. It shows one method of supporting the ceiling joists on the inner tier of bricks. Solid ties between inner and outer brick tiers, intentional

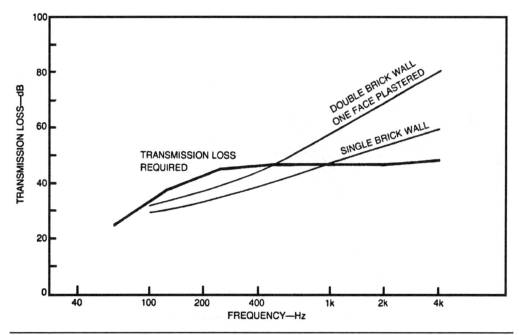

Figure 13.6   The transmission loss required curve is obtained from the number of dB separation between the two curves of Fig. 13.5. The transmission loss offered by different types of wall construction may then be compared to that loss required.

or otherwise, seriously degrade the transmission loss of the wall. Local codes may require such ties, but they should be used as sparingly as possible or not at all.

In building a studio from the ground up there is the opportunity of including flutter echo insurance with no significant increase in cost. The inner brick wall is built into triangular shapes on the south wall and the splayed east wall helps both the studio and the control room.

Sound Lock

The normal studio layout would open both studio and control room doors into the sound lock. In the present case high control room traffic resulted in the decision to open the control room into the entrance lobby (Fig. 13.1). It must be recognized that this somewhat negates the double east wall of the control room as the single control room door will give only some 30 dB isolation, compared to something like 50 dB for the wall. Office noise may be a greater threat than exterior noise.

Studio Floor

The floors of the studio are specified to be covered with wood parquet. This is a beautiful floor covering which is ruled out by high cost in many areas, but not in this one. This is a highly reflective floor as far as sound waves are concerned. It is opposed by a highly absorbent ceiling to be described later. A microphone placed near any reflective surface receives both a direct component of sound from the source and another

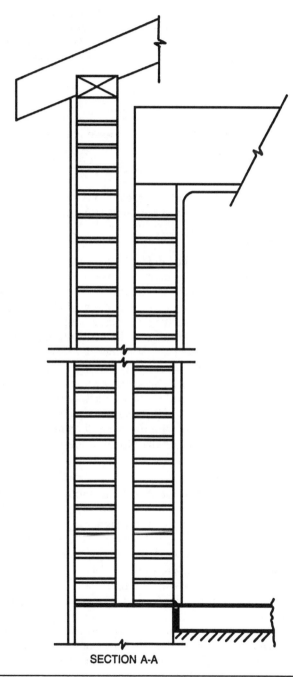

SECTION A-A

FIGURE 13.7   Studio wall construction of double-tier brick, plastered inside and out. Section A-A specified in Fig. 13.1.

reflected from the surface. The latter arrives later and creates comb filter distortion in the electrical signal obtained from the microphone.[29] For a close source, the 10 foot × 12 foot rug under the microphone and source will minimize this form of distortion. For a distant pickup the microphone should be placed very close to the floor so that the two path lengths are approximately equal, or on a stand so that the bounce takes place on the other end of the rug.

Studio Walls

Three types of wall treatment (Fig. 13.8) are used in the studio:

- Wideband (WB) absorbers
- Acoustical tile
- Low frequency (LF) absorbers to compensate for low frequency deficiencies in WB absorbers and acoustical tile, especially the acoustical tile

Wideband Modules

The wideband modules, described in Fig. 13.9, are similar to others in previous chapters. For example, these modules are similar to those of basically 4 inches of 703 type of glass fiber with a suitable mounting structure. Instead of a fabric face cover, a board is used. An ideal cover is Owens-Corning glass cloth covered Fiberglas boards 1 inch thick.

In some foreign countries such a board may not be readily available, in which case a soft wood fiberboard $\frac{1}{2}$-inch thick could be substituted with modest degradation of absorption. Similarly, glass wool or glass fiber of other types could be substituted for the 703 Fiberglas if density approaches 3 pounds per cubic foot.

A degree of variation of studio acoustics can be introduced by mounting the eight wideband modules as shown in Fig. 13.10. Such a mounting allows reversal of each module by the simple expedient of flipping the top latch, lifting the module off the pin, reversing the module, replacing it on the pin and reengaging the top latch. In this way either soft or hard faces of some or all of the modules could be exposed.

Low Frequency Absorbers

Figure 13.8 shows four 4 foot × 8 foot and one 8 foot × 8 foot low frequency absorbers on the studio walls. These can be built with either perforated facings or slat-slit facings, the latter being chosen in this particular design.

The frames of 2 × 8 lumber are first constructed and then the cross pieces of 2 × 8 are added as shown in Fig. 13.11A. At this point the frames are mounted on the wall with suitable angles and expansion screws (Fig. 13.11B). The crack between the frame and wall is then sealed on the inside with running beads of nonhardening acoustical sealant type of mastic so that each smaller section is essentially airtight as far as the wall junction is concerned.

The glass fiber material will probably be stiff enough so that, cut slightly large, the pieces will be held by friction against the backs of the slats without support. If support is needed, zig-zag wires tacked to the 2 × 8s ensure close contact of glass fiber and slats. To avoid stroboscopic optical illusions as eyes are swept horizontally across the vertical slats, the face of the glass fiber should be covered with lightweight black cloth. Instead of relatively light colored 703, Owens-Corning and perhaps other manufacturers, also make a black duct liner which has been used in this service. This eliminates the need for

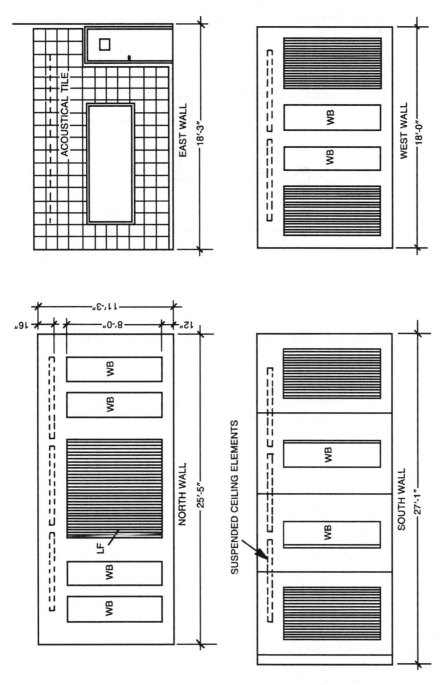

FIGURE 13.8 Wall elevations of studio showing distribution and placement of wideband (WB) panels, low frequency (LF) absorbers, and acoustical tile.

197

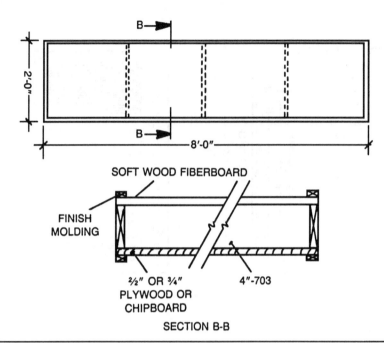

FIGURE 13.9 Details of construction of the wideband wall modules having both a soft and hard face. These are mounted so that they may be reversed to adjust acoustics of the studio.

the black cloth. The slats should be stained and varnished before nailing in place. The spacings should be alternately ⅛ inch and ¼ inch. The use of temporary shims while nailing assures uniform slits from top to bottom.

Acoustical Tile

The entire east wall (with the exception of the observation window and door) is to be covered with ½-inch acoustical tile (Fig. 13.8). These tiles may have perforations or slits or neither if of high quality. They are cemented in place in the usual way. The lower tiles are subjected to considerable abrasion and storing a few dozen matching tiles for later repairs is a good idea.

Studio Ceiling

The studio ceiling has plaster 1 inch thick like the walls. To reduce the reflectivity of the ceiling, six elements are suspended by wires so that the lower edge of the 2 × 6 frames is 16 inches below the plaster surface. These elements hold absorbing material and light fixtures.

Figure 13.12A shows how the 6 foot 6 inch × 6 foot 6 inch frame of 2 × 6 lumber is subdivided into nine smaller sections. The center one holds a specially fabricated sheet metal rectangular box containing four 18-inch fluorescent tubes (remember that the starter reactors should be mounted outside the studio to avoid buzzing noises).

The inside lower edges of this metal box should be painted flat black. The deep recessing of the tubes and the black paint are to remove the tubes and their bright

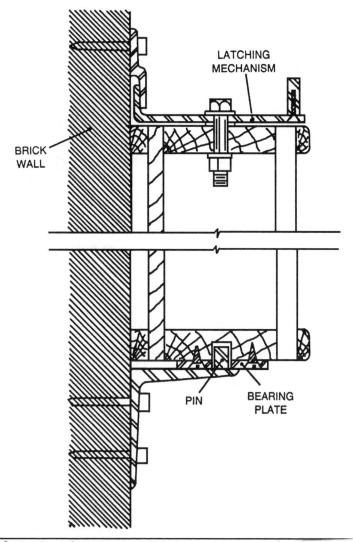

FIGURE 13.10   Hardware for wideband wall modules to allow reversal.

reflections from the view of those in the control room to avoid glare in the eyes and in the observation window.

Section B-B of Fig. 13.12B reveals the basic constructional features of each element. Metal tees and angles, such as used in conventional lay-in suspended ceilings, form the 24 inch × 24 inch sections, eight of which have plastic grilles with honeycomb or square openings resting in them. The 4 inches of 703 is held up by these grilles. A lightweight fabric should be placed between the grille and the 703 it supports to prevent small bits of glass fiber from sifting down the artists' necks. Of the 48 24 inch × 24 inch sections, only 38 are used to hold glass fiber in the design to follow. This means that there will be one or two vacant sections in each element. The vacant sections should be distributed randomly within the element and across the studio.

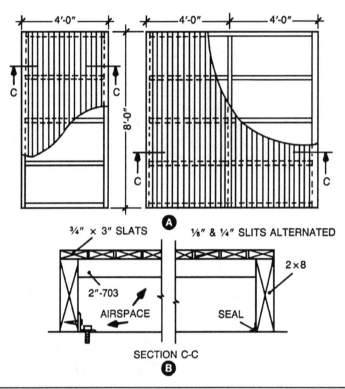

Figure 13.11   Constructional details of slat type of Helmholtz low frequency resonators.

The six ceiling elements are suspended by wires from the ceiling joists according to the projected ceiling plan of Fig. 13.13.

Control Room Treatment

The control room floor is also wood parquet. Opposing it is a conventional lay-in suspended ceiling dropped down 16 inches from the plastered surface. The walls are treated as shown in Fig. 13.14 with the same type of low frequency absorber used in the studio. They are constructed as shown in Fig. 13.11 and use the same acoustical tile.

Studio Computations

The generally accepted optimum reverberation time for a studio of 5160 cubic feet volume is about 0.67 second for music and 0.4 second for speech. The studio is to be used for both, hence some compromise reverberation time must be used. This could be warped upward or downward, depending upon whether music or speech were to be favored. The treatment of this studio has been presented as a fait accompli in previous paragraphs, but retracing the calculation steps should be a profitable exercise as this will provide the basis for the reader who wants to adapt this information to his own studio situation.

Table 13.1 gives reverberation time calculations for two conditions. Condition A is when the soft fiberboard sides of all eight wideband modules are facing the studio.

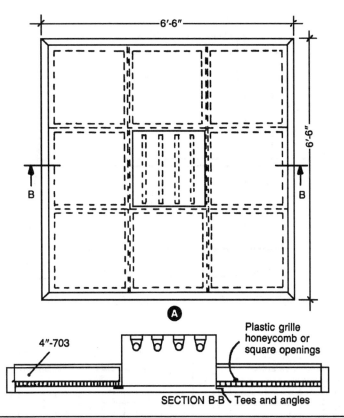

FIGURE 13.12 Details of one of the six suspended ceiling frames in studio which hold the illumination fixtures and wideband glass fiber absorbers. Both faces of the glass fiber are active when so mounted.

Condition B is when the hard faces of all eight wideband modules are exposed. The results of these calculations are presented in graphical form in Fig. 13.15.

The reverberation time can be increased about 30 percent by flipping the eight wideband wall modules from soft to hard side out. Although this must be determined by many listening tests, the best average condition would appear to be Condition A with wall modules soft. This would be essentially a speech condition shifted somewhat in the music direction.

The idea of reversing wall modules for every recording session is too idealistic and just too much work. The type of recording carried out in any studio invariably falls into one, two, or a few categories and usually a single reverberatory condition meets the needs of most of the jobs. It is nice, however, to know that if especially bright conditions are desired for a certain type of music recording, the facilities are available to attain these conditions. The diffusion of sound in the studio is certainly better with soft wall modules than with hard modules. Good diffusion is especially needed for recording of speech.

Even greater flexibility in adjusting studio acoustics is available if the 24 inch × 24 inch pads of 703 in the six ceiling elements are involved. The design represented by Table 13.1 and Fig. 13.15 includes 152 square feet, or only 38 of the 48 possible absorbent

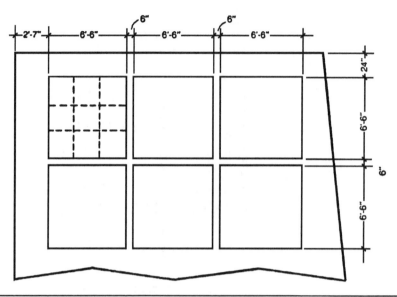

FIGURE **13.13** Placement details of the six suspended frames in the studio.

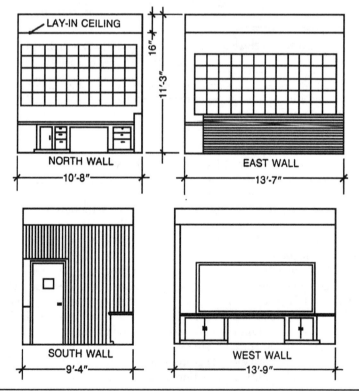

FIGURE **13.14** Control room wall elevations showing distribution of slat resonators and acoustical tile. A conventional suspended ceiling is used in the control room.

SIZE .... 25'5" × 18'3" × 27'1" × 18'0" (two walls splayed)
FLOOR .... Wood parquet
CEILING .... Plastered with 6 suspended elements
WALLS .... 8 Wideband, 5 low frequency modules, acoustical tile
VOLUME .... 5,160 cubic feet

Material	S Area Sq. Ft.	125 Hz a	125 Hz Sa	250 Hz a	250 Hz Sa	500 Hz a	500 Hz Sa	1 kHz a	1 kHz Sa	2 kHz a	2 kHz Sa	4 kHz a	4 kHz Sa
Floor; Parquet	238	0.04	9.5	0.04	9.5	0.07	16.7	0.06	14.3	0.06	14.3	0.07	16.7
Floor; Rug 10' × 12'	120	0.35	6.0	0.10	12.0	0.15	18.0	0.30	36.0	0.50	60.0	0.55	66.0
LF Units	160	0.98	156.8	0.72	115.2	0.33	52.8	0.21	33.6	0.16	25.6	0.14	22.4
Acous. Tile 1/2"	137	0.10	13.7	0.25	34.3	0.65	89.1	0.73	100.0	0.73	100.0	0.68	93.8
Ceiling Elements:													
Lower Surface	152	0.99	150.5	0.99	150.5	0.99	150.5	0.99	150.5	0.99	150.5	0.99	150.5
Upper Surface	152	0.20	30.4	0.20	30.4	0.20	30.4	0.20	30.4	0.20	30.4	0.20	30.4
Total Sabins, Sa without Wall Modules			366.9		351.9		357.5		364.8		380.8		379.8
Condition A Wall Modules, Soft Side Out	128	0.83	112.6	0.99	126.7	0.99	126.7	0.99	126.7	0.99	126.7	0.96	122.9
Total Sabins, Sa			479.5		478.6		484.2		491.5		507.5		502.7
Reverb. Time, sec.			0.52		0.53		0.52		0.51		0.50		0.50
Condition B Wall Modules, Hard Side Out	128	0.28	35.8	0.22	28.2	0.17	21.8	0.09	11.5	0.10	12.8	0.11	14.1
Total Sabins, Sa			402.7		380.1		379.3		376.3		393.6		393.9
Reverb. Time, sec.			0.63		0.66		0.67		0.67		0.64		0.64

TABLE 13.1 Studio Reverberation Time Calculations

sections in the ceiling frames. This would allow adding 10 sections or 40 sabins, or removing 10 sections with only minor sound diffusion effects if a distributed pattern of sections remaining is maintained.

Table 13.1 takes advantage of the fact that the ceiling element pads of 703 are capable of absorbing sound on both the lower and upper faces. As the upper face is somewhat shielded by the framework, an estimated 20 percent absorption is applied to the top surface. Acoustical measurements in the studio could very well demonstrate that the upper surface absorbs more than this.

Control Room Reverberation

The reverberation time goal for the control room is about 0.3 second which is suitably shorter than the studio it serves. The treatment shown in Fig. 13.14 approaches this as the computations of Table 13.2 reveal. Note that the low frequency absorption of the cabinets is a significant contribution. Although specific proprietary materials are listed for both the lay-in ceiling boards and the acoustical tile, many other products of the same type would serve just as well if their different absorption coefficients are known and areas properly adjusted.

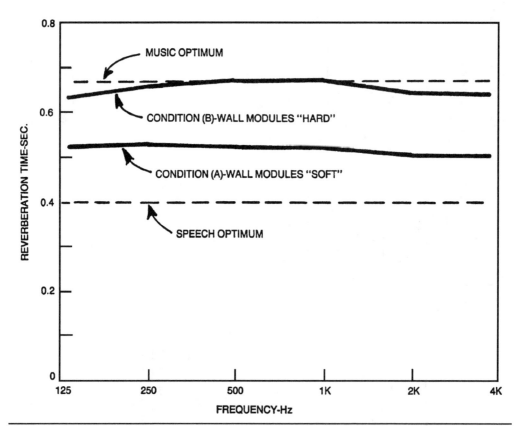

FIGURE 13.15   Degree of reverberation time change made possible with reversible wideband wall modules.

SIZE .... 13'9" × 13'7" × 10'8" × 9'4" (one wall splayed)
FLOOR .... Wood parquet
CEILING .... Lay-in Celotex Safetone Celotone, nat fissured, 3/4" mounting #7
WALLS .... Slat type LF absorbers and Simpson Pyrotect
VOLUME .... 1,530 cubic feet

Material	S Area. Sq. Ft.	125 Hz		250 Hz		500 Hz		1 kHz		2 kHz		4 kHz	
		a	Sa	a	Sa	a	Sa	a	Sa	a	Sa	a	Sa
Floor Parquet	108.	0.04	4.3	0.04	4.3	0.07	7.6	0.06	6.5	0.06	6.5	0.07	7.6
Cabinets-1/2" Plywood	90.	0.30	27.0	0.23	20.7	0.18	16.2	0.14	12.6	0.11	9.9	0.10	9.0
Lay-In Ceiling	136.	0.50	68.0	0.54	73.4	0.55	74.8	0.77	104.7	0.87	118.3	0.88	119.7
LF Slat Absorbers	128.	0.98	125.4	0.72	92.2	0.33	42.2	0.21	26.9	0.16	20.5	0.14	17.9
Acoustical Tile	110.	0.06	6.6	0.27	29.7	0.81	89.1	0.91	100.1	0.68	50.8	0.48	52.8
Total Sabins, Sa			231.3		220.3		229.9		250.8		206.0		207.0
Reverb. Time, sec.			0.32		0.34		0.33		0.30		0.36		0.36

Table 13.2   Control Room Reverberation Time Calculations

205

CHAPTER 14

Studios for a Commercial Radio Station

The client in this example was represented by a very knowledgeable engineer who had already laid out a floor plan suitable for their needs. He wanted it checked out for room resonance distribution, a complete acoustical treatment plan made for each room, and then measurements done to verify the design.

The floor plan already decided upon, shown in Fig. 14.1, includes master control, production control, and a talk booth. The word *booth* infers that it is small and so it is. This one is 805 cubic feet, only half the minimum volume recommended. The other two rooms are somewhat larger than the minimum 1500 cubic feet, 2000 and 1943 cubic feet.

In an operation such as this there is generally a great reluctance to be generous in the area devoted to recording, editing/listening, and live broadcasting. After all, abstract and intangible acoustics are locked in a battle with down-to-earth factors such as space for offices for the station manager, production manager, and engineer, as well as space for library, traffic and continuity, accounting, cafeteria, news and farm department, announcers' work area, and last and most important, ample room for *SALES*.

To keep the studio area under control there are pessimists around who would even question straining for acoustical quality when the audio signal is routinely highly processed to get the effect of greater transmitter power. The signal night not be so good, but signal coverage and resulting sales apparently are best served by such a procedure. Another argument tending to scuttle support for acoustical quality of broadcast studios is that little goes out live anymore, except voice, and the announcer always close-talks the microphone. It is true that room effects are most noticeable for greater source to microphone distances but the imprint of the room is always there, no matter how close the microphone to the source.

Construction

The studio area of Fig. 14.1 is one corner of a 55 foot × 60 foot single story structure. The walls of the studio are of 6-inch concrete block, an air space of 2 inches and an inner wall of 4-inch concrete block as shown in Fig. 14.2. The hollow spaces in the concrete blocks of both tiers were filled with concrete, well-rodded to eliminate air pockets. The external walls have added thermal protection because of the northern location.

The exterior face of the external walls is covered with stucco or siding. The 2-inch air space is filled with thermal type of glass fiber. The inner surface of the 4-inch block

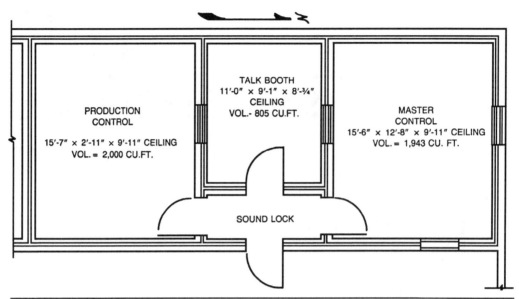

FIGURE 14.1 Floor plan for highly efficient studio and work space for a commercial radio station. The talk booth is substandard in size.

tier of external walls is covered with ⅝-inch gypsum board furred out on 2 × 2s with the space behind also filled with a thermal type of glass fiber.

The interior walls of the studio area are the same basic construction as the external walls as far as the 6-inch and 4-inch concrete block tiers and the 2-inch air space are concerned. The similarity ends there as the air spaces of the interior walls have no glass fiber in them nor is there a gypsum board layer. The inner faces of all walls were to be painted.

An exterior window was insisted upon in the north wall of master control. Another window in the east wall allows other staff workers to see what is going on in master control without entering. The two windows in the talk booth are lined up so that personnel in the three rooms can see their colleagues in the other rooms. This allows the use of hand signals. In fact, someone in production control can look through both the talk booth and master control to the next range of mountains! The production value of this feature is not too evident unless it is to be able to say during a weather broadcast, "It's snowing hard here." From the acoustical standpoint lining up windows is very bad, especially if the glass plates are parallel.

Acoustical Treatment

In this chapter constructional details of wideband wall modules, suspended and other ceiling elements, etc., are passed over in favor of some new and, hopefully, more interesting features. The treatment of master control (Fig. 14.3), production control (Fig. 14.4), and the talk booth (Fig. 14.5) are quite similar to those of other earlier chapters with one simplifying factor: There is no carpet in these rooms. This means no low frequency resonators are required to compensate for the carpet. Presumably, all we need are areas of

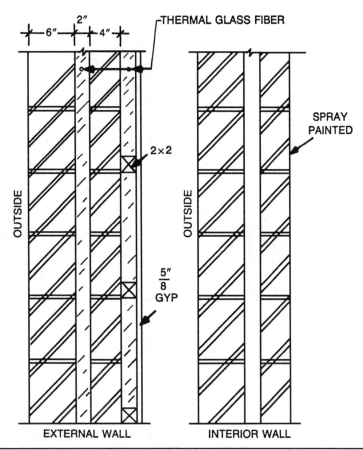

FIGURE 14.2   Wall construction details for studio walls. As this studio is located in a northern area, the external walls are filled with glass fiber insulation between the two concrete block tiers and the inner face is covered with ⅝-inch gypsum board furred out 2 inches with insulation behind. The interior walls are of the same block construction but without the thermal treatment.

4 inches of 703 suitably disposed around each studio to ensure against flutter echo and to give the best sound diffusion. These areas are provided in three forms:

- wideband wall modules
- suspended ceiling elements in master and production control rooms
- an acoustically similar ceiling frame in the talk booth

General Measurements

The three studios treated as shown in Figs. 14.3, 14.4, and 14.5 were subjected to acoustical measurements. Only the reverberation results are discussed in this chapter. These measurements as well as the design were all accomplished by mail. The consultant did not visit the studio site. Working at a distance, the consultant sent a magnetic tape with

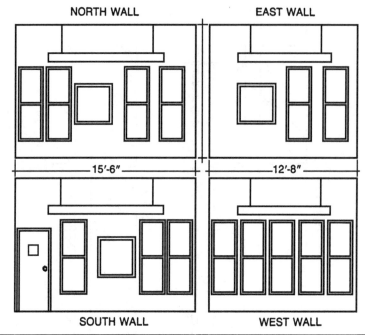

FIGURE 14.3 Wall elevations of master control showing original location of wideband wall modules before adjustments were made.

test signals on it to the engineer who played them according to detailed instructions in each room, picking up each room's response to these signals on a suitable microphone placed as directed and recording the response on another tape recorder. Upon receiving the response tapes in the mail the consultant proceeded to analyze them. In very general terms, reverberation time, swept sine test, and the room's response to impulses were recorded and analyzed.

Reverberation Time

The goal for reverberation time was taken as a nominal 0.3 second, relatively uniform 125 Hz to 4 kHz. The average measured reverberation time of the three studios is shown in Fig. 14.6. All three are substantially below 0.3 second. Well, you win some and you lose some! But the detective in each of us says, "Why?" If this business is not based on rational physical principles as we were led to believe by our physics teacher, perhaps the gnome in the cave up in the mountain might cease his peeping and muttering long enough to design studios for us. Not that the reverberation characteristics of Fig. 14.6 are uniformly low. A studio having a reverberation time of 0.2 second is quite usable, but its sound would tend to be, stated subjectively, dry, dead, and outdoorish in character. The difference between the 0.3 second goal and the 0.2 second reality, however, is a distinct challenge and a problem worth solving.

NORTH WALL — 15·7'

EAST WALL — 12' 11'

SOUTH WALL

WEST WALL

FIGURE 14.4 Wall elevations of production control showing original location of wideband wall modules before adjustments were made.

Theory vs. Practice

The consultant and the radio engineer exchanged views and information by letter, telephone, and one visit of the engineer with the consultant. A number of examples of questionable communication accuracy were disclosed:

- Knowledge of the furred out gypsum board on the inside of exterior walls came after the basic design was completed

- The block surfaces were to be painted. The idea of the original painting specification was to close the tiny surface pores of the coarse concrete blocks which absorb the sound, but the kind of paint to use was not stated clearly by the consultant. The paint used evidently was nonbridging in nature, serving only to stabilize the surface somewhat and to give color, but not to seal off the interstices.

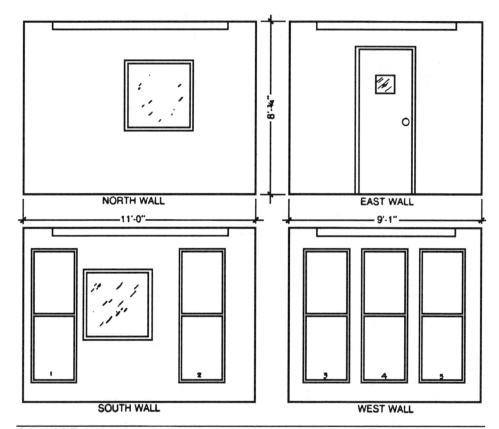

FIGURE 14.5   Wall elevations of talk booth showing original location of wideband wall modules before adjustments were made.

- The third factor was the uncertainty as to how effective the top surface of the 4 inches of 703 in the suspended ceiling would be in absorbing sound. In other words, how well is it shielded from sounds of various frequencies filling the room? The manufacturer's coefficients are not given for such double-sided exposure for this type of product because only one side is normally exposed. In the original design the consultant assumed that the top surface of the 703 would absorb about half as well as the lower surface. The evidence indicates that both sides are fully effective.

The calculations leading to the original design gave essentially 0.3 second reverberation time 125 Hz-4 khz for each of the three studios of Fig. 14.1. But these original calculations did not include the uncertainties introduced later which are listed in the previous paragraph.

The problem now is how to account for the measured reverberation time graphs shown in Fig. 14.6. Our confidence in the whole procedure hinges on our ability to

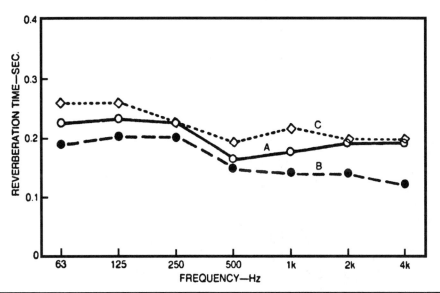

FIGURE 14.6 Measured reverberation time of the three studios; (A) master control,
(B) production control, and (C) talk booth.

account for the actual reverberation time prevailing in these rooms. Accordingly, new
calculations were made in which the absorption coefficients differ in the following ways
from those of the original calculations which predicted about 0.3 second reverberation
time in each room:

1. The area of the 703 used in the suspended ceiling frame was doubled rather than
   increased by 50 percent. This assumes that the top face is just as effective as the
   lower face in absorbing sound.

2. The absorption coefficients for concrete block walls were taken from an obsolete
   source[30] as seen in Table 14.2. The original design, assuming that the paint filled
   the surface pores, used the coefficients in the right column. Learning that the
   walls were spray painted, it was judged that the coefficients of the left column
   applied more closely.

3. The areas of concrete block were reduced by the exterior wall area covered with
   furred out gypsum board. Master control has two such walls and the other rooms
   each have one.

To make a long story short, the recalculated reverberation time for master control,
taking into account the factors in (1) to (3) above, are listed at the foot of Table 14.1.
These calculated points agree reasonably well with the measured values as shown in
Fig. 14.7.

Reverberation time was recalculated for production control using the identical
coefficients and the results are displayed in Fig. 14.8. Here the calculated values are

SIZE .... 15'6" × 12'8", 9'11" Ceiling
FLOOR .... Vinyl tile on concrete
CEILING .... Double ⅝" gypsum board on wood joists.
.... suspended 1 × 6 frame holding 4" of 703 glass fiber
WALLS .... Concrete block exterior walls covered with
.... ⅝" gypsum board furred out on 2 × 2S, space
.... filled with thermal glass fiber (Fig. 14.2)
.... interior concrete book, coarse, spray painted
.... (Fig. 14.2). Wideband modules, 2' × 6', 4" 703 (Fig. 14.3)
VOLUME .... 1,943 cu. ft.

Material	S Area Sq. Ft.	125 Hz a	125 Hz Sa	250 Hz a	250 Hz Sa	500 Hz a	500 Hz Sa	1 kHz a	1 kHz Sa	2 kHz a	2 kHz Sa	4 kHz a	4 kHz Sa
Ceiling 4" 703 69 sq. ft., Both Sides Active	138	0.99	136.6	0.99	136.6	0.99	136.6	0.99	136.6	0.99	136.6	0.99	136.6
N&W Walls ⅝" Gyp Board, 2" Fill	270	0.15	40.5	0.14	37.8	0.12	32.4	0.10	27.0	0.08	16.2	0.05	13.5
E&S Walls Coarse Concrete Block	244	0.36	87.8	0.44	107.4	0.31	75.6	0.29	70.8	0.39	95.2	0.25	61.0
Ceiling: Dbl ⅝" Gypsum Board	196	0.08	15.7	0.05	9.8	0.03	5.9	0.03	5.9	0.03	5.9	0.03	5.9
Wideband Wall Modules (14)	168	0.99	166.3	0.99	166.3	0.99	166.3	0.99	166.3	0.99	166.3	0.99	166.3
Total Sabins, Sa			446.9		457.9		416.8		406.6		420.2		383.3
Reverb. Times, sec.			0.21		0.21		0.23		0.23		0.23		0.25

TABLE **14.1** Master Control Calculations

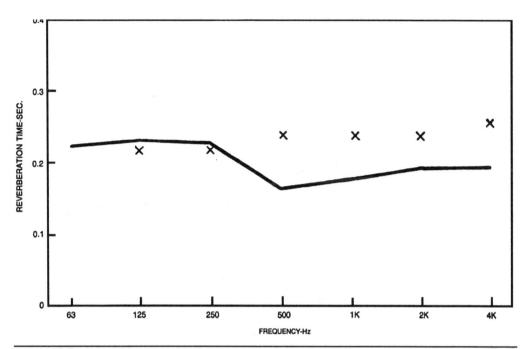

FIGURE 14.7   Measured reverberation time in master control repeated from Fig. 14.6 (solid line). The calculated points (x) are in reasonably close agreement.

very close except at 500 Hz and above. The recalculated values of reverberation time for the talk booth, using the same coefficients, are plotted in Fig. 14.9. The measured and recalculated values of reverberation time, average for six frequencies, are compared in Table 14.3.

These are the types of problems encountered in predicting studio acoustical performance. Figures 14.7, 14.8, and 14.9 show far superior agreement between measured and calculated values, however, than a comparison of the three graphs of Fig. 14.6 with the

Frequency Hz	Coarse, Unpainted	Painted (Bridging Paint)
125	0.36	0.10
250	0.44	0.05
500	0.31	0.06
1k	0.29	0.07
2k	0.39	0.09
4k	0.25	0.08

Source: Ref. 35

TABLE 14.2   Concrete Block Absorbtion Coefficients

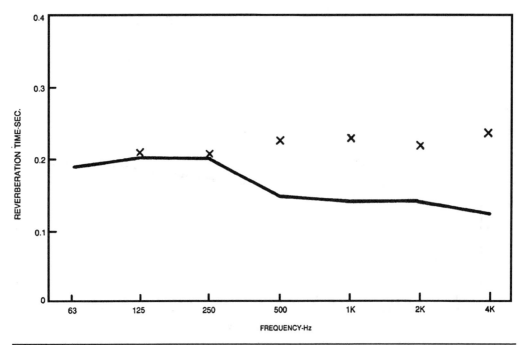

FIGURE 14.8 Measured reverberation time in production control repeated from Fig. 14.6 (solid line). The calculated points (x) show somewhat poorer agreement than Fig. 14.7.

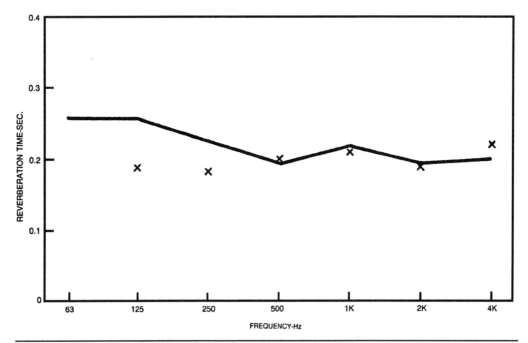

FIGURE 14.9 Measured reverberation time in talk booth repeated from Fig. 14.6 (solid line). The calculated points (x) show reasonably good agreement.

| | Reverberation Time, Seconds | | |
	Measured	Recalculated	
Master Control	0.198	0.237	+20%
Production Control	0.160	0.223	+39%
Talk Booth	0.222	0.198	−11%

TABLE **14.3**   Comparison of Measured and Calculated
Reverberation Times

0.3 second goal of the original design. The agreement would have been even better had the unpainted block coefficients really fitted the type of concrete blocks actually used.

There are a number of lessons here. First, when one of the absorption elements used in the rooms is as unpredictable as concrete block (unpainted or painted with non-bridging paint), calculations are little more than rough guesses. However, concrete block walls, painted in a way that fills the interstices, become predictable; at the same time, they also lose most of their absorption. The absorption of unpainted concrete blocks varies widely because of variations in the density of the materials of which they are made. In the present case, the block walls painted with nonbridging instead of bridging paint introduced significant absorption which resulted in reverberation times near 0.2 second rather than the predicted 0.3 second.

Another lesson is that there is no substitute for measurements to reveal gross problems. If all materials used in a room behave according to the published absorption coefficients, calculated reverberation time can be much closer than Table 14.3 indicates, but if the uncertainty of concrete block absorption is present, accuracies in the −11 percent to +39 percent range must be expected.

A third lesson is that the use of wideband absorption modules allows easy trimming.

Master Control Trimming

To increase the reverberation time of master control from the measured values of graph A of Fig. 14.6 to approximately 0.3 second across the band, considerable reduction in sound absorption is required. The first impulse is to remove wideband modules. There is a problem here. With the concrete block absorption unchanged, almost all the modules must be removed. Bare walls would be a problem. To increase reverberation time to 0.3 second there is only one thing to do: Paint the concrete block walls with a bridging paint and start all over by recalculating the room. This shows that about a dozen wideband modules are required, which must then be positioned carefully to control flutter echoes. The measurements have given a base from which to work. With this added insight a couple of wideband modules can be removed to trim the room.

Production Control Trimming

Painting the concrete block wall with a bridging paint is also recommended for production control. Once this is done, the original design of Fig. 14.4 is quite reasonable. In this case, removing a single wideband module from the room may trim sufficiently.

Talk Booth Trimming

Painting the north, east, and south concrete block walls of the talk booth also brings things close to the original design. This raises the reverberation time to about 0.3 second.

Summary

What a tremendous difference a little thing like the type of paint used on the concrete block walls makes! This is a primary lesson to be learned from this chapter. Unless one has the resources to set up to test a concrete block wall and to measure the absorption coefficients of the specific blocks to be used, it is better to paint the stuff with a heavy paint and depend on known materials. If your heart is set on using the absorption of a particular local concrete block with its natural surface, another approach would be to approach the treatment of the room in increments, measuring reverberation time step by step as materials are added until the desired characteristics are achieved. Most of us do not have the means, the patience, or the love for certain concrete blocks to justify this.

CHAPTER 15

One Control Room
for Two Studios

Whether or not using one control room to serve two studios is satisfactory depends upon the intensity of the recording schedule. If recording sessions are well spaced and the schedule is flexible, having only one control room might work very well. Certainly, the advantages of requiring only one console, one set of recording equipment, and using only one operator are self-evident.

The problem arises when it is necessary to use both studios simultaneously. Even if duplication of equipment and the use of two operators were accepted as the price for a fuller recording schedule, the conflict of monitor loudspeakers becomes apparent. Which recording job gets the loudspeaker(s)? Which operator uses headphones?

For the activity that does not require such overlapped scheduling, however, the single control room for two studios can work out very well. And, as we have seen in Chapter 12, headphone monitoring is becoming an ever more viable alternative for monitoring work as headphone quality is undergoing rapid and dramatic improvement.[12]

The incentive for such improvement, it must be admitted, comes from the hi-fi market, not from recording engineers who still prefer loudspeakers. How much of this is inertia from the recent past when loudspeaker quality far outstripped headphone quality is difficult to say. The point here is that this is a new day and the quality difference is much less. Headphones might very well be used by one operator as he records speech from one studio while another operator records music from another using monitor loudspeaker(s). However, the loudspeaker level would have to be kept down.

Studio Suite Layout

The example to be considered is another of those cases in which the studio suite had to be fitted into space available in an existing building: After the usual battle for space, the prostudio faction came up with the end of the third and top floor of a concrete building overlooking a quiet patio farthest from a busy thoroughfare. The overlooking part was immediately cancelled as all windows in the studio area were bricked up and plastered on both sides.

After the usual preliminary consultation period, the floor plan of Fig. 15.1 was agreed upon. The client was insistent upon one control room for two studios and was willing to go to headphones for a second operator if the recording load increased that much in the future.

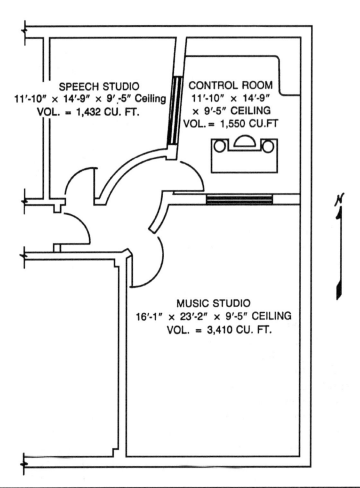

FIGURE 15.1   Floor plan of recording suite in which both a speech studio and music studio are served by a single control room. This arrangement works best if the work load is modest and the schedule flexible. A second operator can handle one studio (e.g., the speech studio) by monitoring with high quality headphones.

The sound lock is a strangely shaped space, but for a good reason. The existing hall required an offset sound lock to reach the control room without robbing the music studio of too much area. The walls of the sound lock, if straight, would cut off corners in both studios with approximately 45-degree walls. Looking into these corners from either studio emphasized that such a straight wall across the corner tends toward a concave effect when considered along with the adjoining walls.

Acousticians always get worried when confronted with concave or quasi-concave surfaces, but nothing makes them happier than convex surfaces. So, why not make convex sound diffusing surfaces out of these sound lock walls and let the concave sides be in the sound lock where they can do no harm? Walls of brick are quite amenable to shaping in this fashion. In fact, leaving the rough brick texture aids diffusion just that much more and provides an attractive visual feature in each room.

The volumes of the control room (1550 cubic feet) and the speech studio (1432 cubic feet) are very close to the 1500 cubic foot minimum, yet are adequate for their intended purposes. The music studio at 3410 cubic feet provides reasonably adequate space for the largest music group contemplated.

Acoustical Treatment

The acoustical treatment of the three rooms is strongly based on a modular plan. There are three types of modules of 24-inch × 24-inch outside dimensions. Each has a different absorption characteristic:

- A wideband (WB) module which absorbs equally well over the frequency range 125 Hz-4 kHz.
- A low peak (LP) module having a peak absorption at about 125 Hz.
- A midpeak (MP) module which peaks at about 800 Hz.

The absorption of these three modules is constrasted in Fig. 15.2. These are three basic building blocks capable of compensating for the absorption of carpets, drapes,

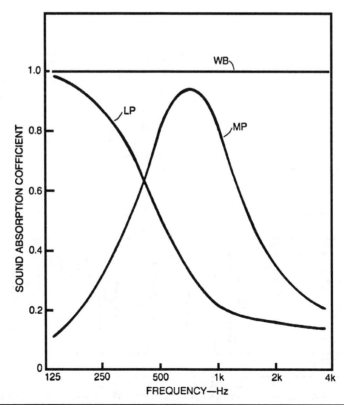

FIGURE 15.2 Comparison of absorption vs. frequency characteristics of wideband (WB), low peak (LP), and midpeak (MP) types of sound absorbers.

acoustical tile, or other materials having unbalanced absorption and performing other tasks in the acoustical treatment of a room. As there is no reason to be limited to these three, versatile as they are, other materials such as cork and convex panels will be used.

Music Studio Treatment

The wall treatment of the music studio is shown in Fig. 15.3, the floor and ceiling treatment in Fig. 15.4.

The north wall leaves little space for acoustical elements because of the observation window and the convex brick wall in the northwest corner. The east and south walls are practically covered with 24-inch × 24-inch modules of two types, low peak and wideband. The west wall is dominated by three large and three smaller convex panels.

The ceiling (Fig. 15.4) has 96 cork tiles 12 inches × 12 inches × 1 inch cemented to it in an irregular pattern built up of groups of four cork tiles. It was not possible to attain the desired reverberation time if the floor was covered with carpet wall to wall as requested, hence a compromise of an 8-foot × 10-foot rug is specified. Each absorptive element will be discussed individually in connection with the music studio, although the same elements are used in other rooms as well.

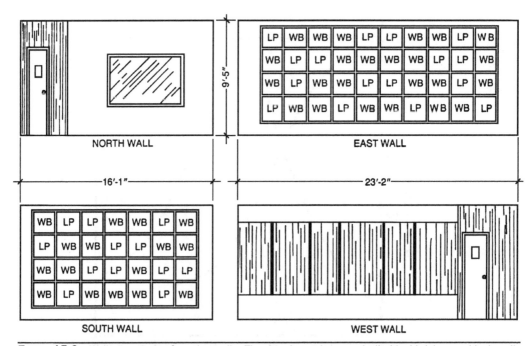

FIGURE 15.3   Wall elevations of music studio. The door is set in a semicylindrical brick sound lock wall. Plywood polycylindrical diffuser/absorbers cover the west wall and wideband (WB) and low peak (LP) modules cover the east and south walls.

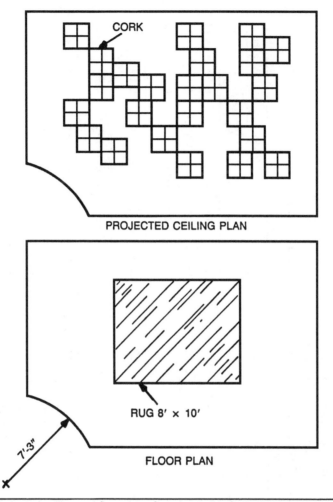

PROJECTED CEILING PLAN

FLOOR PLAN

Figure 15.4 Projected ceiling and floor plan of music studio. Cork tiles 1 inch thick are arranged in a pattern on the ceiling and an 8-foot × 10-foot rug is placed anywhere on the floor dictated by microphone placement.

Ceiling Treatment

The cork tiles of 12 inches × 12 inches × 1 inch were selected because they were readily available in this particular foreign country and are considered very attractive by some people. Absorption coefficients for cork may be difficult to locate. As it was a popular acoustical material 50 years ago before the acoustical industry sprouted, coefficients can be found in old textbooks.

Inspecting these coefficients (Table 15.1), low absorption at low frequencies and better absorption at higher frequencies are noted. The absorption varies much the same as the rug on the floor and about the same order of magnitude. If cork is to be replaced by

SIZE .... 16'1" × 23'2" × 9'5" Ceiling
FLOOR .... Wood parquet, 8' × 10' rug
CEILING .... 96 Cork tiles 12" × 12" × 1" random pattern
WALLS .... 40 Wideband, 28 low peek modules, convex panels
VOLUME .... 3.410 cu. ft.

Material	S Area Sq. Ft.	125 Hz a	125 Hz Sa	250 Hz a	250 Hz Sa	500 Hz a	500 Hz Sa	1 kHz a	1 kHz Sa	2 kHz a	2 kHz Sa	4 kHz a	4 kHz Sa
Floor: 8' × 10' Rug	80.	0.05	4.0	0.10	8.0	0.15	12.0	0.30	24.0	0.50	40.0	0.55	44.0
Floor: Wood Parquet	282.	0.04	11.3	0.04	11.3	0.07	19.7	0.06	16.9	0.06	16.9	0.07	19.7
Cork Tiles: 12" × 12" × 1"	96.	0.05	4.8	0.10	9.6	0.20	19.2	0.55	52.8	0.60	57.6	0.55	52.8
Walls: Low Peak 28 3.62 Sq. Ft.	101.	0.98	99.0	0.88	88.9	0.52	52.5	0.21	21.2	0.16	16.2	0.14	14.1
Walls: Wideband (40)	145.	0.99	143.6	0.99	143.6	0.99	143.6	0.99	143.6	0.99	143.6	0.99	143.6
Convex Panels: 3 Large, Empty	66.	0.41	27.1	0.40	26.4	0.33	218	0.25	16.5	0.20	13.2	0.22	14.5
Convex panels: 3 Small, Empty	44.	0.32	14.1	0.35	15.4	0.30	13.2	0.25	11.0	0.20	8.8	0.23	10.1
Total Sabins, Sa			303.9		303.2		282.0		286.0		296.3		298.8
Reverb. Time, sec.			0.55		0.55		0.59		0.58		0.56		0.56

TABLE 15.1 Music Studio Calculations

acoustical tile, it should be the thinnest and cheapest tile available because ¾-inch modern acoustical tile is a far more efficient absorber than 1-inch cork.

Actually, it would be better to have a compensating type of absorber on the ceiling opposite the rug, but there is the potential problem of not knowing just where the rug will be in the studio and mounting compensating (LP) modules on the wall is easier than on the ceiling.

Wall Treatment

The modules on the east and south walls are constructed according to the plan of Fig. 15.5. The wideband (WB) module is simply a frame holding 4 inches of 703 (or two layers of 2 inches) or similar glass fiberboards of about 3 pounds per cubic foot density.

This glass fiber is held in place by zigzag wires at the appropriate level creating an air space of approximately 1¼ inches. This air space helps to extend good absorption to lower frequencies, or, looking at it another way, it helps us come closer to realizing in practice the 0.99 coefficient at 125 Hz and 250 Hz supplied by the manufacturer.

The face of the 703 is held in place by a lightweight expanded metal lath appropriately spray painted before installation. It may be advisable to cover the 703 with a stretched lightweight fabric before the expanded metal is applied. The backs of both of these modules are uncritical; ¼-inch hardboard or plywood is adequate.

The LP module having an absorption peak at about 125 Hz is housed in a frame identical to that of the wideband absorber. Only 2 inches of 703 is required in this case and the facing is ³⁄₁₆-inch hardboard or plywood drilled with ³⁄₁₆-inch diameter holes

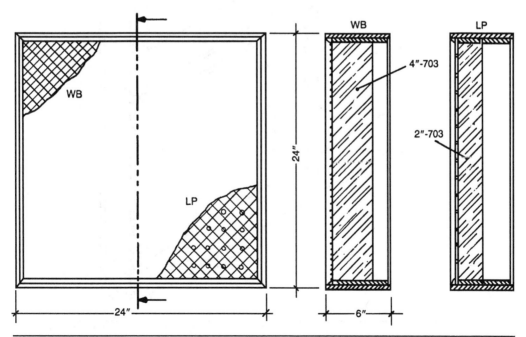

FIGURE 15.5 Constructional details of wideband (WB) and low peak (LP) absorbing modules which utilize identical boxes.

spaced 1%₁₆ inches center to center. By stacking the panels a number can be drilled simultaneously, reducing the tedium somewhat. A loosely-woven fabric cover over this perforated cover for esthetic reasons is optional.

One of the weightier problems of treating these studios is how to mount the 24-inch × 24-inch modules on walls and ceilings. One of the advantages of the modular approach is to facilitate trimming room acoustics if measurements indicate the need.

Therefore, ideally, the best type of mounting is the one which allows easy removal or interchange of modules. This is best accomplished by building a frame into which the individual modules may be inserted or removed at will. A frame of this type is described in Fig. 15.6.

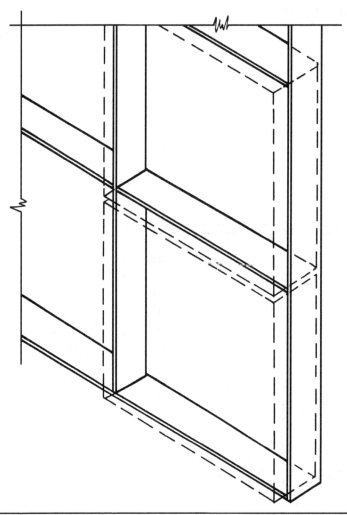

FIGURE 15.6   Wideband (WB), low peak (LP) and midpeak (MP) absorbing modules can be conveniently slid in and out of a simple wall frame of 1 × 4 lumber.

Another less desirable approach is to affix the box itself firmly to the wall and to consider the contents interchangeable, but not the external box. If the contents were removed, a fabric face would preserve visual continuity, but its acoustical effect would be essentially removed.

On the other hand, a WB absorber could be changed to an LP by changing glass fiber depth and cover. Making a midpeak (MP) absorber, to be described later, out of a WB or LP would require using the bottom of the box, leaving some empty space behind the grille cloth cover, but it would serve its acoustical function satisfactorily. It is also possible to stack the boxes like cord wood on a low support, making the stack stable with well-placed cleats or other ties.

Each WB and LP box weighs about 20 pounds. With the adhesives available today the boxes could be cemented directly to the wall. With the acoustical guidance in this and other chapters, it is not too much to expect the mounting problem to be solved, along with a host of other problems, by the ingenuity in residence.

West Wall Polys

The polycylindrical or convex acoustical elements on the west wall of Fig. 15.3 have two things going for them: bass absorption and excellent diffusing characteristics.[3] They require only modest skill in construction.

The details of Fig. 15.7 show a skin of $3/16$-inch hardboard or plywood stretched over bulkheads previously cut on the arc of a circle with a band saw. The larger unit requires a sheet about 48-inches wide and the smaller one about 32 inches. No acoustical filling is required inside.

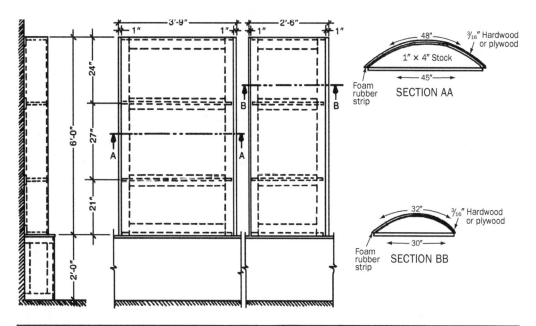

FIGURE 15.7 Constructional details of polycylindrical diffusers/absorbers on west wall of music studio. They are fastened to the wall and rest on a simple base enclosed with plywood.

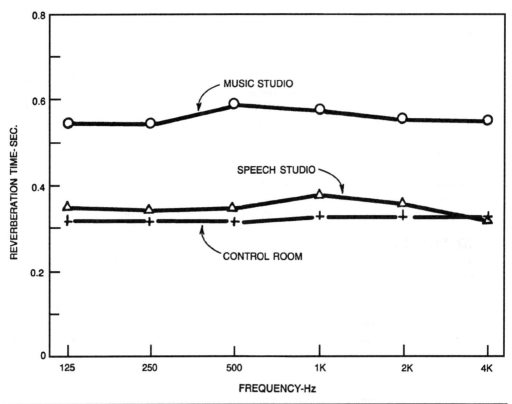

FIGURE 15.8   Calculated reverberation time vs. frequency characteristics of music and speech studios and control room.

When music fills the room the vibration of the covering skin is easily felt by placing the fingers on it. It is important that rattles be prevented. This can best be done by placing a thin foam rubber strip with a self-adhesive backing along the edge of each bulkhead before screwing down the skin. The bulkheads divide the space within each convex element into three unequal, essentially airtight, compartments. All three large and three small convex elements rest on a low, enclosed shelf. As this is covered with ½-inch plywood, it also is a fair bass absorber of the flat panel type. Because this 18-foot shelf contributes only 2.5 percent of the absorption of the room at the most, its effect is minor and is not included in Table 15.1.

With the treatment described the reverberation time of the music studio hovers around the desired 0.55 second level as shown numerically in Table 15.1 and graphically in Fig. 15.8.

Speech Studio Treatment

A reverberation time of approximately 0.3 second is the goal for the speech studio. Carpet was specified in this room. In fact, an indoor-outdoor type of carpet was specifically requested for economy. This is included wall-to-wall. Opposing the carpet, 27 low peak (LP) modules are mounted to the ceiling, arranged as shown in Fig. 15.9. The matter of attachment to the ceiling, admittedly somewhat more complicated than to the wall, is left to local ingenuity.

The treatment of the walls of the speech studio is shown in Fig. 15.10. The east and south walls are dominated by the convex masonry wall shared with the sound lock. The north and west walls each have 20 modules, most of them (25) are the wideband (WB) modules as used in the music studio. A third type of 24-inch × 24-inch module, the mid-peak (MP), is used only in this studio.

Combining the absorption of the indoor-outdoor carpet with the low peak (LP) ceiling modules, a deficiency of absorption around 800 Hz results. One method of boosting this sagging midband absorption is to introduce 13 modules having peak absorption in this frequency region as shown in Fig. 15.2. The net result is a calculated 0.35 second reverberation time for the speech studio which is a quite uniform 125 Hz to 4 kHz as shown in Fig. 15.8. The computations and other details are found in Table 15.2.

The construction of the midband (MB) modules is detailed in Fig. 15.11. It is a very shallow module with a 1-inch cavity filled with 703 or other type of glass fiber material covered with a perforated panel. Like the low peak (LP) module, it is a tuned Helmholtz type of resonator. The resonance frequency is determined by such things as the percentage perforation (the percentage of the hole area to the entire cover area), the thickness of the panel, and the depth of the enclosed cavity.

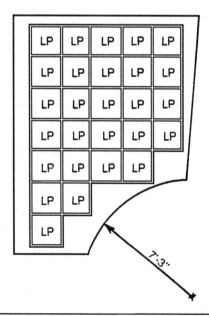

Figure 15.9   Projected ceiling treatment plan for speech studio. The low peak (LP) absorber modules are the same as those used on the wall of the music studio.

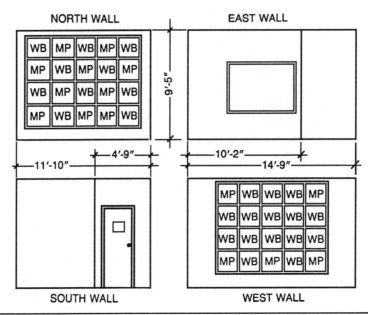

FIGURE 15.10 Wall elevations of speech studio showing placement of wideband (WB) and midpeak (MP) absorber modules.

The glass fiber broadens the absorption peak. The cover is common stock pegboard. Different types of pegboard have different hole configurations and hence different perforation percentages. The diamond hole configuration is square when the panel is rotated 45 degrees. A common diamond type of pegboard comes with 5/32-inch holes spaced 3/4 inches center to center on the square.

The perforation percentage of this type is about 3.4 percent. A perforation percentage between 3 and 6 percent is required. Figure 15.11 includes a sketch showing how the perforation percentage for the entire cover of any type of pegboard can be readily calculated on a unit basis knowing only hole diameter and spacing.

Control Room Treatment

The goal is a reverberation time in the control room somewhat shorter than either of the studios it serves. Making it shorter than the 0.35 second of the speech studio, however, runs head-on into another request. The control room acoustics has to be suitable for the occasional recording of interviews, etc., in that room. A satisfactory compromise value of reverberation time for the control room is 0.32 second as shown in Fig. 15.8.

The treatment of the control room with its hard, reflective floor is a very straightforward task as only wideband (WB) modules are required. With a volume of 1550 cubic feet and surface area of 818 square feet it is easy to plug these figures and a reverberation time of 0.28 into Sabine's equation and come up with a 0.32 second reverberation time.

SIZE .... 11'10" × 14'9", East wall splayed
FLOOR .... Indoor-outdoor carpet wall to wall
CEILING .... 27 Low peak modules
WALLS .... 27 Wideband and 13 midband modules
VOLUME .... 1,432 cu. ft.

Material	S Area Sq. Ft.	125 Hz a	125 Hz Sa	250 Hz a	250 Hz Sa	500 Hz a	500 Hz Sa	1 kHz a	1 kHz Sa	2 kHz a	2 kHz Sa	4 kHz a	4 kHz Sa
Floor: Indoor/ Outdoor Carpet	152.	0.01	1.5	0.05	7.6	0.10	15.2	0.20	30.4	0.45	68.4	0.65	98.8
Ceiling: 27 Low Peak 3.62 Sq. Ft.	96.	0.98	96.0	0.88	86.2	0.52	51.0	0.21	20.6	0.16	15.7	0.14	13.7
Walls: 27 Wideband	95.	0.99	97.0	0.99	97.0	0.99	97.0	0.99	97.0	0.99	97.0	0.99	97.0
Walls: 13 Midband	47.	0.09	4.2	0.30	14.1	0.80	37.6	0.80	37.6	0.35	16.5	0.20	9.4
Total Sabins, Sa			196.7		204.9		200.8		185.6		197.6		218.9
Reverb. Time, sec.			0.35		0.34		0.35		0.38		0.36		0.32

TABLE 15.2 Speech Studio Calculations

231

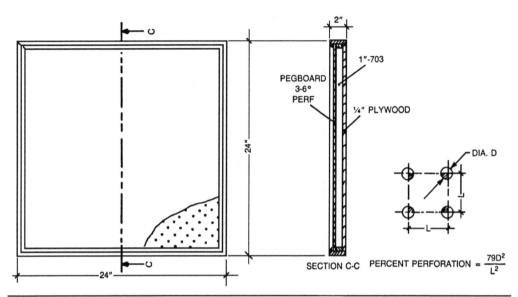

FIGURE 15.11   Constructional details of midpeak (MP) absorber module used in speech studio.

As shown in Fig. 15.12, 24 WB modules are mounted on the ceiling. Figure 15.13 shows 15 WB units on the north and 21 on the east walls of the control room. The enclosed cabinets under the work table, which are of ½-inch plywood enclosing an air space, must be considered panel absorbers. However, the 61 square feet of such a surface yields only about 17 sabins; hence, the reverberation time is pulled down only a slight amount at the low end of the spectrum.

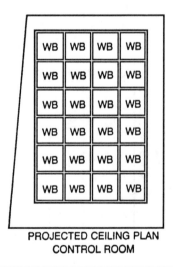

PROJECTED CEILING PLAN
CONTROL ROOM

FIGURE 15.12   Projected ceiling treatment plan of control room showing positions of 24 wideband (WB) modules.

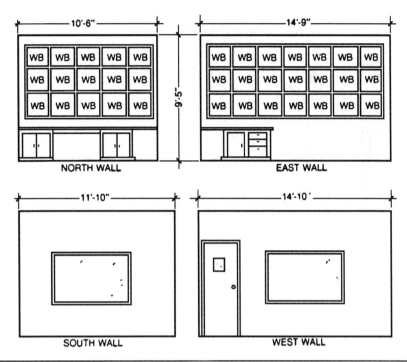

FIGURE 15.13   Wall elevations of control room showing locations of wideband (WB) modules on north and east walls.

As all walls are masonry, their absorptive effect is negligible. Squeezing out a few sabins from the wood parquet floor and glass surface the total 231 total is approached. If the walls were of drywall, the low frequency absorption could be significant. Absorption of walls and cabinets provide the only unbalanced absorptive effect in a room such as this, and it is usually of nominal magnitude.

Air Conditioning

Tearing holes in studio walls comes under the heading of bad news to acoustical consultants. The client in the present case considered a central air conditioning system too expensive and elected to use wall-mounted room units and live with the resulting inconvenience, discomfort, and increase in background noise.

The sectional view of Fig. 15.14 shows how an acoustical shield can be built over the face of the air conditioning unit to be used when quiet conditions are required. Two layers of $\frac{3}{4}$-inch particleboard are used because its density is about 30 percent greater than plywood. Such a double panel offers 35 dB attenuation at 500 Hz on a mass law basis. The trick is to seal this lid around the edges in a way which matches or exceeds this. Figure 15.14 shows an arrangement of a double 2 × 4 or 2 × 3 frame which can accomplish this if carefully constructed. The secret is well-caulked joints and use of

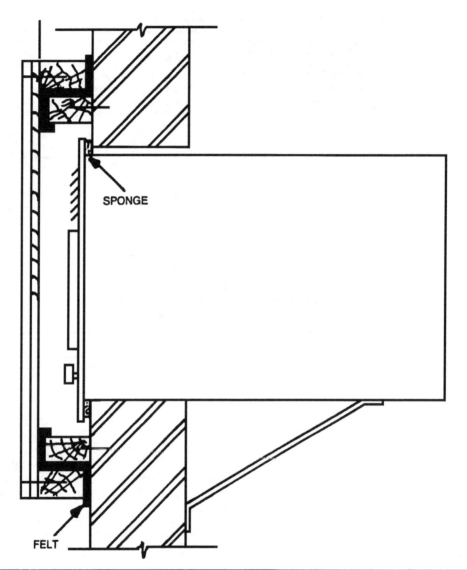

FIGURE 15.14   Room type air conditioners are not recommended for sound sensitive areas, but if they must be used, the above plan provides reasonable protection against intrusion of outside noise. The A/C unit is turned off and the door closed during actual recording.

heavy felt for a snug wiping fit. The lid can be hinged if desired and some device to clamp it in place is needed. This is not a very convenient solution to the air conditioning problem as the lid must be continually closed for a recording and opened to cool things off between takes.

CHAPTER 16

A Video Mini-Studio

Recording studios of various types have been covered in previous chapters and it is time to consider one for video and television production. The client was not interested in commercial television production but instead wanted to train television technicians to operate the camera, lights, and control console as well as writers, producers and directors. This calls for a rather generous control room to accommodate the many observers in addition to those actually engaged in the task at hand. As there are no special requirements for the acoustics of the control room, this chapter deals only with the studio. Not only was television training stipulated, the studio also had to be suitable for recording musical groups and for speech in the form of dramatics, single narrator, and interviews.

The acoustical requirements for TV, music, and speech recording in the studio of Fig. 16.1 are divergent. The optimum reverberation time for speech for a studio of this size (7,418 cubic feet) is about 0.5 second; for music about 0.7 second. The requirements for television are less well defined as both speech and music are involved.

Considerable noise is produced in a TV studio as cameras are rolled around, cables dragged, and people move behind the camera. It is customary to use a boom-mounted microphone which can be brought only so close to those talking or it will dip into the picture. Even though the microphone is highly directional, the greater average source to microphone distance means that studio noise becomes a significant problem.

All of these things taken together have led the television production people to demand very absorbent walls and ceiling, the floor surface remaining hard to make rolling camera dollies and pedestals that much easier.

The requirements are much like the motion picture soundstage—make it as dead as possible. For both motion picture and television production, however, there is a saving grace in that the local acoustics are influenced strongly by reflections of sound from the walls of the setting. For example, if in the TV picture we see people in a library setting, reflections from the *flats* making up the visual bounds of the picture make the sound more or less what one would expect in a library in spite of one or two *open walls*. Movable absorptive *flats* are used to correct acoustical flaws which occur in such a fluid situation. A law of TV and motion picture production seems to be, "The next shot is entirely different!" Our goal, then, for television training work is to provide as dead a studio as is feasible.

Louvered Absorbers

There are three acoustical conditions required: reverberation times of 0.5 and 0.7 second, and a third even more dead than the 0.5 second speech condition. To meet all of these requirements in a single room demands some method of adjusting room acoustics. There are numerous ways the acoustics of a room can be varied.[4]

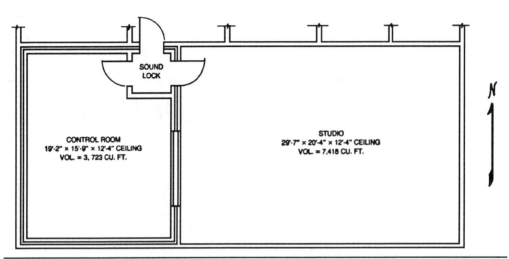

FIGURE 16.1   Floor plan of studio suitable for television instruction and recording speech, drama, and music. The divergent requirements of these varied uses are met by adjustable acoustical elements in the studio.

In Chapter 10 the possibility of flipping panels having one reflective side and one absorptive side was explored. Hinged panels can expose deep absorptive layers when open, cover them when closed and so on. The application of adjustable louvered panels as described in Fig. 16.2 seems to fit the present case best of all. The hardware of the louvered windows commonly used in homes in the more temperate climates can easily

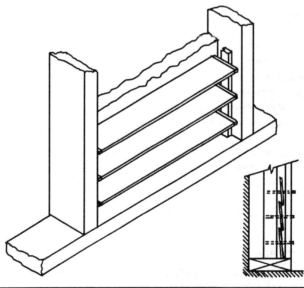

FIGURE 16.2   Louver controlled adjustable absorber. The position of the panels (open or closed) controls the effectiveness of the 703 type of glass fiber behind. Partial closing of panels creates a Helmholtz resonator type of low peak absorber.

be adapted by using ³⁄₁₆-inch tempered hardboard in place of the usual glass plates. Glass would be quite suitable acoustically, but far more expensive than hardboard. The simple movement of a lever opens and closes the panels of one segment, exposing the glass fiber behind when the panels are open (horizontal) and shielding it acoustically from the room when the panels are tightly closed. Adopting this means of adjustment of room absorption, the calculation of acoustical parameters of the room may proceed.

Cyclorama Curtain

To provide the neutral background so often demanded in television work, a cyclorama curtain arranged along the east and south walls is used as shown in Fig. 16.3. This curtain, supported from a curved track, may be moved or retracted at will. Assuming that this curtain is made of 14-ounce cotton material, it is a significant absorber, especially at the higher audio frequencies. Besides being active as a visual background for TV, its deployment as shown in Fig. 16.3 also hides the low peak absorbers which are something less than beautiful. This does not cause an acoustical problem because the low frequencies upon which the low peak absorber acts are attenuated very little by the fabric.

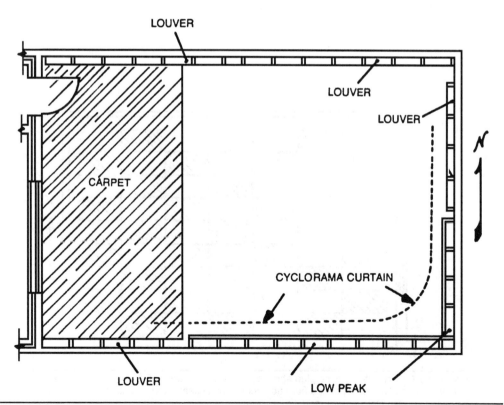

FIGURE 16.3   Floor plan of video and multipurpose studio showing positions of carpet, cyclorama curtain, low peak and adjustable louver absorbers.

Floor Covering

A carpet, 200 square feet in area, is placed at the west end of the studio next to the observation window. Speech and drama activities can take place at this end of the studio. If the adjacent louvers on the north and south walls are open, something approaching the dead end of a live-end-dead-end studio could be arranged.

The floor not covered with carpet is covered with vinyl tile which is excellent for rolling camera dollies. In fact, the entire floor should be covered with vinyl tile, even under the carpet. For unusual situations the carpet can be rolled up or its location shifted at will.

Ceiling Treatment

To keep the cost of acoustical treatment as low as possible a rather inelegant ceiling treatment is specified. A pattern of 31 panels of Owens-Corning Type 703 Fiberglas, 2 feet × 4 feet × 2 inches is cemented to the ceiling.

A suggested arrangement is given in the projected ceiling plan of Fig. 16.4. These semi-rigid boards (3 pounds per cubic foot density) should be relatively free from the sloughing off of troublesome glass fibers. If this is a problem, each panel could be wrapped in lightweight cloth before cementing in place.

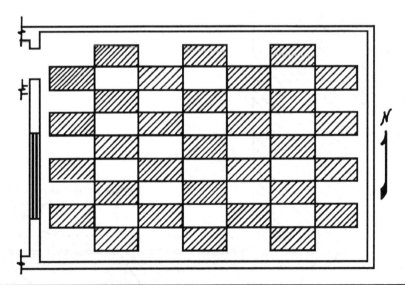

FIGURE 16.4 Projected ceiling plan showing suggested pattern of 24-inch × 48-inch × 2-inch panels of glass fiber of 3 pounds per cubic foot density cemented to ceiling. These may be covered with thin cloth before mounting if desired.

Louver Absorbers

The basic section of louver absorber is shown in Fig. 16.5. The frame is of 2 × 6 lumber. The width of each section is 24 inches inside. The height of the two lower segments of each section is 48 inches inside. This allows mounting the 2 inch 703 glass fiber with no cutting in the lower two segments and only a single cut for the top segment. The louvers

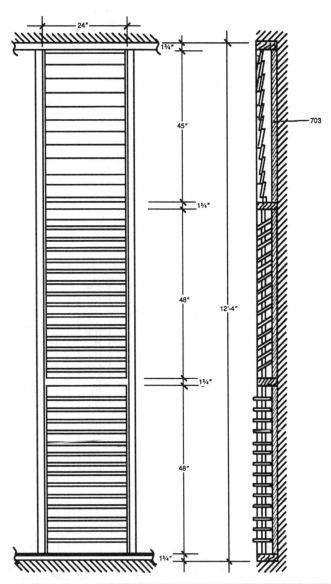

FIGURE 16.5 Typical floor-to-ceiling louvered section. Construction of frame of 2 × 6s allows depth for only 2 inches of glass fiber. Using 2 × 8s would provide space for 4 inches of glass fiber.

can be adapted to the width of section selected (24 inches) by cutting the louver panels of $\frac{3}{16}$-inch tempered hardboard the right length.

The louver sections are located in the studio as shown in Fig. 16.3. Thirteen louver sections are fixed to the north wall. By splitting the 13 into six and seven section groups a space is left near the center of the north wall for power and microphone outlets. Another group of four sections is similarly mounted on the east wall so that some wall surface is left exposed for electrical services in the NE corner of the studio. A fourth group of four louver sections is affixed to the west end of the south wall. A total of 21 floor-to-ceiling louver sections are thus distributed across three of the walls of the studio for a total of about 489 square feet.

Low Peak Absorbers

In certain foreign countries the specification of slat absorbers causes no problem because beautiful hardwoods are cheap and plentiful. Lumber costs in the United States have gone out of sight and buying straight-grained, good quality lumber is often ruled out by its cost.

To avoid such expense a low peak absorber using sheets of $\frac{3}{4}$-inch plywood with saw slots has been designed. This form of the familar Helmholtz low peak absorber is shown in Fig. 16.6. The framework is made of 2×10 lumber spaced 24 inches center to center to accommodate the 4-foot $\times$ 8-foot sheet of plywood without waste. The cross-bracing serves the double purpose of strengthening the structure and breaking the air space

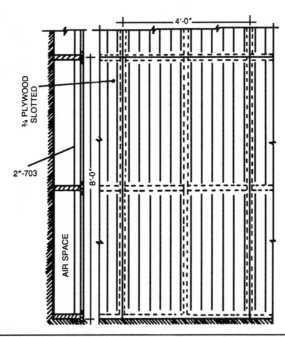

FIGURE 16.6 Inexpensive low peak Helmholtz resonators on south and east walls made of $\frac{3}{4}$-inch plywood panels with saw slots cut in them.

behind a single sheet of plywood into four cavities somewhat less than 2 feet × 4 feet each. This is important to discourage resonances in the cavity parallel to the plywood face. The slots are 4 inches center to center, arranged with respect to the 2 × 10 frame as shown in Fig. 16.6.

An uncertain factor in the design of this low peak absorber is the width of the saw slot, which affects tuning. A width of $\frac{1}{8}$ inch is assumed in the present case, although increasing this to $\frac{3}{16}$ inch increases the frequency of resonance about 20 percent (from 144 Hz to 176 Hz for 4-inch spacing of slots). This is not too serious as the top of the resonance peak (the absorbence peak) is fairly broad if the 2 inches of 703 glass fiber is placed against the back of the slotted panel where the air particle velocity is great.

Keeping the saw slot width between $\frac{1}{8}$ inch and $\frac{3}{16}$ inch should keep this shift of resonance peak within usable limits. The saw slots stop at the cross-bracing to keep the 4-foot × 8-foot panel strong.

The glass fiber may be cemented to the back of the panel, spots of cement being placed between, but not close to, the saw slots. The saw slots should be cleaned out as much as possible, removing slivers and rough protrusions in the slots. A stain and varnish finish for the slotted panels is recommended, taking care not to clog the saw slots in the process.

Figure 16.7 summarized the placement of acoustical materials and devices on the four studio walls. Because the cyclorama curtain hides most of the slotted panel absorbers, the visual impression on entering the room is that louvers cover most of the walls. This will, at least, give an impression of functional novelty, if not a thing of beauty and a joy forever.

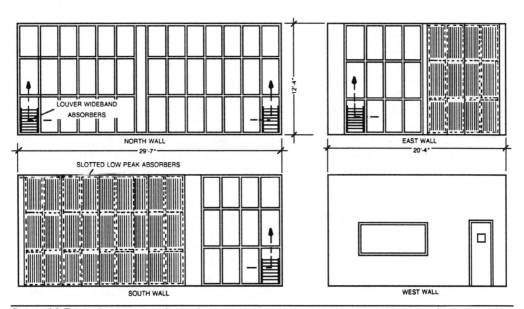

FIGURE 16.7 Wall elevations of video and general-purpose studio showing placement of adjustable louvered absorbers and slotted low peak absorbers. The slotted absorbers are largely hidden by the cyclorama curtain.

Louver Complications

The practical aspects of louver-controlled wideband absorbers have been discussed, but there is more to this type of structure than meets the casual eye. The simplistic view is that the louver is only a cover for the glass fiberboards which may be closed or opened. But what happens when the louver slats are not tightly closed, when a narrow slit remains? The slit makes that segment into a low peak absorber of the Helmholtz type. This is a very indeterminate condition. How wide is the slit?

A slit formed by two louver slats approaching each other is a far more complex problem, mathematically, than having a slit of fixed width in a cover of fixed thickness. Therefore, the frequency at which the absorption peak occurs is uncertain in using the louver elements as low peak absorbers, but the possibility is intriguing.

Computations

Table 16.1 gives the step-by-step calculations to the following three conditions:

1. TV condition (all louvers open)
2. Voice condition ($\frac{1}{2}$ of louvers open)
3. Music condition (all louvers closed)

The resulting reverberation times for the three conditions of Table 16.1 are shown graphically in Fig. 16.8. The unbalanced absorption provided by the carpet, cyclorama curtain, and the low frequency deficiency of the 2-inch glass fiber ceiling boards is compensated by the saw slot low peak absorbers in varying degrees for the above three conditions. Consequently, the bass rise of the reverberation time of Fig. 16.8 is least for the TV condition and greatest when all louvers are closed for the music condition.

Now a confession: the construction of the louver sections detailed in Figs. 16.2 and 16.5 for 2 × 6 lumber allows only room enough for 2 inches of 703. The computations of Table 16.1 and the heavy line graphs of Fig. 16.8 are for 4 inches of 703 in all louver segments and sections. If calculations are carried through for 2 inches of 703 in all louver sections, the bass rise of reverberation time is much greater, as shown by the broken lines of Fig. 16.8. Of course, changing the thickness of 703 has no effect on the music condition because all louvers are closed.

What is needed at this point is a series of detailed reverberation measurements for the specific studio in question. These should include measurements of reverberation time with different percentages of louvers in a narrow slit (low peak) and open condition.

Why go to the expense of 4 inches of 703 if 2 inches of 703 will work just as well using normally closed louvers as low peak slit resonators to compensate for the bass rise in reverberation time? This approach, obviously, requires some careful measuring and planning. It also requires some method, such as the use of a shim to set the slit width accurately in each segment. And it may be more straightforward to pay for 4 inches of 703 for the louver sections and live with the heavy graphs of Fig. 16.8. Of course, this would require a 2 × 8 framework instead of the 2 × 6 frame of Fig. 16.5.

However, the whole approach of using the louver sections as slit resonators could be reduced to simple operational procedures. If not followed, or carelessly followed, such operational procedures could result in some weird unbalanced acoustical conditions. To calibrate judgment, a rough calculation has been carried through to help evaluate the use of louvers as slit resonators for supplying low frequency absorption.

SIZE .... 29'7" × 20'4" × 12'4" Ceiling
FLOOR .... Vinyl tile, 200 sq. ft. carpet, 3/16" pile, foam underlay
CEILING .... 31 pieces 703 Fiberglas 2' × 4' × 2" cemented to ceiling
WALLS .... Low peak absorbers, 278 sq. ft.
.... Louver adjustable wideband, 489 sq. ft. 4" 703
.... Cyclorama curtain 35 lin. ft. 10' high, 14 oz mil.
VOLUME .... 7,418 cu. ft.

Material	S Area Sq. Ft.	125 Hz a	125 Hz Sa	250 Hz a	250 Hz Sa	500 Hz a	500 Hz Sa	1 kHz a	1 kHz Sa	2 kHz a	2 kHz Sa	4 kHz a	4 kHz Sa
Carpet	200	0.05	10.0	0.10	20.0	0.10	20.0	0.30	60.0	0.40	80.0	0.50	100.0
Floor: Vinyl Tile	401	0.02	8.0	0.03	12.0	0.03	12.0	0.03	12.0	0.03	12.0	0.02	8.0
Ceiling: 31/703	248	0.18	44.6	0.76	188.5	0.99	245.5	0.99	245.5	0.99	245.5	0.99	245.5
Cyclorama	350	0.03	10.5	0.12	42.0	0.15	52.5	0.27	94.5	0.37	129.5	0.42	147.0
Low Peak Absorbers	278	0.90	250.2	0.84	233.5	0.64	177.9	0.36	100.1	0.17	47.3	0.06	16.7
Sa Subtotal			323.3		496.0		507.9		512.1		514.3		517.2
TV Condition All Louvers Open 4" 703	489	0.99	484.1	0.99	484.1	0.99	484.1	0.99	484.1	0.99	484.1	0.99	484.1
Total Sabins, Sa			807.4		980.1		992.0		996.2		996.4		1001.3
Reverb. Time, sec.			0.45		0.37		0.37		0.36		0.36		0.36
Voice Condition 33.6" Louvers Open	164	0.99	162.0	0.99	162.0	0.99	162.0	0.99	162.0	0.99	162.0	0.99	162.0
Total Sabins, Sa			485.3		658.0		669.9		674.1		676.3		679.2
Reverb. Time, sec.			0.75		0.55		0.54		0.54		0.54		0.54
Music Condition All Louvers Closed													
Total Sabins, Sa			323.3		496.0		507.9		512.1		514.3		517.2
Reverb. Time, sec.			1.1		0.73		0.72		0.71		0.71		0.70

TABLE 16.1 Television Studio Calculations

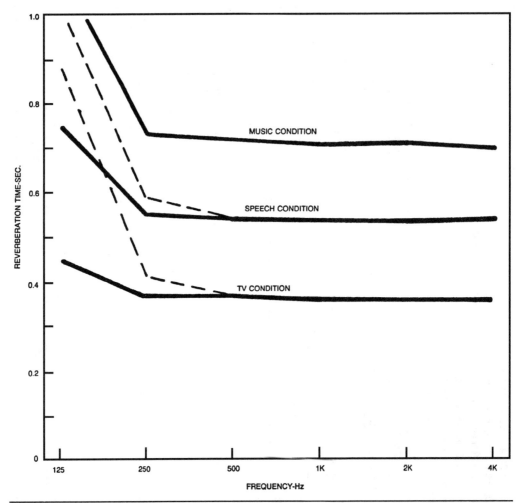

FIGURE 16.8 Calculated reverberation time vs. frequency characteristic of studio for three conditions of adjustable louvers: video condition (all louvers open), voice condition (¹/₃ of louvers open) and music condition (all louvers closed). The heavy lines apply to louvers with 4 inches of glass fiber, the broken lines apply to louvers with 2 inches of glass fiber.

In Fig. 16.9 the heavy graph A represents the reverberation time of the studio for voice condition with 2 inches of 703 in all louver sections. In this voice condition ¹/₃ of the louvers are open and ²/₃ closed. Assuming that the slit width and depth (the overlap of the louvers), the effective cavity depth and all other parameters result in a low peak resonator comparable to the saw slot low peak absorbers, the absorption coefficients of the saw slot units can be used in this rough calculation (see Table 16.1). This is not a very secure assumption; hence the results may be off a considerable amount, but still be of help in a qualitative way.

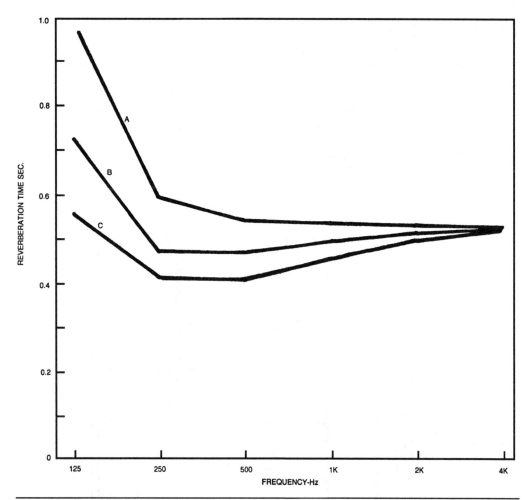

Figure 16.9 Graph A is the reverberation time for the voice condition of Fig. 16.8 ($^1/_3$ louvers open with 2 inches of glass fiber) repeated for reference. Opening some of the closed louvers slightly converts them to low frequency slit absorbers. Graph B is for $^1/_3$ of the louvers open, $^1/_3$ closed, and $^1/_3$ in low peak slit condition. Graph C is for $^1/_3$ of the louvers open and $^2/_3$ in the slit condition. This illustrates the possibility of reducing the bass rise by using louvers in the slit condition.

With this assumption graphs B and C have been computed. Graph B is for $^1/_3$ of the louvers open, $^1/_3$ closed, and $^1/_3$ in low peak slit condition. Graph C is for $^1/_3$ of the louvers open and $^2/_3$ in slit condition. Graph C flattens out fairly well, but falls to far below the 0.46 second goal. Having $^1/_3$ of the louvers open is common to all three graphs. Other combinations of open and slit louvers would undoubtedly lift and straighten graph C but, it is emphasized, this should be done on the basis of measurements, not calculations with uncertain coefficients.

Television Facilities

The cyclorama certain has been discussed, perhaps a bit prematurely, in connection with its acoustical absorption effect. The relationship of this curtain to an overhead pipe grid for supporting lamps is shown in Figs. 16.10 and 16.11A.

The advantage of supporting lamps from above is that the floor is kept free of most lamp stands and cables. For full lighting it is sometimes necessary to provide a suitable *scoop* or *broad* on a stand, but a grid can care for most of the lighting equipment required for a small stage such as this.

The pipes, often 1½ inches inside diameter, are secured at intersections with Ubolts as shown in Fig. 16.11B. The individual lamp units are suspended below the grid at an

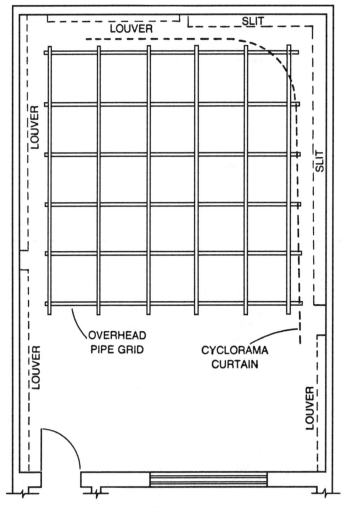

FIGURE 16.10   Projected ceiling plan of video general-purpose studio showing relationship of overhead pipe grid for supporting lamps and the cyclorama curtain.

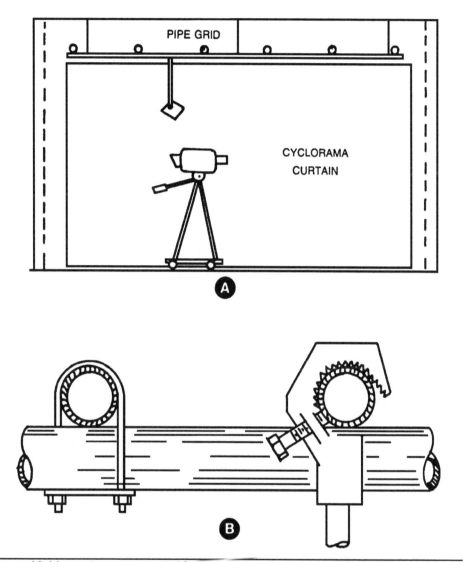

FIGURE 16.11   (A) Overhead pipe grid for supporting lamps and its relationship to cyclorama curtain. (B) U-bolt clamp method of securing pipe grid and conventional hook for lamp hangers.

adjustable height on a vertical pipe secured to the grid by the toothed clamp also depicted in Fig. 16.11B. A safety chain from the pipe grid to the lamp unit is necessary as the adjustable devices sometimes slip.

Lamps are mounted, positioned, adjusted, and pointed from a ladder. In larger studios this is done from catwalks above. With one or more cameras, conventional acoustical treatment in the control room, videotape recorder, monitor bank, video and sound consoles, and the required supporting gear, the TV mini-studio described should be quite adequate for a meaningful training program. An illustrated trade reference catalog is helpful in setting up such a television training facility.[35]

CHAPTER 17
A Video and Multitrack Studio

W hat do video studios and multitrack recording studios have in common? Off-hand, TV and multitrack seem like quite divergent activities. They both need ample space. In multitrack recording the acoustical separation between sources obtainable by screens and other devices is limited and it is soon discovered that a certain minimum of physical separation is necessary, which adds up to requiring a certain amount of floor space for each performer. Television's need for elbow room is more obvious, what with several cameras rolling around, overhead space for lights and multiple sets.

The studio selected for this chapter is, according to the general plan of this book, a budget project. This means compromises in many areas; the challenge is in keeping the magnitude of these compromises small. For example, the walls of this studio are 8-inch concrete block rather than double concrete with an air space. By plastering both faces and filling the block voids with concrete, a respectable STC 56 can be obtained which is within range of, say, STC 62. This illustrates a physical principle not too generally appreciated—that it is the last ounce that boosts the cost to excessive levels.

Put another way, the curve flattens out so that much more effort and money must be put in to get that last ounce than the earlier ounces. The message is simply that work of quality can be done with facilities somewhat less than the world's best. The best studio in the world comes far from guaranteeing the best product; skill and resourceful-ness are still the indispensable ingredients.

Studio Plans

The layout of the studio to be studied is shown in Fig. 17.1. The plan includes two studio-control room suites. The large Studio A and its control room are for video and multitrack recording. The smaller Studio B and its control room are for general speech work.

Studio B serves also as an isolation booth for vocals or drums when complete separa-tion from what is happening in Studio A is required. A window in Control Room A lines up with a window in Studio B for coordination of activities via eye contact. Another window in Studio B looking into Studio A does the same between those two rooms. These incidental windows for eye contact are small but are built to the same double-glass standards as the larger observation windows.

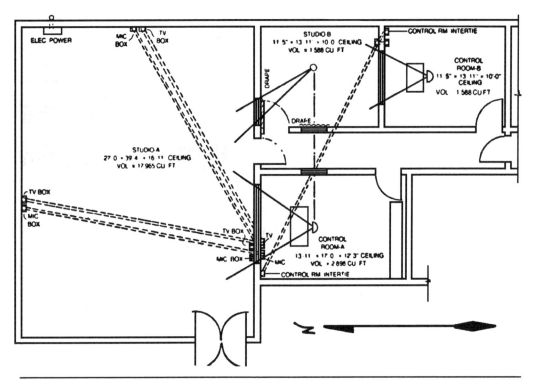

FIGURE 17.1   Floor plan of larger studio complex adapted for both television and multitrack work. The use of Studio B for drum or vocal isolation booth is aided by small windows for visual contact and intertie lines between Control Rooms A and B.

Conduits

A suggested plan for conduit runs is indicated in Fig. 17.1. These conduits are laid before the floor slab is poured as shown in Fig. 17.2A. This gives the most direct path which pays off later in ease of snaking in the cables. Conduit terminal boxes are on the north and east walls. The boxes on the south wall under the control room window are fed directly through the wall from the control room as indicated in Fig. 17.2A. To avoid acoustical leaks these short conduits through the observation window wall must be very carefully caulked and after the cables are pulled through, glass fiber should be packed in the conduit from both ends.

One conduit serves video equipment and another holds the audio lines for each of the three terminal positions in Studio A. These conduits are fitted with boxes of the type shown in Fig. 17.3. Each audio pair terminates on a barrier strip from which flexible leads run to the professional type microphone connectors mounted on the lid. The video box must be adapted to the type of camera equipment to be used, but this is straightforward if the conduit and boxes are provided.

In the control room boxes larger than those in Studio A are required because they must terminate three conduits and the one going through the wall coming into the back of the box, if desired. Figure 17.2A and Fig. 17.4 illustrate one satisfactory method of arranging both video and microphone boxes and conduits in Control Room A.

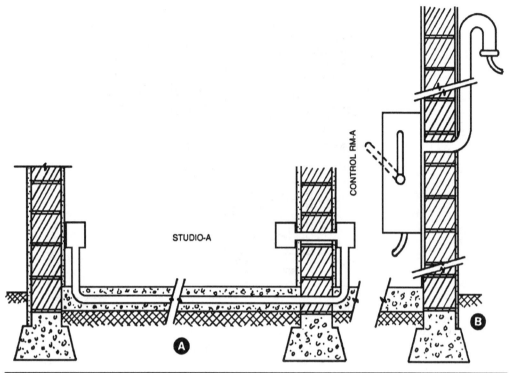

FIGURE 17.2 (A) Duct runs laid before the concrete floor is poured provide separate ducts for television and audio use. (B) Electrical switch box arrangement for energizing portable cable runs for all set lighting. House lights are handled in the conventional way.

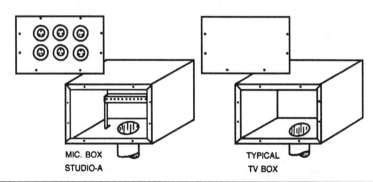

FIGURE 17.3 Conduits are terminated in boxes adapted to audio or television use.

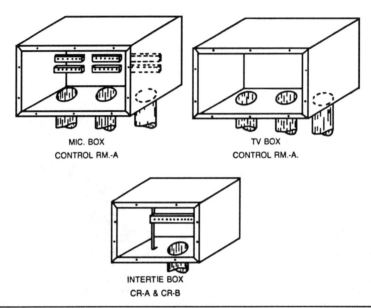

MIC. BOX
CONTROL RM.-A

TV BOX
CONTROL RM.-A.

INTERTIE BOX
CR-A & CR-B

FIGURE 17.4 In Control Room A the microphone box must terminate three conduits; the television box is the same. An intertie box terminates the conduit connecting Control Room A and Control Room B.

A conduit should always tie control rooms together in a studio complex. By running six audio pairs in this conduit and terminating at both ends on a jackstrip, great flexibility results. Equipment in Control Room B can be used for a big job in Control Room A, or vice versa, without moving the equipment physically. Using the lines for microphone, cue foldback or talkback is also made easy with these intertie lines. The intertie box is shown in Fig. 17.4.

Power Facilities

Figure 17.2B gives details of the heavy duty electrical power switch and breaker box on the east wall of Studio A. This is to provide power only for set illumination. The general house lights are handled independently in the conventional manner. From this box portable cables run to spider distribution boxes and then to the individual lamp circuits on the pipe grid and on the floor. When this switch is open, all portable circuits are dead. For safety, great care must be exercised around these heavy duty lamp circuits and only experienced personnel should be allowed to work with them.

Studio Treatment

It has been pointed out that both television and multitrack recording require relatively *dead acoustics*. The client also expressed a desire to have Studio A acoustics suitable for recording musical groups in the more conventional manner.

In fact, this use was in immediate demand while both video and multitrack were activities they hoped to enter soon. To minimize the degree of compromise between the three types of work a certain amount of adjustability has been built into Studio A acoustical treatment. This is accomplished by seven swinging panels on the north wall and five (rather, four full panels plus a half panel) on the east wall as shown in Fig. 17.5. By closing all panels approximately 334 square feet of 2-inch 703 are, in effect, removed from the room and replaced by about half that area of plywood. The closed panels also are physical protrusions which assist in diffusing sound.

Swinging Panels

The swinging panel construction is detailed in Fig. 17.6A and 17.6B. It is a simple frame of 1 × 4 lumber, with a plywood back (½-inch or ⅝-inch) for stiffening, which holds and protects 2-inch Type 703 Fiberglas absorbent. For efficient plywood cutting the frame is made of a size to accept 2-foot × 8-foot pieces of plywood. A cross-member at the mid-point adds strength, breaks the 8-foot length visually, and supplies a logical position for a third hinge.

Some acoustically transparent protective cover is desirable for the 703. This could be perforated metal sheets having at least 25 percent of the area in holes. It could be expanded metal, such as metal lath, or wire screen. Perhaps the simplest and most attractive, if not the most resistant to mechanical damage, is colorful cloth such as burlap or other loosely-woven fabric.

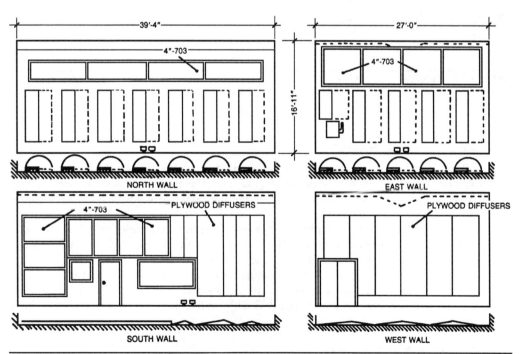

FIGURE 17.5 Wall elevations showing treatment plan in large Studio A. Swinging panels on the north and east walls allow considerable range in adjustment of acoustics, supplementing plywood diffusers/absorbers and wideband elements.

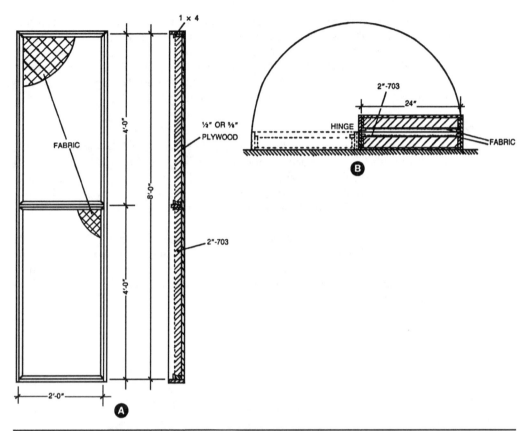

FIGURE 17.6 Constructional details of 2-foot × 8-foot hinged wall panels. When open, double width of 2-inch 703 is exposed. When closed, plywood protuberances contribute to diffusion and low frequency absorption.

Plywood Wall Diffusers

Diffusers of ¾-inch plywood of triangular construction are used on the west wall and a portion of the south wall as shown in Fig. 17.5. On this figure the cross-sectional shapes are associated with each elevation. To utilize the plywood without waste each face is either 2 feet or 4 feet in width.

The 12-foot length uses plywood of that length, if available, or 8-foot plywood sheets may be extended because of the placement of the section dividers. A 2 × 4 ridge and frame provide the basic stiffening and nailing facilities required.

Figure 17.7 gives basic details of construction which apply to diffusers having either the 2-foot or 4-foot face width. These plywood diffusers, which are also fairly low frequency absorbers, can be painted without affecting their acoustical properties significantly.

Wideband Wall Absorbers

The east end of the south wall (Fig. 17.5) is largely covered with wideband modules containing 4-inch thicknesses of Type 703 Fiberglas. Modules of similar construction

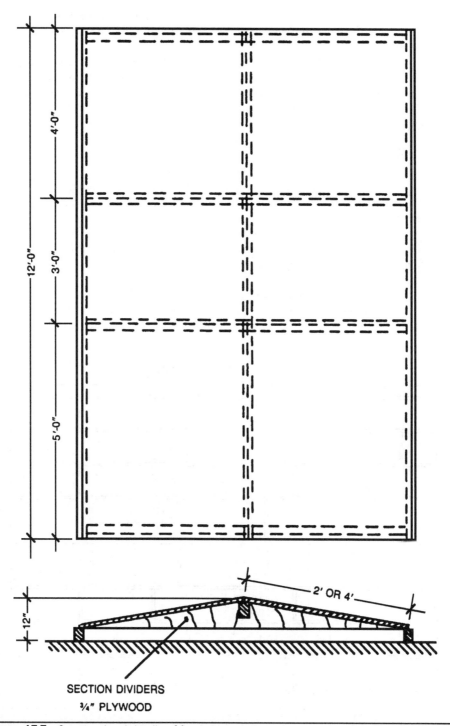

FIGURE 17.7   Constructional details of fixed plywood elements on west and south walls. Panels of 2-foot and 4-foot width are utilized.

but of different shape are also mounted above the plywood diffusers on the north and east walls. The exact dimensions of the individual modules are not significant. The total effective area is significant.

As similar units have been described in earlier chapters, no space is given to their construction other than to mention the need for a frame of 1-inch lumber (plywood backs are optional) and a lightweight, loosely woven fabric for appearance and control of glass fiber particles. A perforated metal sheet (25 percent perforation minimum), screen, or expanded metal cover may be added for protection against physical abuse if desired.

Ceiling Treatment

The Studio A projected ceiling plan of Fig. 17.8 reveals a triangular plywood diffuser approximately 8-feet wide running the length of the room down the center. The remainder of the ceiling surface is covered with a wideband absorber composed of the usual 4 inches of 703 Fiberglas.

A 2 × 4 gridwork holds the glass fiber semirigid boards which may be covered as discussed previously. In this case, expanded metal lath without the usual fabric is sufficient because of the height of the ceiling and the resulting distance from critical eyes and probing fingers.

The triangular plywood ridge on the ceiling is constructed as shown in Fig. 17.9. The entire structure is built down from the 2 × 4 gridwork fastened securely to the roof slab. There is a significant amount of open space between the plywood skin and the

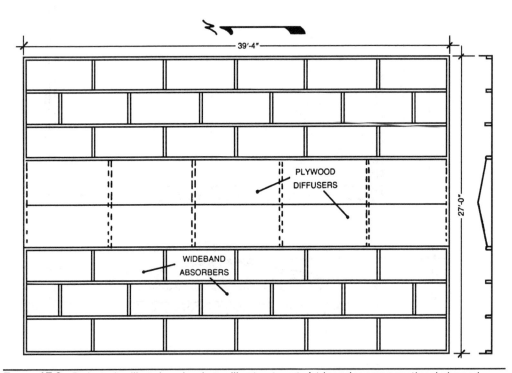

FIGURE 17.8 Projected ceiling plan showing ceiling treatment. A triangular cross-sectional plywood element about 8-feet wide runs the full length of the room. The remainder of the ceiling area is covered with wideband sections of 4-inch 703 glass fiber.

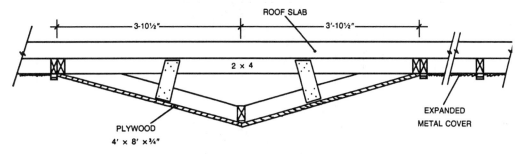

FIGURE 17.9   Details of construction of ceiling plywood diffuser/absorber.

concrete roof slab. This space can be put to good use for conduit runs or, possibly even as a duct for air circulation.

Reverberation Time

Table 17.1 provides a summary of pertinent acoustical data and computations of reverberation time. This table has been simplified and abbreviated by neglecting the sound absorbed by the vinyl tile-covered concrete floor, the plastered concrete block wall areas not covered by plywood or wideband units, windows, doors, and air contained in the room. Admittedly, the absorbence of each of these is minor, yet overall accuracy would be improved by their inclusion. But the price paid in confusion and complication is too high. Measurements should be relied upon for final evaluation in any event.

Table 17.1 includes reverberation time calculations for two conditions:

- All swinging panels open exposing the 2-inch 703.
- All swinging panels closed, covering the 703 and exposing the plywood box of half width and double depth.

The values of reverberation time from the table are plotted in Fig. 17.10 to give a graphic picture. It is interesting that at 125 Hz the reverberation time is essentially the same whether the panels are open or closed.

The 2 inches of 703 has an absorption coefficient of only 0.18 at 125 Hz, effective when the panels are open. When closed, the plywood shows a coefficient of 0.38, but this applies only to about half the area of the 703.

For frequencies of 250 Hz and above at which the 703 becomes essentially a perfect absorber there is a significant spread between the *all panels closed* and the *all panels open* graphs. When the panels are open a reverberation time of 0.58 second is obtained over most of the band. When closed, this increases to about 0.74 second.

Traditional music recording techniques normally require a reverberation time of about 0.9 second for a studio of this size (almost 18,000 cubic feet). However, 0.74 second constitutes a reasonable and usable compromise. For television and multitrack work the 0.58 second should be very favorable. Placing some of the musicians near the highly absorbent opened swinging panels, with adequate screens between, should result in excellent separation for multitrack work. Further, the thousand square feet of floor space should accommodate something like 20 musicians of either the traditional or multitrack types.

SIZE .... 27'0" × 39'4" × 16'11" Ceiling
FLOOR .... Vinyl tile on concrete
CEILING .... 3/4" Plywood diffuser 304 sq. ft.
WALLS .... Wideband 4' 703-722 sq. ft.
.... Swinging panels 2" 703-334 sq. ft.
.... Wideband 4' 703-449 sq. ft.
.... 3/4" plywood diffuser 297 sq. ft.
VOLUME .... 17,965 cu. ft.

Material	S Area Sq. Ft.	125 Hz		250 Hz		500 Hz		1 kHz		2 kHz		4 kHz	
		a	Sa	a	Sa	a	Sa	a	Sa	a	Sa	a	Sa
Plywood: Wall-Ceiling	601	0.38	228.4	0.19	114.2	0.06	36.1	0.05	30.1	0.04	24.0	0.04	24.0
Wideband 4 703 Walls–Ceiling	1171	0.99	1159.3	0.99	1159.3	0.99	1159.3	0.99	1159.3	0.99	1159.3	0.99	1159.3
Swing Panels OPEN 2 703	334	0.18	60.1	0.76	253.8	0.99	330.7	0.99	330.7	0.99	330.7	0.99	330.7
Total Sabins, Sa			1447.8		1527.3		1526.1		1520.1		1514.0		1514.0
Reverb. Time, sec.			0.61		0.58		0.58		0.58		0.58		0.58
Swinging Panels CLOSED Wideband, 4 703			1159.3		1159.3		1159.3		1159.3		1159.3		1159.3
Plywood: Walls–Ceiling + Panels	768	0.38	291.8	0.19	145.9	0.06	46.1	0.05	38.4	0.04	30.7	0.04	30.7
Total Sabins, Sa			1451.1		1305.2		1205.4		1197.7		1190.0		1190.0
Reverb. Time, sec.			0.60		0.67		0.73		0.73		0.74		0.74

TABLE 17.1 Studio A Acoustical Data

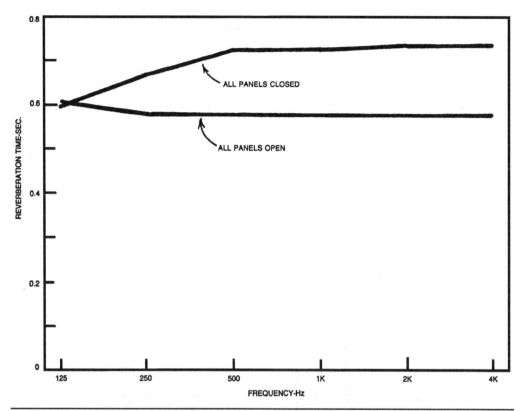

FIGURE 17.10 Studio A calculated reverberation time characteristics for all panels closed (conventional music condition) and all panels open (multitrack and television condition). Swinging the panels has practically no effect at 125 Hz because the lower absorption of the 2-inch thickness of 703 is offset by the plywood absorption of the closed boxes.

    For practically infinite separation, sometimes required for soloist or drums, the use of Studio B as an isolation booth is possible. The arrangement of Fig. 17.1 makes this both possible and convenient, sending the microphone outputs from Studio B to Control Room A by way of the intertie lines between Control Rooms A and B. Studio B, in the present case, has a heavy work load in conventional recording of radio programs with a single narrator.

Control Room Treatment

The wall treatment of Control Room A to achieve a reverberation time of about 0.3 second is described in Fig. 17.11. By plugging 0.3 second reverberation time into Sabine's equation we find that about 473 sabins (absorption units) are required. How shall this be distributed between the three pairs of room surfaces? This is an excellent opportunity for illustrating the good acoustical practice of distributing each type of absorber between the N-S, E-W, and vertical pairs of surfaces in proportion to the areas of these

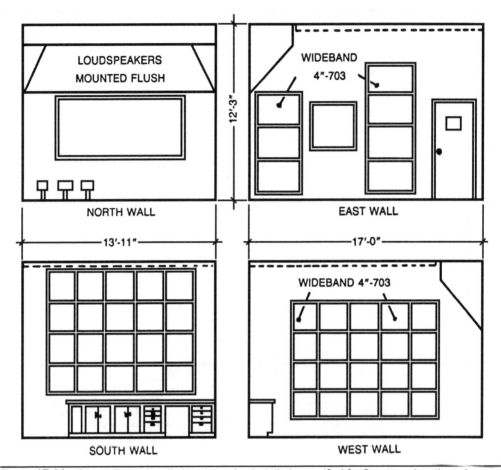

FIGURE 17.11 Control Room A wall elevations. As vinyl tile is specified for floor covering, the only material required for acoustical treatment is 4-inch 703 glass fiber.

pairs. In this control room with no carpet, we have only one kind of absorber-areas of 4-inch 703. In the case of Control Room A this distribution is shown in Table 17.2.

For the frequency range 125 Hz-4 kHz we can take the absorption coefficient of 4 inches of 703 as essentially unity. This means that one sabin of absorption is given by each square foot of 703. For Control Room A we then need 473 square feet of 703. This should be distributed approximately as shown in column (a) of Table 17.2. The distribution of the 703 shown in Fig. 17.11 accommodates only 390 square feet which yields a reverberation time of 0.36 second, which is acceptably close to 0.3 second. About 160 square feet of 703 should then be applied to north and/or south walls, but the north wall is almost filled with the observation window. The south wall has the work table and built-in drawers and cabinets. So, we do what we can and mount a frame of 20 sections, each 24 inches × 24 inches inside, on the south wall, totaling only 80 square feet.

Axial Modes	Area Sq. Ft.	% Total Area	(a) Exact Distribution of 473 Sabins	(b) Practical Distribution of 390 Sabins
N-S walls	416.5	33.8	160	80
E-W walls	341.0	27.7	131	122
Floor-ceiling	473.2	38.5	182	188
Totals	1230.7	100.0%	473	390

TABLE 17.2   Control Room A Distribution of Absorption

The other 80 square feet must be placed elsewhere. This is about all that can be done for the N-S mode.

How about the east and west pair of walls? On the west wall (Fig. 17.11) there are 20 sections, each 24 inches × 24 inches inside, yielding 80 square feet, and seven sections 24 inches × 36 inches inside on the east wall. This gives an effective area of 42 square feet for a total of 122 square feet. The E-W walls plus the N-S walls then offer a total or 80 plus 122 or 202 square feet of 4-inch 703.

Ceiling Treatment

A total of 473 square feet of 703 is required and 202 square feet are applied to the walls; therefore, 271 square feet should be applied to the floor-ceiling pair of surfaces. Because 703 is not a very satisfactory floor covering, 188 square feet are placed on the ceiling. With a width of 13 feet 11 inches (13.9 feet), this means that a length of about 13.5 feet of 703 area yields an area of 188 square feet.

The frame takes up a respectable portion of this area; hence a length of 14 feet 6 inches is actually required to give the 188 square foot net.

Figure 17.12 shows a grid of sections mounted on the ceiling, leaving some bare ceiling for the loudspeakers on the north end. The 24-inch inside dimension of each section is for the purpose of efficient cutting of the 24-inch × 48-inch sheets of 703. Of course, 24-inch × 48-inch sections require cutting only in the odd sections. It is the 703 area that counts. There are many ways to handle the mechanical mounting.

The rationale of Table 17.2 is based on distributing the 4-inch 703 glass fiber material in proportion to the areas of the surfaces associated with the three axial modes of Control Room A. Column (a) of this table gives the exact number of sabins for each pair of surfaces based on this premise. Column (b) lists the practical distribution of 703 areas of Figs. 17.11 and 17.12 yielding a reverberation time of 0.36 second. There are some unavoidable slips betwixt theory and practice. Observation windows, doors, hard-surfaced floor and work table have brought compromise. The principle is still a good initial guide, even if certain departures from it are necessary.

Floor Treatment

As indicated in the floor plan of Fig. 17.12, the floor covering specified is vinyl tile. Linoleum, wood parquet, or other hard surface is acceptable. By avoiding carpet with its unbalanced absorption, only 4-inch 703 is required in the acoustical treatment.

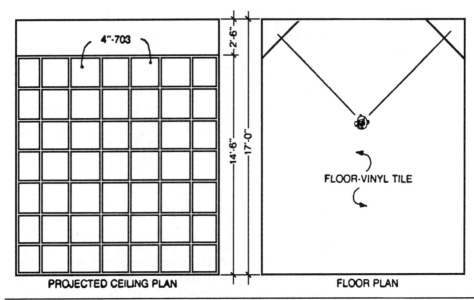

FIGURE 17.12 Projected ceiling plan and floor plan of Control Room B.

In the context of this chapter no detailed discussion of the treatment of Studio B and Control Room B is required because of their similarity to suites covered in previous chapters. For normal use of Studio B a reverberation time of the order of 0.3 second is required. The suitability of 0.3 second when used as an isolation booth depends on the type of sound source placed in it. As a vocal booth 0.3 second may be a bit low, but reverberation can always be added electronically. As a drum booth it should be satisfactory as it is, although idiosyncrasies of individual drummers often require temporary adjustments. For normal use of Studio B, short drapes can be drawn over the two unused windows to minimize reflection defects.

CHAPTER 18

A Screening Facility for Film and Video

Favorable conditions for viewing motion pictures must include both visual and aural factors. Visual quality depends on screen diffusion, screen brightness, and the steadiness of the projected image. Aural quality in regard to room acoustics is much the same problem as covered in previous discussions of studio and monitoring room quality with some relaxation of tolerances. From the viewpoint of sound reproduction, the motion picture projector sound head quality is vitally important, along with amplifiers and loudspeakers. Both the visual and the acoustical factors are treated in this chapter, as they are inseparably bound together in projecting motion pictures and in viewing video presentations effectively.

Floor Plan

For the best impression of a projected motion picture, the projection machinery must be in a separate room. This not only reduces the noise, it eliminates one more potential diversion of attention of those who should be paying full attention to the screen.

In Fig. 18.1 a very practical, low cost, high quality screening facility is described. It is small but effective accommodating up to 20 people very comfortably, a few more in a pinch. The individual seats may be upholstered, swivel-type, or simply canvas director's chairs. Arranged on carpeted wood risers of 4-foot width, a feeling of comfort and luxury can be imparted with modest outlay.

Door Arrangement

The background noise level standard adopted for this small theater (the NC 25 contour is commonly used, Fig. 7.5), and the environmental noise level outside the theater determine the type of walls required. This has been covered in Chapter 7.

Doors A and B of Fig. 18.1 provide entrance and egress for the screening room. If these are single doors, even though they have a solid core and are well weatherstripped, their insulation against noise from the outside is quite limited. If the external noise is low level, single doors may be adequate. Adding door C makes a sound lock between doors C and B. This would make little sense unless something similar is done to door A. In addition to the noise, another potential problem associated with door A is that, if

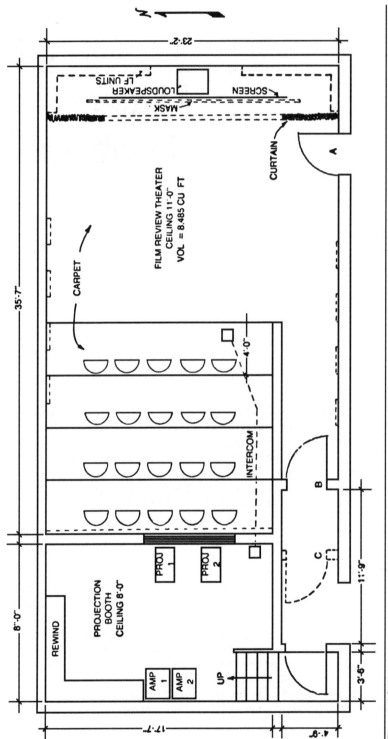

Figure 18.1 Floor plan of film and video review theater seating about 20 people on a stepped riser. Precautions must be exercised to control projector noise and noise from outside as well as distracting light falling on the screen through open door A.

opened while the theater is darkened, it may be very disruptive as light from the outside falls on the projected image, disturbing the viewers.

Screen

This theater should be equally valuable for projection of 35 mm or 16 mm film prints, or 70 mm or super 8 mm prints for that matter. The same screen can serve them all equally well. Aspect ratio is defined as the ratio of width to height of the film image and the screen proportions are determined by the proportions of the image on the film. Historically, an aspect ratio of four units wide by three units high (1.33:1) has been used.

When sound was introduced to 35 mm film, less room was available for the image of each frame, but a smaller 4 × 3 proportion persisted. Starting with the early introduction of *CinemaScope* and progressing on through many stages, the aspect ratio varied from 2.66:1 on down.

It is now general practice to provide for projection at proportions between 1.33:1 (old 35 mm film and normal 16 mm films) to 1.65 and 1.95:1 for flat, widescreen films and 2.35:1 for anamorphic projection. This means that ample screen area must be provided and that black masks are to be used to shape the screen to the format to be used, cropping the edges neatly. In Fig. 18.2 a screen 6-feet high and 14-feet wide meets the maximum proportions required, 2.35:1.

The masks are simply black cloth-covered frames of light wood or metal. These are arranged as a top and bottom horizontal pair, and a left and right pair for cropping the

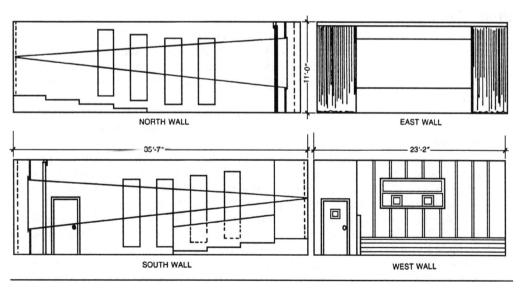

FIGURE 18.2 Wall elevations showing placement of screen and its associated elements as well as acoustical treatment. The rear wall and side wall panels are basically 4-inch 703 glass fiber covered with a perforated vinyl fabric.

vertical edges. A fancy (and very convenient) installation would have these pairs adjustable by motor drive and remotely controlled from the projection booth so that one button would be pushed for 1.33:1, another for 1.65:1, etc. Other controls for highly professional projection would include motor-driven curtains and light dimmers, preferably synchronized.

The screen must be of the perforated type if the loudspeaker is positioned behind it as indicated in Fig. 18.1. The small perforations make the screen essentially transparent to sound in the audible band, yet are not visible at normal viewing distances.

Three types of screens are in common use, distinguished from each other by their surface.

- matte
- beaded
- metallized

A picture projected on a screen with a matte surface appears to have much the same brightness when viewed off to one side as when viewed from directly in front. The general screen brightness level of the matte screen is, however, quite low. Beaded and metallized screens have a more pronounced directional characteristic, throwing most of the light directly back toward the projector and giving a much dimmer picture off to the side. For quality projection giving a picture of equal brightness over the seating area of the small screening room, a perforated screen with a matte surface is probably the more suitable. It should be mounted in a frame with elastic cord so that its surface is very flat.

Projection Booth

The floor of the projection booth is at least 3 feet above the main floor of the screening room so that the projector beam clears the heads of those seated on the top riser. Actually, this beam should not be interrupted by persons walking anywhere in the theater, but with the limited 11-foot ceiling height this ideal cannot be attained with the riser plan shown.

The projection room is reached by five steps up from the alcove shared with door B. The projectionist's observation window is 7-feet long and about 18-inches wide and is of customary double-glass construction.

Glass in the two projector ports is quite a different problem because of possible color tint and refraction affecting the projected image. From the standpoint of theater noise, however, it is imperative that these small ports be fitted with at least one good thickness of glass. A work table is suggested for film rewinding and similar tasks.

The projection booth should be acoustically treated to reduce the effect of projector noise both in the booth and in the screening room. The surface area of the projection booth is about 691 square feet with an 8-foot ceiling. If the floor is vinyl tile and the walls and ceiling are bare gypsum board the average absorption coefficient might be about 0.05, giving a total absorption of $691 \times 0.05 = 34.55$ sabins at midband.

If the absorption coefficient was increased to 0.30 by the acoustical treatment of room surfaces, the total absorption in the room would be increased to $691 \times 0.30 = 207.3$ sabins.

This would result in a decrease in projector noise level of 10 log 34.55/207.3 = 7.78 dB. This means that anywhere in the room, except in the immediate vicinity of the projector, the noise level is thus reduced almost 8 dB by the introduction of the absorbing material. This would make it much more comfortable for the projectionist and would reduce projector noise in the screening room as well.

Fire regulations pertaining to projection rooms must be determined before actual construction and acoustical treatment are begun, but acoustically the only requirement is to make the room as highly absorbent as feasible over the audible frequency range with no worry about uniformity.

Another detail which can contribute to smooth operation of the projection booth is some form of intercommunication with a control seat in the audience area. This link provides a professional touch to a projection event in such things as when to roll it, the adjustment of sound level, etc.

Theater Treatment

A screening theater of this type is basically a listening room and should be acoustically treated as such. There is the question of whether speech or music should be favored, but the understanding of narration and dialog is taken as the most fundamental requirement. For a room with an 8,485 cubic foot volume, the optimum reverberation time for speech is close to 0.5 second. The goal is 0.5 second, uniform 125 Hz-4 kHz, recognizing in advance that tolerances are not as tight for film or video viewing as they are for recording studios or critical monitoring rooms.

The entire floor area of the film review theater, including the riser seating area, is carpeted with a heavy carpet and pad as indicated in Table 18.1. This is in deference to the comfort and enjoyment of guests and in spite of the acoustical compensation required.

Another component of acoustical absorption having characteristics similar to carpet is the curtain. For this the area entered in the computation is that which the curtains offer when retracted to reveal the 14-foot width of screen (the 2.35:1 aspect ratio). The question arises, "should both sides of the partially retracted curtain be considered active absorbers?" This depends on how the curtain was placed when the coefficients were measured. The book (Appendix A) says, "Medium velour, 14 ounces per square yard draped to half area."

Although the inference is that the drape is not far from a wall (these are about 3 feet from a wall), the area of only one side of the curtain has been entered in Table 18.1 and the possible error of a few dozen sabins registered in the inner consciousness.

Wideband absorption, again with 4 inches of Owens-Corning 703 Fiberglas as the basic dissipative element, is applied to the north and south walls in the form of 2 foot × 8 foot wall panels and to the entire rear (west) wall. In both cases the 703 is covered with a decorative vinyl fabric perforated for a reasonable degree of sound transparency. The vinyl fabric supplied by L.E. Carpenter and Company (Appendix D), called *Vicrtex*, comes in many attractive patterns and colors, but is perforated at the factory only on order. The perforation percentage of the Vicrtex is estimated to be between 12 percent and 15 percent. This vinyl covering can be a main contributor to the decor of the room if carefully chosen.

SIZE .... 23'2" × 35'7", ceiling 11'0"
FLOOR & RISERS .... Heavy carpet and pad
WALLS .... Side walls 8 wideband panels 2'—8'
.... 4" 703 covered with perforated vinyl fabric (Fig. 18.3)
.... rear 4" 703, perforated vinyl cover (Fig. 18.4)
.... behind screen low peak absorbers in corners.
.... 0.31" perf., 8" deep 4" 703 (Fig. 18.5, 6)
VOLUME .... 8,485 cu. ft.

Material	S Area Sq. Ft.	125 Hz		250 Hz		500 Hz		1 kHz		2 kHz		4 kHz	
		a	Sa	a	Sa	a	Sa	a	Sa	a	Sa	a	Sa
Carpet	841	0.05	42.1	0.15	126.2	0.30	252.3	0.40	336.4	0.50	420.5	0.60	504.6
Curtain, Open	100	0.07	7.0	0.31	31.0	0.49	49.0	0.75	75.0	0.70	70.0	0.60	60.0
Wideband Panels 8 at 14.8 sq. ft.	119	0.99	118.0	0.99	118.0	0.99	118.0	0.99	118.0	0.99	118.0	0.99	118.00
Wideband East Wall	120	0.99	119.0	0.99	119.0	0.99	119.0	0.99	119.0	0.99	119.0	0.99	119.0
Riser, ¾" Ply	324	0.38	123.1	0.19	61.6	0.06	19.4	0.05	16.2	0.04	13.0	0.04	13.0
Gypsum Bd. ½"	1660	0.10	166.0	0.05	83.0	0.04	66.4	0.03	49.8	0.03	49.8	0.03	49.8
Low Peak	190	1.0	190.0	0.83	157.7	0.44	83.6	0.29	55.1	0.24	45.6	0.20	38.0
Total Sabins, Sa			765.2		696.5		707.7		769.5		835.9		902.4
Reverb. Time, sec.			0.54		0.60		0.59		0.54		0.50		0.46

TABLE 18.1 Film and Video Review Theater Calculations

The wideband wall panels are located as shown in Fig. 18.2, four to each side wall and positioned so that a panel on one wall faces bare wall between panels on the opposite wall. This should provide reasonable control of flutter echoes in the north-south mode, even as the carpet does for the vertical mode.

Figure 18.3 gives necessary details of construction of the wall panels. The perforated vinyl fabric comes in a 54-inch width which will provide two 24-inch panels. If

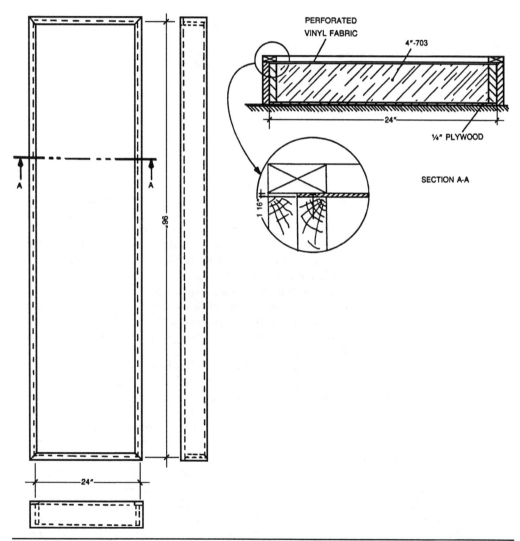

FIGURE 18.3   Constructional details of 2-foot × 8-foot wall absorbing panels covered with vinyl fabric. The frame and spacer width must be adjusted to accommodate the 4-inch semirigid glass board without bulging of vinyl cover.

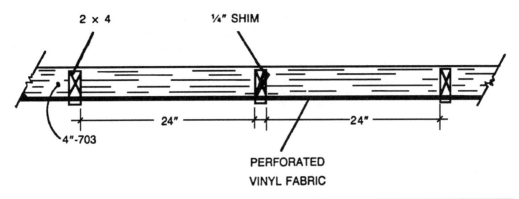

FIGURE 18.4   Detail of rear (west) wall treatment. The 2 × 4 frame is shimmed out about $\frac{1}{4}$ inch to avoid bulging of perforated vinyl fabric face by the glass fiberboard.

desired, 1 × 6 lumber can be used for the external frame which results in an air space between the 703 and the back board. This increases low frequency absorption.

Figure 18.4 shows how the rear (west) wall is framed with 2 × 4s making spaces 24 inches wide to accommodate both the 24-inch width of the 703 and the 54-inch width of Vicrtex cover. The 2 × 4s are only about $3\frac{3}{4}$ inches and by shimming them out from the wall about $\frac{1}{4}$ inch the 703 does not tend to bulge the Vicrtex. A finish strip is nailed to each 2 × 4, covering the Vicrtex edges and overlap.

Low frequency absorption is required to compensate for low frequency deficiencies of carpet and curtains. The risers, constructed of $\frac{3}{4}$ inch plywood on a wood frame, contribute a significant amount of absorption, peaking about 125 Hz. In addition, the gypsum plaster board on walls and ceiling, even though covered by other elements, also absorbs well at low frequencies.

In Table 18.1 gypsum board of $\frac{1}{2}$-inch thickness has been assumed, although relatively small changes would be expected if it were $\frac{5}{8}$-inch or even double thickness.

As the riser and wall/ceiling surfaces do not give quite enough low frequency compensation, some perforated panel Helmholtz resonators are introduced to the room. These are not the most beautiful things in the world; hence they are hidden behind curtains and screen. By placing them in the corners, they take advantage of the fact that all modes terminate in corners.

The positions of these Helmholtz resonators are shown in Fig. 18.5 and their construction is detailed in Fig. 18.6. The $\frac{3}{16}$-inch diameter holes on a square pattern 3 inches on centers turns out to be a perforation percentage of about 0.31 percent.

Other configurations of hole diameter and spacing yielding perforation percentages of about 0.31 percent ±10 percent are acceptable. Care should be exercised to assure that supporting 2 × 8s and dividing 1 × 8s fall between rows of holes. The $\frac{3}{16}$-inch Masonite sheets can be stacked for drilling to facilitate this chore.

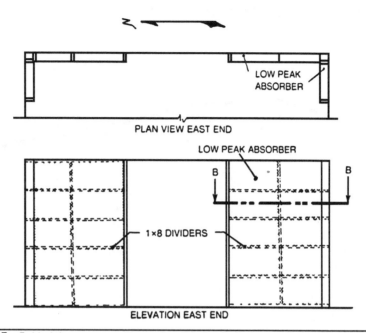

FIGURE **18.5**   Behind the screen on the east wall is located a low peak absorber in each corner, floor to ceiling. Dividers of 1 × 8 lumber break up the air space to discourage modes of vibration parallel to the face of the absorber.

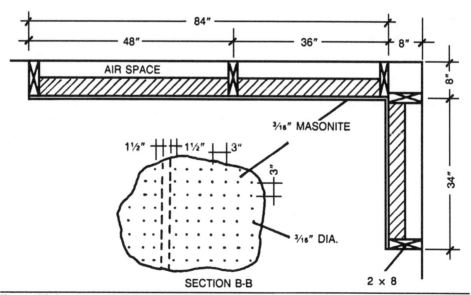

FIGURE **18.6**   Section B-B from Fig. 18.5 of a typical low peak corner absorber behind the screen. The perforation percentage is about 0.31 percent.

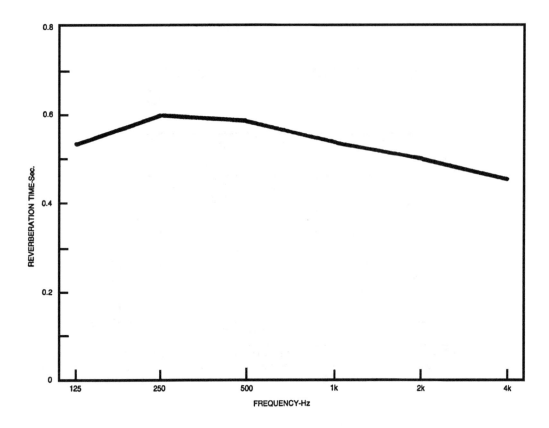

FIGURE 18.7   Calculated reverberation time of film review theater. This characteristic is satisfactory for this room which is considered primarily a listening room for speech.

Reverberation Time

Following through on the calculations of Table 18.1 the reverberation time at each of the six frequencies is determined. The values from the table are plotted in Fig. 18.7. This graph varies from 0.46 to 0.60 second. However, as previously stated, the listening conditions should be quite acceptable.

CHAPTER 19

Multiple Studios

All the good things should be maximized, all the bad things minimized. This message is repeated frequently enough to encourage maximum recall with minimum effort. A new building was to be built, but many activities were in competition for the maximum space a minimum budget would allow.

As for recording studios, talk at first swung toward a large number of very small studios. This was parried by solid information on the adverse effect of small studio spaces on sound quality and it was agreed to reduce the number of studios so that at least the 1500 cubic foot minimum volume could be realized. Most of the work is the recording of voice programs, but numerous languages are involved, necessitating a multiplicity of studio-control room suites.

It is desirable to make all these suites very similar so that language teams can go from one to another with no delay or inconvenience caused by lack of familiarity. Essentially identical acoustics of all speech studios would also allow complete freedom in intercutting.

Music recording of fairly large vocal and instrumental groups requires one music recording studio, but this larger studio must also be capable of being pressed into service for voice recording at times. This is a very challenging and interesting set of requirements, the solution of which poses some technical problems of general interest.

Heretofore in this book splayed walls have been conspicuous by their scarcity. One reason for this is that many of the studios studied here were located in existing buildings. In such cases splaying of walls required either reconstruction of great sections of the building or losing precious studio volume, or both. This is costly financially and acoustically. Splayed walls do reduce the chance of flutter echoes being produced, even as well-distributed absorbing materials do. In a new building, however, splaying represents little or no additional cost and no loss of room volume. Under such conditions it is most logical to include it.

Typical Recording Suite

Figure 19.1 represents a typical speech studio-control room suite having:

- Rooms of minimum volume (about 1500 cubic feet)
- Two splayed walls in each room
- Sound lock space shared by two or more suites
- An equipment storage space for each two suites

This is certainly maximizing function in limited space as two speech studios, their associated control rooms, an adequate sound lock and a shared storeroom are obtained

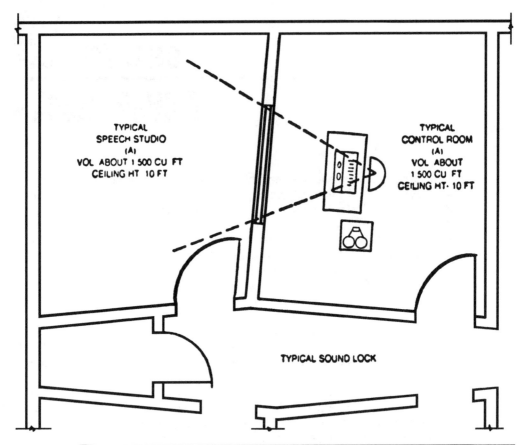

FIGURE 19.1 Typical speech recording suite featuring rooms of about 1,500 cubic feet volume and splayed walls coordinated with a sound lock corridor. This is the basic unit of which the larger studio complex is composed.

in a rectangular area about 17 feet × 24 feet with a ceiling height of 10 feet. The plan of the single recording suite of Fig. 19.1 thus becomes an elemental building block of the larger studio complex. This sets the pattern for all control rooms and all speech studios; in fact, everything but the music studio.

Splaying Plan

Normally walls are splayed 1:10 or 1:5. Ratios in this form are readily understood by construction workers but they can also be expressed in degrees (5.7 degrees or 11.3 degrees) by plugging 0.1 or 0.2 into the trusty calculator and punching *arc-tan*. Due to a trigonometric inclination and a love of round figures an even angle of 5 degrees was adopted for the splay of these studio walls which was later realized to be a very odd ratio for the construction people, $2\frac{1}{8}$:24.

The scheme of splaying all speech studios and control rooms is illustrated graphically in Fig. 19.2. The two walls to be splayed are simply rotated 5 degrees about their

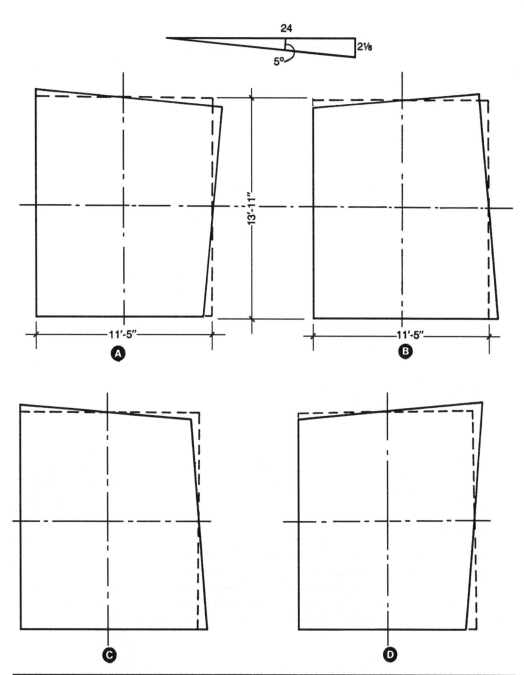

FIGURE 19.2   Splaying plan for the speech recording studios and associated control rooms. The two walls to be splayed in each room are rotated 5 degrees about their midpoints.

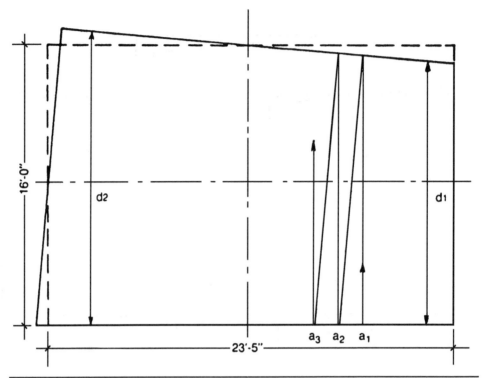

Figure 19.3   Splaying plan for the larger music recording studio is also based on 5-degree rotation about the midpoint of each of the two walls.

midpoint. As two walls are involved, and each wall can be rotated clockwise or counterclockwise about its midpoint, there are four possible combinations of room shapes with the 90-degree corner held in the same position. This 90-degree corner can be placed in other positions by rotating or flopping the sketches. Three of these four possibilities account for all the room shapes to be included.

The larger music studio splaying plan is shown in Fig. 19.3. Actually, it is the plan of Fig. 19.2A adapted to the different proportions and dimensions of the music studio, Studio C. It should be emphasized at this point that the four splaying plans of Fig. 19.2 have somewhat different areas, even though based upon identical rectangles. It should also be remembered that only the N-S and E-W modes are touched with the above wall splays. The vertical flutter echo must be cared for in some other way.

Room Proportions

To a first approximation the dimensions of the basic rectangle from which the splay pattern is derived can be used to establish proportions for the best distribution of axial modes. The actual modal frequencies, of course, differ slightly from these. One way of looking at it (Fig. 19.3) is that the sound energy reflected to a splayed wall from an opposing, but unsplayed, wall does not return to the same spot on the originating wall, a_1, but returns to a_2, a_3, etc. It "walks the slope" and tends toward becoming a tangential mode.

Another approach is to consider dimension d_1 to give one axial mode frequency and dimension d_2 another one slightly lower. Both outlooks are based on geometrical acoustics which fail miserably in the low frequency region giving us the most trouble. It is really a very complex problem and the mathematical tool of wave acoustics is a more powerful approach. Although only an approximation, establishing favorable room proportions according to the basic presplaying rectangle is a logical and practical approach and the only simple alternative.

Floor Plan

Using the typical speech studio-control room layout of Fig. 19.1 and the splaying plans of Figs. 19.2 and 19.3, the floor plan of Fig. 19.4 has been derived. It includes three speech studio suites, A, B, and D, and one music studio, Studio C, with its control room. Control Room C, serving the music studio, is comparable to the other smaller rooms. One sound lock corridor serves three studios and three control rooms. Were it not for the stairwell, all eight rooms might well have been served by a single sound lock. These studios are located in one corner of the top floor of the two-story building.

Traffic Noise

As the building housing the studios is on a well-traveled boulevard in a major city, traffic noise must be considered. Traffic noise varies greatly with the time of day and the only way to evaluate it properly is to run at least a 24-hour noise survey on the proposed site. Obviously, there is less traffic at night and noise conditions are at their lowest point in the early morning hours, but the 24-hour survey makes possible such statements as "The

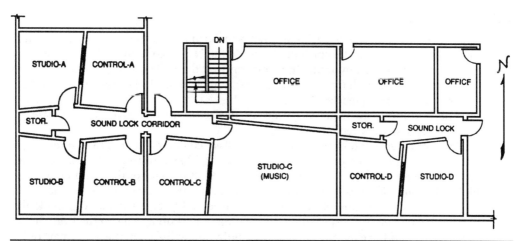

FIGURE 19.4 Floor plan incorporating three speech recording suites such as shown in Fig. 19.1 and a larger music recording studio. The control room for the music studio (Control Room C) is comparable to the speech studios and their control rooms.

noise level exceeds 75 dB(A) only 4 percent of the time." This is the sort of data required to support various types of decisions.

This ideal approach was not possible in this instance, therefore an octave analysis of peak boulevard traffic noise was made at the curb closest to the building site with the results shown in Fig. 19.5. This was done before the exact location of the building was known. Once the building location was set, the measured values at the curb were extrapolated to the nearest face of the building by assuming spherical divergence of sound with its resulting 6 dB reduction with each doubling of the distance from the line of traffic. This procedure yielded the broken line spectrum of Fig. 19.5, the estimated noise spectrum just outside the building.

The NC-15 contour is our goal for noise level within studios and control rooms which, of course, could very well have noise contributions from other activity within the building, air-conditioning equipment, etc., in addition to traffic noise. At the moment, however, only traffic noise is under consideration.

Special urging resulted in the placement of the studio corner of the building on the side away from the boulevard. This offers some protection from traffic noise. The building code requires ventilation of the attic space above the studio area which means that traffic noise of considerable magnitude pervades the attic space immediately above the studio ceilings.

With the wall construction to be described later, the greatest prospect for a traffic noise problem in the studios turns out to be via this ceiling path. If sound level measurements within the studios reveal traffic noise levels appreciably above the NC-15 contour, a layer of sand will be added between the ceiling joists above the double drywall ceiling. A 1-inch layer of sand "beefs up" the ceiling, acoustically speaking, 36 dB at 500 Hz on a mass basis and weighs only about 8 pounds per square foot. Each doubling of the sand thickness adds 3 dB transmission loss, but doubles the weight.

It is advisable to stop short of a thickness at which sand would break through the ceiling and pour down on unsuspecting personnel below. A modest amount of sand could add very substantially to the insulation strength of the ceiling against external noise.

External Walls

There is information in Fig. 19.5 which helps in deciding how heavy to make the external walls of the studio. The distance between the NC-15 contour and the broken line graph represents the minimum transmission loss the walls must provide. This loss requirement varies with frequency. It is maximum at about 1 kHz, decreasing for both lower and higher frequencies.

This is unfortunate in one sense because the sensitivity of human ears is greatest in this general frequency region.

It is fortunate in another sense because walls of common materials and normal construction can offer quite good transmission loss at 1 kHz, much greater than at much lower frequencies. The octave noise level of 64 dB at the face of the building is 47 dB higher than the NC-15 contour at 1 kHz which we are striving for within the studio. The external wall construction of Fig. 19.6 was considered to be adequate to provide this much transmission loss, especially knowing that the 64 dB applies to the front face of the building and the external walls of the studios are in the rear. The lower part of the wall of Fig. 19.6 is the top of the tilt-up panels. The upper part of frame construction is plastered outside and covered with double $\frac{5}{8}$-inch gypsum drywall panels inside. The

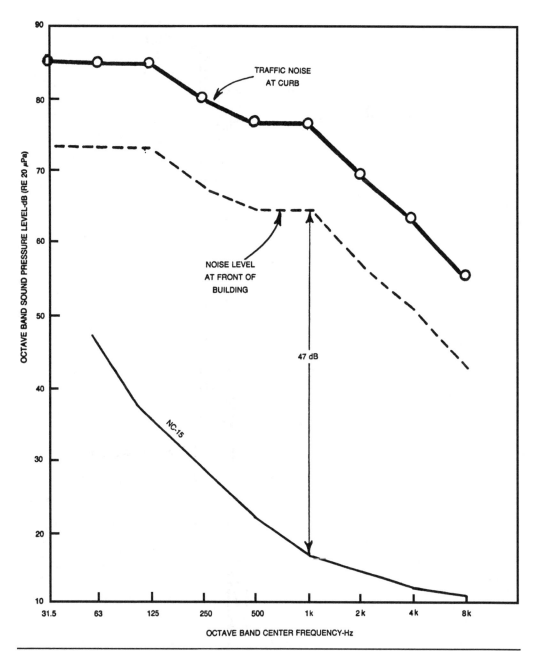

FIGURE 19.5 Traffic noise peaks were measured at the boulevard curb and later (when the building position was established) extrapolated to the face of the building. The NC-15 contour is the goal for noise within the studio. The difference between these two graphs is the transmission loss the studio external walls must provide.

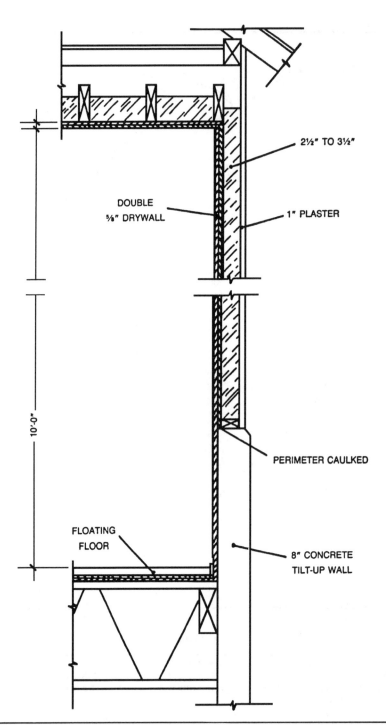

2½" TO 3½"

DOUBLE
⅝" DRYWALL

1" PLASTER

PERIMETER CAULKED

FLOATING
FLOOR

8" CONCRETE
TILT-UP WALL

10'-0"

FIGURE 19.6   Plan of external wall of studios and control rooms. Tight sealing of the base layer of gypsum board around its periphery contributes significantly to wall performance in protection of studios against external noise.

insulation between wall studs serves the double purpose of thermal insulation and discouraging acoustical resonances in the cavity which degrade the effectiveness of the wall in attenuating external noise.

Ceiling construction is similar, except for the plaster. Should sand be added to the upper surface of the ceiling drywall at a later time, the insulation would first be removed, the sand applied and then the insulation would be replaced on top of the sand.

Internal Walls

The construction of typical internal walls is specified in Fig. 19.7. Double layers of ⅝-inch gypsum board are standard on every wall or ceiling separating sound sensitive areas from outside noise. The gypsum panels of all studio walls (except external) are not nailed to the studs but are supported resiliently. The base layer board is secured to the resilient channels with screws. These resilient channels, U.S. Gypsum Rc-1 or equal, are first nailed horizontally to the studs, spaced 24 inches.

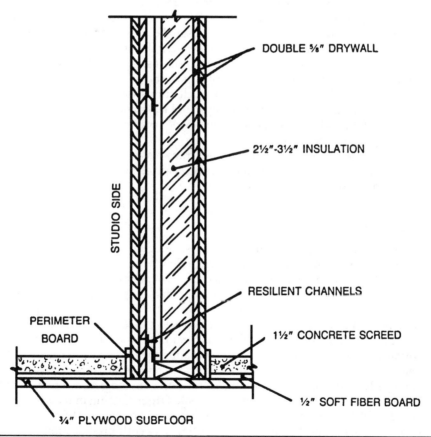

FIGURE 19.7   Plan of interior walls of studios and control rooms. On the studio side the base layer of gypsum board is screwed to resilient channels and the face layer is cemented to the base layer. Such a resilient mounting makes the resonance frequency of the wall diaphragm on the studio side different from that of nailed panels on the other side, preventing coincidence and thus improving transmission loss of the wall.

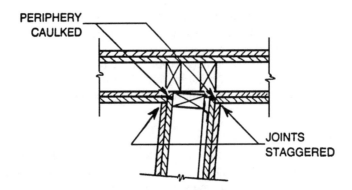

FIGURE 19.8   Plan for staggering gypsum layers at corner joints to reduce noise leaks.

The vertical base layer board is then secured to the resilient channels with special screws of proper length so that the flexible action of the channel is not destroyed by screws hitting studs. The face layer of gypsum board is then applied horizontally with adhesive. All joints are then finished in the normal way.

Figure 19.8 illustrates the preferred method of staggering layers of gypsum board at corner intersections. The entire periphery of the base layer should be carefully caulked with nonhardening acoustical sealant. Such efforts toward hermetically sealing each room pays great dividends in reducing sound leaks and assuring maximum transmission loss of the wall. The resilient studio face of a wall resonates at a different frequency than the opposite nailed face.

Such resonance effects tend to reduce transmission loss at the resonance frequency and different handling of the two wall faces displaces one resonance point from the other, improving overall wall performance.

Floating Floor

To obtain sufficient protection against noise of other activities in ground floor rooms below the studio area, a floating concrete floor was required in studios and control rooms. There are numerous fancy mechanical ways of supporting floating floors on springs and rubber devices as well as proprietary impregnated glass fiberboards and strips which are excellent in floating floors, but expensive. An inexpensive method used in Germany a quarter of a century ago was pressed into service.[32]

A soft fiberboard is laid on the structural floor as a support for the concrete. The stiffness of the fiberboard is reduced by coating the underside of it with cork granules before laying it on the structural floor. The fiberboard is then covered with plastic sheets and overlapped at least 3 inches. It also runs up over the perimeter board indicated in Fig. 19.7. The concrete screed is thus prevented from running into cracks between fiber boards which would form solid bridges between the structural floor and the floating floor, destroying the floating characteristics. The cork granules improve the impact sound insulation about 16 dB over the fiberboard alone. The 1½-inch concrete thickness is certainly minimum.

To reinforce such a floor (wise precaution) the thickness of the concrete must be 3 inches to 4 inches. In thinner layers it is almost impossible to keep the reinforcing screen in the center of the concrete layer during the pouring. If the reinforcing wires are on the bottom of the layer, little reinforcing results. The danger is that concentrated loads, such as the legs of a grand piano, may crack the concrete, reducing its sound insulating value.

Treatment of Studio A

The specification of carpet for all studios dominates their acoustical treatment. The drywall wall and ceiling surfaces provide a modest, but insufficient, amount of low frequency compensation for the carpet. Helmholtz resonators can easily supply the remainder of compensation required, but the problem is, where to put them? Thick boxes on the walls and ceilings are not an esthetic delight. It was decided to use a frame suspended from the ceiling to hold the low frequency boxes and illumination fixtures and to shield both from view with a lower frame face of plastic louver panels of either egg crate or honeycomb type openings.

Should the usual panels of 4 inches of 703 Fiberglas adorn the walls? In the interests of appearance it was decided to employ proprietary panels of glass fiber covered with decorative, perforated vinyl wallcovering manufactured by L.E. Carpenter and Company (Appendix D). *Vicracoustic panels* 2 feet × 8 feet × 2 inches were selected. The core of semirigid glass fiber, which does the absorbing, can be covered on one or both sides with ⅛-inch high density glass fiber substrate if required for protection against impact.

In this studio application the less expensive Type 80 panel, which consists only of the absorbing core wrapped on face and edges with perforated vinyl, was considered adequate. Panels (Fig. 19.9) are mounted on the walls by the use of Z-clips, one

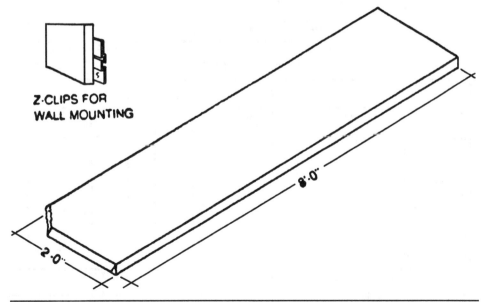

Z-CLIPS FOR WALL MOUNTING

8'-0"

2'-0"

FIGURE 19.9 The panels selected for use in all studios and control rooms are 2-foot × 8-foot Vicracoustic panels composed of 2 inches of dense glass fiber covered with an attractive perforated vinyl fabric. Mounting to walls is by use of Z-clips.

part of which is cemented to the backs of the Vicracoustic panel, the other screwed to the wall.

The low frequency absorption of these panels is increased from 0.47 to 0.57 at 125 Hz if the Z-clips are mounted on 1×3 strips, but the advantage of doing so at other frequencies is almost nil. The treatment of Studios A, B, C, and D is accomplished with carpet, low frequency Helmholtz resonators, and Vicracoustic panels added to the built-in absorption of the drywall surfaces.

The placement of the Vicracoustic wall panels in Studio-A is shown in Fig. 19.10. Even though the east and south walls are splayed, an attempt is made to place panels on one wall to oppose bare wall (or window, or door) on the opposite wall. This can best be judged in the projected ceiling plan of Suite A in Fig. 19.11.

The constructional details of the low frequency absorbing boxes are given in Fig. 19.12. Similar to others considered in other chapters, the frame is of 1×8 lumber with a

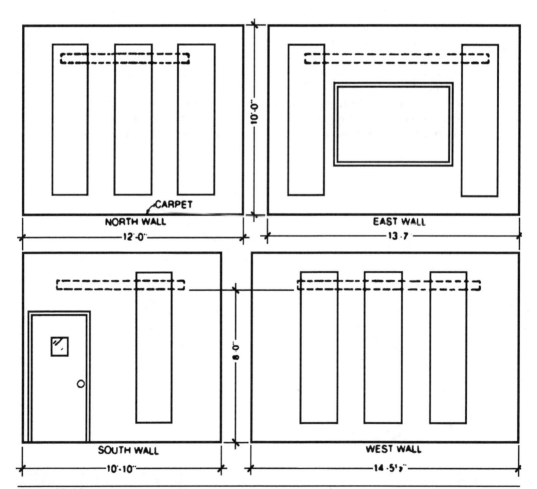

FIGURE 19.10 Wall elevations of Studio A showing placement of Vicracoustic panels. The broken lines indicate relative position of the suspended frame holding low peak absorbing boxes.

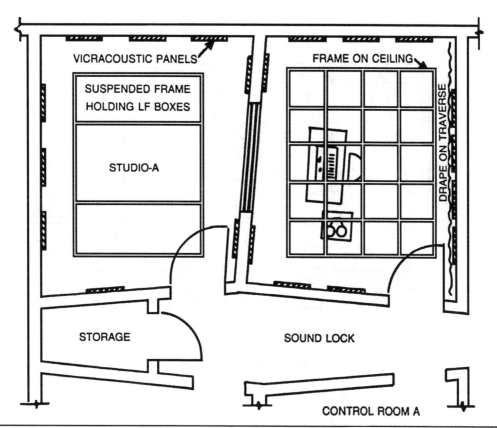

FIGURE 19.11   Projected ceiling plan of Studio A and its control room showing position of the following acoustical elements: Vicracoustic wall panels, frame suspended from ceiling in Studio A, frame fastened to ceiling in Control Room A and drapery on traverse in Control Room A.

back of ½-inch plywood or particleboard. The face of ³⁄₁₆-inch masonite is filled with ³⁄₁₆-inch holes drilled on 3-inch centers. This gives a perforation percentage of about 0.3 percent and a resonance peak in the vicinity of 100 Hz.

The 4 inches of 703 glass fiber broadens this peak. The boxes should be spray painted with flat black paint to reduce their visibility.

The 7 foot × 10 foot suspended frame is placed in Studio A approximately as indicated in the projected ceiling plan of Fig. 19.11. Figure 19.13 shows the relationship of the 1 × 6 frame and the plastic louver layer. The three fluorescent fixtures assure that the plastic louver plane is the dominant visual feature of the room.

The 13 black low frequency boxes the frame contains will scarcely be visible. The placement of the low frequency boxes in the frame is important. Boxes 1, 2, 3, and 4 have their faces downward, resting on the open cells of the plastic louvers. Boxes 5 and 6 rest on their long edges and point north; 7 and 8 point south; 9 faces west; and 10 faces east. The three low frequency boxes 11, 12, and 13 resting on the fluorescent fixtures must, of course, be directed upward.

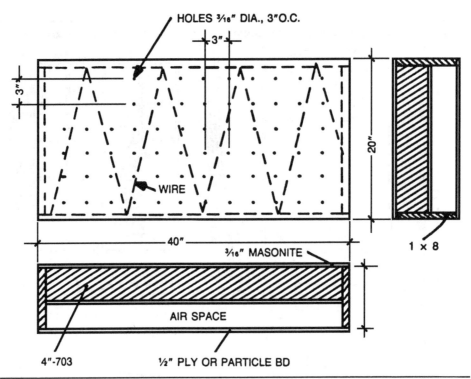

FIGURE 19.12 Constructional details of Helmholtz type perforated face resonator which provides low frequency absorption to compensate for carpet deficiency in the studios.

It is well that these last three have some soft material between the boxes and the metal reflectors to avoid sympathetic rattles when the room is filled with sound. In fact, an awareness of the possibility of rattles in the entire assembly is advised.

Reverberation Time of Studio A

Table 19.1 lays out the details of calculating (estimating) the reverberation time of Studio A. A reverberation time of 0.35 second was the goal and the plotted graph for Studio A in Fig. 19.14 shows that this goal has been approximated. The exceptionally high absorption of the Vicracoustic panels at 250 Hz and 500 Hz generates the characteristic dip noticed in the control rooms and Studio C as well. An irregularity of this magnitude in the calculation stage has little significance unless confirmed by subsequent measurements. At that time, and not before, some trimming might be necessary and justified. The beauty of the modular acoustical treatment is that such trimming, if required, can be easily carried out.

Acoustical Treatment of Control Room A

The speech studios and control rooms are very similar in size and differ chiefly in the somewhat lower reverberation time goal of about 0.3 second for control rooms. Acoustically, a very major difference between the two types of rooms is that the floors

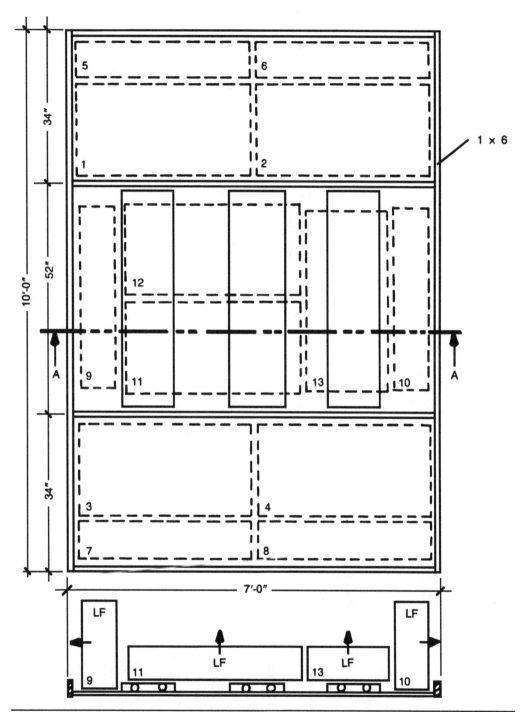

Figure 19.13 Frame of 1 × 6 lumber suspended from the ceiling of Studio A which holds illumination fixtures and the 13 low frequency boxes required in the room. Some boxes face downward, some point upward, and others point in the four horizontal directions.

SIZE .... Two walls splayed from basic rectangle 11' 5" × 13' 11" ceiling 10'.
FLOOR .... Carpet, heavy with pad, on floated floor.
CEILING .... Suspended frame holding 13 low-frequency Helmholtz absorbers 4.75 sq ft each.
WALLS .... Vicracoustic type 80 panels (9) 2' × 8' × 2', perforated vinyl covering 2" glass fiber, furred out 1".
VOLUME .... 1,598 cubic feet

Material	S Area Sq. Ft.	125 Hz		250 Hz		500 Hz		1 kHz		2 kHz		4 kHz	
		a	Sa	a	Sa	a	Sa	a	Sa	a	Sa	a	Sa
Carpet	160	0.05	8.0	0.15	24.0	0.30	48.0	0.40	64.0	0.50	80.0	0.60	96.0
Drywall	668	0.08	53.4	0.05	33.4	0.03	20.0	0.03	20.0	0.03	20.0	0.03	20.0
(LF) Low Frequency Absorbers	62	1.0	62.0	0.68	42.2	0.39	24.2	0.17	10.5	0.13	8.1	0.10	6.2
Vicracoustic Panels	144	0.57	82.1	0.98	141.1	0.92	132.5	0.76	109.4	0.71	102.2	0.78	112.3
Total Sabins, Sa			205.5		240.7		224.7		203.9		210.3		234.5
Reverberation Time, seconds			0.38		0.33		0.35		0.38		0.37		0.33

TABLE 19.1 Speech Studio Calculations for Typical Studio A

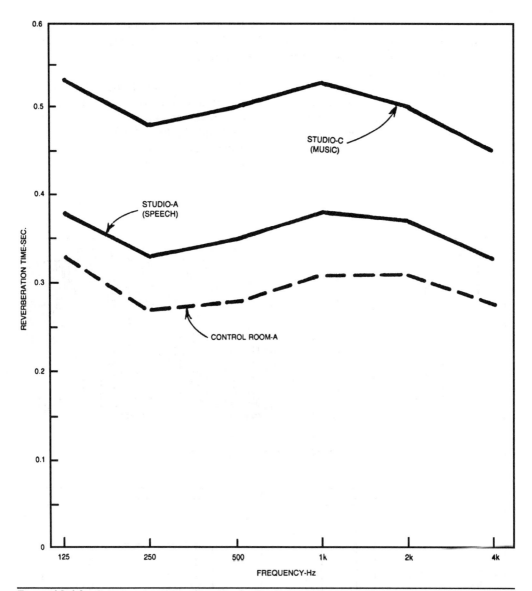

FIGURE 19.14   Calculated reverberation time for Studio A as shown which compares to the goal of 0.35 second. The goal for Control Room A is 0.3 second and for Studio C, 0.5 second. Measurements must verify such calculated estimates. The modular treatment plan allows trimming if required.

of the control rooms are covered with vinyl tile instead of carpet. This factor means that the basic treatment will be accomplished primarily with wideband, or quasi-wideband, materials.

The grid of 1 × 6 lumber attached to control room ceilings makes 20 squares having inside dimensions of 24 inches × 24 inches which hold pads of 4 inches of 703 giving a total area of 80 square feet. The actual configuration is not too important, the effective area is.

Construction can follow similar frames described in earlier chapters with a fabric or screen facing. The air space between the 703 and the ceiling aids low frequency absorption of the material. The positioning of the ceiling frame is not critical, the position of Fig. 19.11 is suggested.

The placement of the Vicracoustic panels in typical Control Room A is shown in Fig. 19.15. The four panels on the east wall are normally hidden behind a drapery which is retractable. This drape is included to flatten the reverberation time at 1 kHz and above is very close to 0.3 second as shown in Table 19.2 and Fig. 19.14. If the drape is retracted the reverberation time in the same high frequency region is close to 0.35 second, making the 250 Hz dip stand out a bit more. This drape may be considered an approved variable acoustical element, if desired.

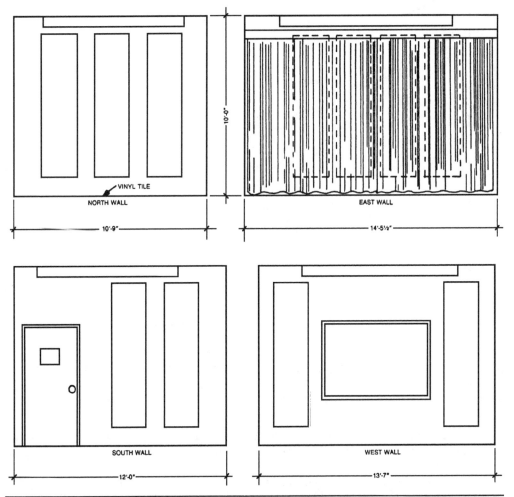

FIGURE 19.15 Wall elevations of Control Room A showing placement of Vicracoustic wall panels. The drape, normally covering the panels, is included to flatten the reverberation characteristic slightly. When retracted, the reverberation time in Control Room (A) is suitable for speech recording.

		125 Hz		250 Hz		500 Hz		1 kHz		2 kHz		4 kHz	
Material	S Area Sq. Ft.	a	Sa	a	Sa	a	Sa	a	Sa	a	Sa	a	Sa
Drywall	668	0.08	53.4	0.05	33.4	0.03	20.0	0.03	20.0	0.03	20.0	0.03	20.0
Wideband Ceiling	80	0.99	79.2	0.99	79.2	0.99	79.2	0.99	79.2	0.99	79.2	0.99	79.2
Vicracoustic Panels	176	0.57	100.3	0.98	172.5	0.92	161.9	0.76	133.8	0.71	125.0	0.78	137.3
Drapery	126	0.03	3.8	0.04	5.0	0.11	13.9	0.17	21.4	0.24	30.2	0.35	44.1
Total Sabins, Sa			236.7		290.1		275.0		254.4		254.4		280.6
Reverberation Time, seconds			0.33		0.27		0.28		0.31		0.31		0.28

SIZE .... Two walls splayed from basic rectangle 11' 5" × 13' 11", ceiling 10'.
FLOOR .... Vinyl tile over floated concrete.
CEILING .... Wood frame holding 80 sq. ft. of 4" 703.
WALLS .... Vicracoustic Type 80 panels (11) 2' × 8' × 2", perforated vinyl covering. 2" glass fiber, furred out 1". Retractable drapery 9' × 14' 10 oz sq. yd. east wall
VOLUME .... 1,598 cubic feet.

TABLE 19.2 Typical Control Room Calculations

With the drapes extended over the east wall, conditions are proper for listening to sounds from the speech studio with its reverberation time of 0.35 second. With the drapes retracted, the control room becomes more adaptable acoustically for recording an interview, for instance. The drapery material should not be too heavy (10 ounces per square yard material was used in the calculations).

Ordinary monk's cloth, running about 8 ounces per square yard, is acceptable. It should be hung as close to the wall as possible and still clear the panels. Only enough material should be hung to result in the fold almost disappearing when the drapes are extended. Having deeply folded drapes would introduce too much absorption.

Music Studio Treatment

Music Studio C should have a longer reverberation time than the speech studios for two reasons, its greater volume and the fact that the music is better served by a longer reverberation time than used for speech. A goal of about 0.5 second was selected, compromising somewhat toward a speech requirement because the studio will be used for speech recording at times. In spite of its greater size, the music studio uses the same acoustical elements as the speech studios: carpet, Vicracoustic panels, and peaked low frequency absorbers.

Figure 19.16 shows the placement of the Vicracoustic panels on the walls of Music Studio C. The suspended ceiling frames are shown with broken lines, the lower edge

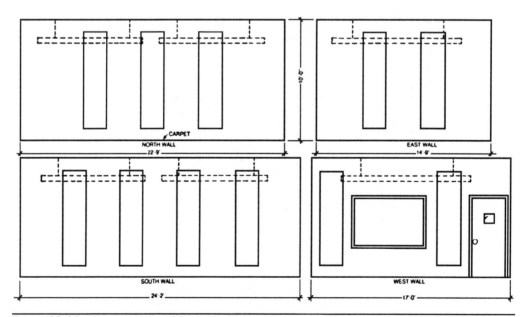

FIGURE 19.16 Wall elevations of Studio C showing placement of Vicracoustic panels in this music recording studio. Two suspended frames are required for illumination and to hold low frequency boxes.

being about 8 feet above the floor. These frames each hold 14 of the same low frequency resonator boxes used in the speech studios and described in Fig. 19.12.

Figure 19.17 shows the configuration of the 14 low frequency boxes in one of the ceiling frames. In general, the plan is the same as for the speech studio frame: the perforated side of those standing on edge facing outward (5, 6, 7, 8, 9,10): those resting on the fluorescent fixtures (11,12,13,14) facing upward; and the rest (1, 2, 3, 4) facing downward.

Table 19.3 reveals the details of the calculation of reverberation time for the music studio. The calculated values of reverberation time are plotted in Fig. 19.14. The deviations from the goal of 0.5 second are not significant, but measurements should verify the actual shape which will then be the basis of any trimming adjustments considered necessary.

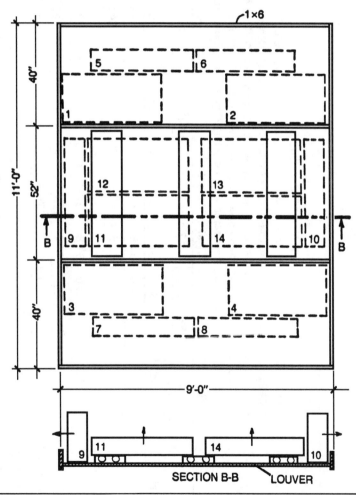

FIGURE 19.17 Detail of one of the suspended frames in the music studio. The positions of the 14 low frequency boxes are indicated.

SIZE .... Two walls splayed 5° from basic rectangle
16' 0" × 23' 5" ceiling 10'
FLOOR .... Carpet, heavy with pad on floated concrete.
CEILING .... Two suspended frames 9' × 11", each holding
14 low frequency Helmholtz absorbers, 4.75 sq ft each
WALLS .... Vicracoustic type 80 panels (11). 2' × 8' × 2"
perforated vinyl covering 2" glass fiber. furred out 1"
VOLUME .... 3,700 cubic feet

Material	S Area Sq. Ft.	125 Hz		250 Hz		500 Hz		1 kHz		2 kHz		4 kHz	
		a	Sa	a	Sa	a	Sa	a	Sa	a	Sa	a	Sa
Drywall	1,157	0.08	92.6	0.05	57.9	0.03	34.7	0.03	34.7	0.03	34.7	0.03	34.7
Carpet	370	0.05	18.5	0.15	55.5	0.30	111.0	0.40	148.0	0.50	185.0	0.60	222.0
(LF) Low Frequency Absorbers	133	1.0	133.0	0.68	90.4	0.39	51.9	0.17	22.6	0.13	17.3	0.10	13.3
Vicracoustic Panels	176	0.57	100.3	0.98	172.5	0.92	161.9	0.76	133.8	0.71	125.0	0.78	137.3
Total Sabins, Sa			344.4		376.3		359.5		339.1		362.0		407.3
Reverberation Time, seconds			0.53		0.48		0.50		0.53		0.50		0.45

TABLE 19.3   Calculations for the Music Studio C

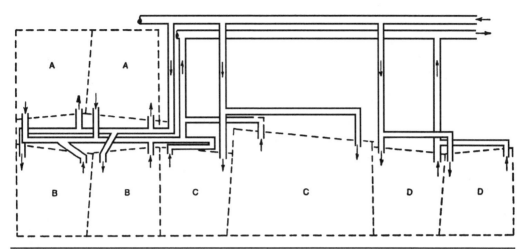

FIGURE 19.18 Air handling ducting plan for the studio complex. This plan places a maximum length of ducting between grilles of adjacent rooms to prevent crosstalk from room to room via the duct. Lined ducts attenuate sound in the ducts.

A/C Duct Routing

In Chapter 7 the several basic principles concerning air conditioning ducts in studios were elucidated. Applying these principles to the present case of multiple studio suites, the plan of Fig. 19.18 resulted. Note that the maximum length of duct is placed between grilles of adjacent rooms or even rooms on the opposite side of the sound lock. Supply and return grilles in a given room should not be too close together to assure adequate circulation. In the Control Room C return duct, a U-shaped section was inserted to avoid a short path to the adjoining room.

CHAPTER 20

Diffusion Confusion

While we touched upon modern acoustic diffusers in Part I, in order to understand everything in the next section, a more thorough explanation is needed. One of the best basic explanations I know of was written by F. Alton Everest more than twenty five years ago. While there have been numerous advancements in this area, he not only clarified the need for diffusers but shed light on the theory behind their operation.

F. Alton Everest was a man of science. He held a B. Sc. (EE) degree from Oregon State University, an EE degree from Stanford University, and did graduate work in physics at the University of California/Los Angeles. Mr. Everest was involved in the early stages of television. He did underwater sound research during WW II and cofounded the Moody Institute in Los Angeles, where he was director of science and produced science films. He spent several years as senior lecturer on communications at the Hong Kong Baptist College in Hong Kong. He published many books and papers and was a member of SMPTE, ASA, Institute of Electrical and Electronic Engineers, and AES. He was cofounder and past president of the American Scientific Affiliation and is listed by American Men of Science.

It should be no surprise that he chose a scientific approach to unravel the mystery of diffusion. What follows represents some of what Mr. Everest had to say on the subject.

All that is required of the acoustical consultant is the design of studios and other rooms that engineers, musicians and the general public consider "good." This is a difficult and subjective evaluation and the job definitely does not fall into the neat categories of definite black-white, go-no-go things of this world. If a room is too reverberant or too dead it is judged "bad" and adjusting reverberation within relatively close limits is probably the greatest single factor in elevating a poor room to a good or at least a better condition. However, reverberation time is not the only factor involved. Another factor is the diffusion of sound in the room. Often two studios very similar in size with the same reverberation time have a very different sound. This difference can possibly be traced to diffusion of sound in the room.

The relationship of diffusion of sound in a studio to the general acoustical quality of that studio is something of a mystery that has baffled studio designers for the last half century. What is diffusion? The sound field in a studio is diffuse if at any given instant the intensity of sound is uniform everywhere in that room and at every point sound energy flows equally in all directions. It has to do with homogeneity of sound in a room. Such a diffuse condition is a basic assumption in the derivation of the reverberation time equations of Sabine and Eyring. It is apparent that a dominant standing wave condition or knowledge that sound conditions vary throughout a studio means that a diffuse condition does not exist.

Diffusion is not the problem in large rooms such as auditoriums as it is in small rooms such as recording studios and listening rooms. This is the result of the fact that the dimensions of the smaller rooms are comparable to the wavelength of sound to be recorded or reproduced in them.

In Chapter 6 the effect of room size was considered along with room proportions. It was noted that the more uniform the distribution of room resonance modes the better. This procedure contributes to the diffusion of sound in the room. Selecting a cubical space in which all axial modes pile up at certain frequencies with great empty spaces between these pile-up frequencies is a move away from reasonably diffuse conditions. It is impossible, by traditional methods at least, to attain truly diffuse sound conditions in a small space, but approaching it as closely as possible is a major goal in studio design.

Sound Decay Irregularity

In the measurement of reverberation time the modes of the room are excited, say, with high intensity random noise from a loudspeaker. When the loudspeaker sound is suddenly terminated, these room modal frequencies die away, each at its own frequency and own rate.

Figure 20.1 shows tracings from graphic level recorder records of five successive decays of an octave band of random noise centered at 125 Hz. The loudspeaker and

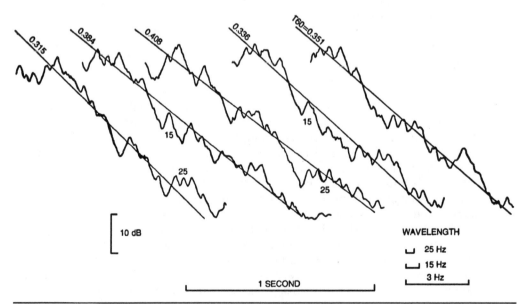

FIGURE 20.1   Graphic level recorder tracings of successive reverberatory decays under identical conditions of an octave band of random noise centered on 125 Hz. Evidence of beats between axial mode resonances is apparent.

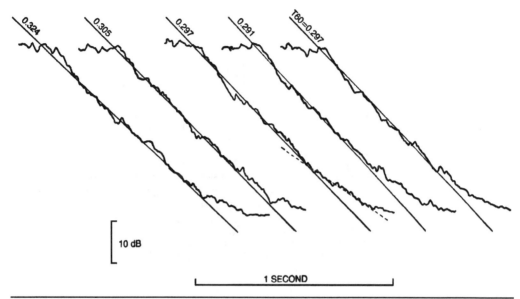

FIGURE 20.2   Successive graphic level recorder tracings or decays under the same conditions as Fig. 20.1 except for an octave band centered on 4 kHz. An octave at this frequency contains so many modal frequencies that the decay is much smoother than an octave at 125 Hz. A second decay slope at low levels gives evidence of a slower rate of decay of certain room modes.

microphone positions remained fixed. Figure 20.2 shows similar five successive decays under identical conditions except it is for an octave band of random noise centered on 4 kHz. The contrast in smoothness of decay is striking, yet these are typical of small studio decays at these frequencies. It is instructive to dig into this a bit further.

The 125 Hz and 4 kHz decays of Figs. 20.1 and 20.2 were made in a small multitrack studio 13 feet 5 inches × 18 feet 5 inches with a ceiling height of 7 feet 6 inches, volume 1,853 cubic feet. The object of the measurement was basically the determination of the reverberation time of the studio.

Establishing a best fit straight line average slope to the erratic decays at 125 Hz is far less precise than for the 4 kHz decays. To illustrate this, the *squint-eye slope* lines are included in Figs. 20.1 and 20.2 with the reverberation time (T_{60} seconds included at the top of each slope.

Of course, different observers establish slightly different slope fits, but the averaging of five slopes for each frequency and each microphone position gives a statistically significant mean value.

The mean value for the five 125 Hz decays for each of three microphone positions in this studio is 0.291 second with a standard deviation of 0.025 second. The same for the 4 kHz octave is 0.311 second with a standard deviation of 0.013 second. The standard deviation (the plus and minus deviations from the mean value which includes 67

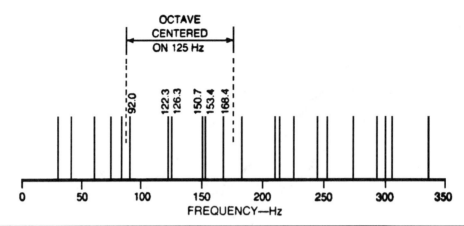

FIGURE 20.3 The studio in which the decays of Figs. 20.1 and 20.2 were taken has axial modal frequencies as shown. The octave centered on 125 Hz passes only the six indicated. The close pairs within this octave tend to beat with each other causing fluctuations in the decay trace at the difference frequency.

percent of the measurements) for 125 Hz is twice that for 4 kHz which reflects the greater fluctuations in the 125 Hz measurements.

Of special interest in this chapter, the reverberation decays of Figs. 20.1 and 20.2 also reveal something of the sound diffusion conditions in this studio. Octave bands of random noise were used in both the 125 Hz and 4 kHz cases. An octave band centered on 125 Hz is considered to include energy from 88 to 177 Hz, the half-power (3 dB down) points. The 4 kHz octave covers 2,828 to 5,656 Hz, the one spanning 89 Hz, the other 2,828 Hz. The 125 Hz octave band includes relatively few modal frequencies of the room, the octave at 4 kHz many.

The axial mode frequencies for this small multitrack studio below 250 Hz are shown graphically in Fig. 20.3. Although there are no pile-ups several pairs are very close together and wide gaps (compared to the approximately 5 Hz bandwidth of each mode) occur.

In Fig. 20.3 the span of the 125 Hz octave includes six modal frequencies. Each of these modes has its own decay rate determined by the absorption material in the room involved in that particular mode. A single mode, if excited and allowed to decay without influence of any other mode, would decay exponentially which gives a nice straight line decay on a dB scale as shown in Fig. 20.4A. The octave containing the six axial modes of Fig. 20.3 might be considered a combination of B and C, as seen in Fig. 20.1.

Then how is the smoothness of the decay curves of the 4 kHz octave (Fig. 20.2) explained? The 125 Hz octave is only 89 Hz wide while the octave centered on 4 kHz is 2,828 Hz wide. The greater smoothness is explained by the greater width and thus the greater number of modal frequencies included in the 4 kHz band. Only the axial modes are plotted in Fig. 20.3. It would be impossible to show graphically even the numerous axial modes within the 4 kHz octave, let alone the tangential and oblique modes. In fact, considering all three types of natural frequencies of this multitrack studio, something of

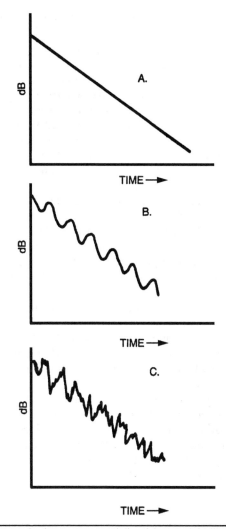

FIGURE 20.4   (A) A single mode decay exponentially, giving straight line decay on a logarithmic scale. (B) Two closely spaced modes, each having the same decay rate, beat with each other causing the decay to vary at the difference frequency. (C) Many closely spaced modes result in an erratic decay, the more modes the smoother the decay.

the order of 800,000 modal frequencies exist in the 4 kHz octave band while only 328 exist in the 125 Hz octave band.

Of course, as pointed out in Chapter 6, the tangential and oblique modes have less influence than the axial, but they do have some effect and this effect would be in the direction of smoother decays and better diffusion.

The effect of *difference beat frequencies* between axial modes of Fig. 20.3 can be detected in the decays of Fig. 20.1. The graphic level recorder paper speed for both Figs. 20.1 and 2 was 100 mm per second and a one second scale is indicated on both of these figures.

If the 153.4 Hz and the 168.4 Hz modes beat together, a difference frequency of 15 Hz is produced.

One cycle of a 15 Hz signal is represented by the length of the line so indicated in the lower right-hand corner of Fig. 20.1. In the second and fourth 125 Hz decays there are fluctuations closely matching this frequency. The 126.3 Hz and 150.7 Hz modes beating together would yield a difference beat frequency of 24.4 Hz.

There are fluctuations in decay one and three which are close to 25 Hz. The closely spaced modes near 125 Hz and 150 Hz (Fig. 20.3) produce beats of 3.6 and 2.7 Hz. Variations corresponding to the more slowly varying beats near 3 Hz are more difficult to pinpoint, but there are even suggestions of these. In other words, the modal frequencies within the 125 Hz octave band account for the relatively great fluctuations of the 125 Hz decays.

The reason the five decays are not identical or similar can be explained by the fact that the different modal frequencies were not all excited to the same level. The random noise signal constantly changes in amplitude and frequency (within the octave limits). It is entirely fortuitous as to what instantaneous amplitude and frequency were at the time the sound was interrupted to begin the decay. A very smooth low frequency decay could result from a dominant single mode, although with octave bands this is unlikely.

An important indication of the diffusion of sound in this small multitrack studio is given in the low frequency reverberation decays as in Fig. 20.1. If the fluctuations are very great, the diffusion is poor. The smoother the decays, the better the diffusion.

Quantitative evaluation of diffusion conditions are not yet available from such decays, but good qualitative comparisons are not only possible, but part of the arsenal of informed workers in studios.

Diffusion information may also be gleaned from decay curves at higher frequencies. In Fig. 20.2 the broken line indicates a fairly definite suggestion of a second slope. This is probably the result of certain modal frequencies having less contact with the absorbing material in the room (i.e., modes that are less damped) or modal frequencies not fully excited as the decay begins. In this particular case, these modes do not affect things until the sound has decayed 30 dB, hence their effect would probably not be detectable in normal program material.

Variation of T$_{60}$ with Position

Measuring reverberation time at different locations in a studio often reveals small but significant differences in reverberation time. These are usually averaged together for a better statistical description of conditions in the studio.

For example, Fig. 20.5 shows the reverberation time measured at three different microphone positions in the small multitrack studio mentioned in the previous section. It is noticed that the three graphs tend to draw together as frequency is increased. This suggests that such changes in reverberation time at different locations in the same studio are the result of a certain degree of nondiffuse conditions because we know that diffusion is better at high frequencies. Can this method then be used to evaluate the sound diffusion condition in a room?

The Engineering Research Department of the British Broadcasting Corporation asked the same question.[33] In their characteristically thorough way they measured

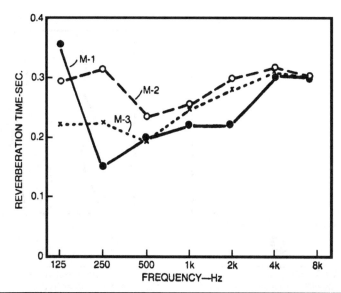

FIGURE 20.5 Variation of reverberation time/frequency graphs with position in a small studio. The modal content of the octaves at different frequencies varies, and the decay rate of the different modes varies. To obtain a statistical picture of the sound field in the room, it is customary to average the measured reverberation times for each frequency at each position.

reverberation time at 100 microphone positions in a 10-foot × 10-foot room. With no absorption material in the room (Fig. 20.6A), very diffuse conditions resulted in reverberation time long, but essentially constant throughout the room.

In Fig. 20.6B the reverberation time contours are shown when one wall was treated. The reverberation time is lower nearer the absorbing surface. They then demonstrated that geometrical diffusing elements on the untreated walls plus one absorbing wall resulted in quite complex contours. The laborious nature of this approach discourages further exploration of the method, although it shows some promise if special instrumentation were devised.

Directional Microphone Method

The signal output of a highly directional microphone in a perfectly diffuse room should be the same no matter where it is pointed, except when pointed at the source of sound. Ribbon microphones, with their figure-8 pattern, have been tried. Parabolic reflectors with a microphone at the focal point have been tried, as well as the line array type of directional microphone.

All of these methods have proved to be rather awkward to use and the results difficult to interpret. The greatest shortcoming of this method, from the point of view of small studios, is that sharp microphone directivity is hard to get at the low frequencies at which diffusion is the greatest problem. The prospect of this method of appraising diffusion in small studios is poor.

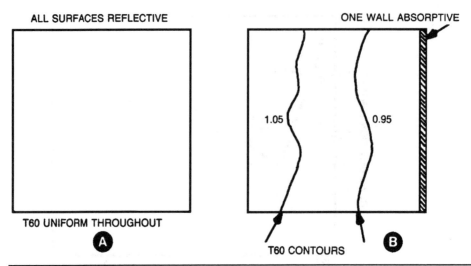

FIGURE 20.6   (A) If all surfaces of a room are 100 percent reflective, the sound field is completely diffuse and the decay rate is the same at every point in the room. (B) If one wall is absorptive, the decay rate varies from point to point in the room. The contours of decay rate tend to be parallel to the treated wall.

Frequency Irregularity

For the last half century there have been many serious efforts to evaluate diffusion in rooms by steady-state transmission measurements.[34, 36] Microphone and loudspeaker positions remain fixed. The constant amplitude swept sine wave signal radiated from the loudspeaker and picked up by the microphone has the room effect impressed upon it. This room influence should reveal something about the room.

Figure 20.7 shows typical frequency response records taken in a music studio of 16,000 cubic foot volume having a reverberation time of about 0.6 second at the 100–300 Hz frequency region under investigation. Two things are very striking:

- The magnitude of the variations
- The differences from position to position in the room

The amazing thing is that fluctuations in point-to-point response of such magnitude occur in studios having the best acoustical treatment and those considered excellent in subjective evaluations. The loudspeaker response in the 100–300 Hz region is included in these recordings, but it remains constant through all the tests.

If such wild fluctuations are to be of any help in evaluating studios, it is necessary to find some method of reducing them to numbers. Bolt has suggested the term *frequency irregularity factor* obtained by adding all the peak levels, subtracting the sum of all the corresponding dip levels, and dividing the difference dB by the number of hertz swept. This frequency irregularity factor, or simply *FI factor*, is in dB/Hz.

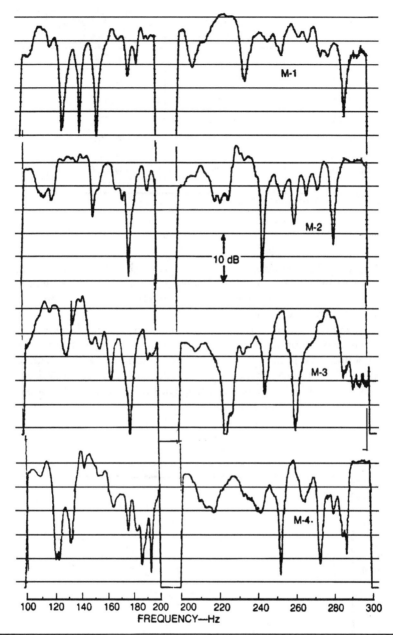

FIGURE 20.7   Typical steady state swept sine frequency response records taken at different positions in a music recording studio having a volume of 16,000 cubic feet and a reverberation time of 0.6 second. The striking variations are an indication that even a well-treated studio falls far short of truly diffuse conditions. At frequencies above 300 Hz, the curves becomes progressively smoother.

Microphone Position	Frequency, 100–200 Hz	Irregularity Factor, 200–300 Hz	dB/Hz 100–300 Hz
M-1	0.777	0.556	0.667
M-2	0.497	0.775	0.626
M-3	0.663	0.852	0.758
M-4	0.764	0.771	0.768
Mean	0.675	0.739	0.705
Standard Deviation	0.129	0.127	0.069

TABLE 20.1   Frequency Irregularity

Applying this procedure to Fig. 20.7 yields the FI factors tabulated in Table 20.1. Comparison of Fl factors for the different microphone positions would seem to tell us that conditions at M-2 are the best, considering the 100–300 Hz range, and that M-1 and M-2 are superior to M-3 and M-4.

Glancing back to Fig. 20.7 would seem to support this. What the FI factors of Table 20.1 tell us about sound diffusion in the room is not so clear. In measuring many studios it was noticed that larger FI factors were commonly associated with longer reverberation times. The 100–300 Hz FI factor for a dozen studios is plotted in Fig. 20.8 against their corresponding reverberation times. The broken line (not a least-squares fit) would seem to indicate a definite relationship.

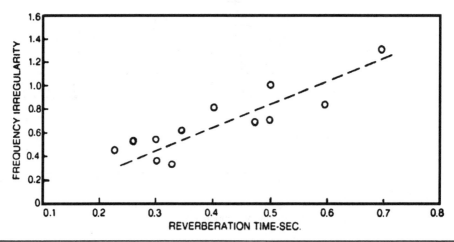

FIGURE 20.8   The variation of measured frequency irregularity factor with studio reverberation time.

In fact, theoretical studies and experimental results have shown that at high frequencies *frequency irregularity* is related only to reverberation time and that it gives no additional information on diffusion of sound in the room. Whether or not this is true in the 100–300 Hz region remains to be seen.

One thing seems to be clear, if at a certain microphone position the swept frequency response is within, say, ±5 dB, that would be a good spot for a narrator to sit.

A corollary to that observation is that it is possible to compare microphone positions by a swept frequency signal test. For such a test a good place for the loudspeaker would be in a corner of the room. If standing waves such as indicated by the runs of Fig. 20.7 exist in what are called well-treated studios, conditions may be less bad in some spots than in others. This is of operational value only if recording in a studio can be done with a single microphone in a fixed position.

Size and Proportions of Room

A minimum studio or control room volume of 1,500 cubic feet has been urged. This is step one toward better diffusion. Rooms smaller than this are often plagued with coloration problems, impossible or impractical to correct. Of course, rooms having substantially greater volumes but still in the general small room category have plenty of diffusion problems also, but the chances of achieving satisfactory conditions by the application of the methods to be described are better.

By making a room large in terms of the wavelength of the sound to be recorded in it means that the modal frequencies will be closer together which means improved diffusion. Most of the methods of diffusing sound in a room to be considered later are most effective at the higher audio frequencies. Optimizing the proportions of the room is one of the most effective ways to improve diffusion at the low end of the audible band. There are a number of steps in the acoustical treatment of a room which tend toward better diffusion of sound in the room, once the major basic matter of room size and proportions are set (review Chapter 6 in this regard).

Distribution of Absorbing Materials

Numerous controlled experiments and practical experience have demonstrated that concentrating the required absorbing material on one or two surfaces of a room is an acoustical abomination. Common sense emphasizes that this procedure often leaves some opposing or parallel walls untreated, producing some axial resonances.

The application of absorbing material in patches has been established as far superior to application in fewer large areas. This accounts for the proliferation of wall modules and sectionalized ceilings in studio designs in this book. The patches of absorbing material may be distributed by determining the areas of the N-S, E-W, and vertical pairs of surfaces and dividing the material between the three axial modes proportionally. At least, this is a respectable criterion to use as a rough guide, even though practical considerations like doors, observation windows, and floor coverings demand compromise.

Another important contribution of patches of absorbing material is that diffraction of sound, especially at the higher frequencies, takes place at the edges of each patch.

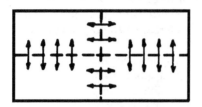

Figure 20.9   The broken lines indicate the sound pressure modal contours for a simple mode in rectangular and trapezoidal rooms. The small arrows represent the direction of particle motion. It is obvious that the trapezoidal room shape contributes something to diffusion.

Such diffraction contributes helpfully to diffusion. Placing absorbent where it acts on every axial, tangential and oblique mode is smart, remembering that all modes terminate in corners. Distributing absorbent in patches contributes to absorption efficiency as well as diffusion. Diffraction effects act as though the absorbent "sucks" sound energy from the surrounding reflective area which, in effect, increases its absorption coefficient.

Splayed Walls

The conventional wisdom among studio people has long held that splayed walls aid in diffusion of sound. Splayed walls do have the ability to help in the control of flutter echoes between opposite reflective surfaces, but do they really contribute significantly to diffusion? Model experiments have shown that the frequency irregularity factor is reduced with walls canted 5 percent but there is some question of this applying to practical studios with less smooth walls. The BBC made subjective tests in which experienced listeners listened with and without splayed walls with patches of absorbent on them and the results were inconclusive.

Rectangular and trapezoidal room shapes are illustrated in Fig. 20.9. The broken lines represent the sound pressure modal contours for a simple mode for both shapes. The small arrows represent the directions of particle motion in the two cases. The trapezoidal room shape most certainly has contributed something to diffusion, but the magnitude of the contribution is small for walls splayed the usual one part in 10.

It appears that justification of splayed walls must come from reduction of flutter echoes rather than improvement of diffusion. As there are other ways to prevent flutter echoes (such as patches of absorbent) it would seem that tearing an existing building apart to cant walls might be ill-advised. In new construction, however, inclining the walls might cost very little.

Resonator Diffusion

In low frequency Helmholtz resonators, what happens to incident sound energy that is not absorbed by the system? It is scattered and scattering contributes to diffusion of sound energy in the room. This is not true of the porous type of absorbers in which energy not absorbed is reflected from the backing surface. Remember, however, that high frequency energy not affected by a perforated or slat low frequency absorber can be reflected and contribute to a flutter echo problem. This would suggest facing such units with a high frequency absorber or inclining its surface.

Geometrical Diffusers

There have been numerous, and presumably effective, geometrical protuberances employed in studios to diffuse sound. Common among these are semicylindrical (poly) diffusers[37] and diffusers of rectangular and triangular cross section. The polycylindrical surface has been widely applied in studios, not only for its low frequency absorption, but for its ability to take sound arriving from a given direction and reradiate it through an angle of 100 degrees to 120 degrees. This contributes positively to diffusion in the room.

In spite of the salutary effect of polycylindrical elements, both controlled experiments[38] and theoretical studies[39] have demonstrated the marked superiority of rectangular protrusions over both cylindrical and triangular. The rectangular protrusions produce some effect when their depth is as shallow as one-seventh of the wavelength. Thus a rectangular element 6-inches deep has some effect down to approximately 325 Hz. This effect works to change the normal modes of a smooth-walled room. The cylindrical and triangular projections also do this, but to a lesser degree.

The acoustical distinguishing feature that sets the rectangular apart is the fact that it has finite portions perpendicular to the wall on which it is mounted. It has the ability of breaking up concentrations of modal frequencies better than cylindrical or triangular protrusions, of reducing the magnitude of dominant modes, and lowering the frequency irregularity in swept sine transmission. This provides some support for the proliferation of the not-too-beautiful wall modules in studio designs already considered.

Other diffusing elements found in every studio, control room, and listening room are people, tables, chairs, door and window frames, and equipment of every sort.

Diffraction Grating Diffusers

The optical diffraction grating can break down a beam of sunlight into all the colors of the rainbow. The type of grating used in optical studies has microscopically fine, parallel lines ruled on glass. Inexpensive plastic replicas are available as toys and the principle is applied in colorful advertising signs.

What has this to do with acoustics? Dr. Manfred Schroeder of the AT&T Bell Laboratories has made the connection in a very positive way.[44] Imagine a surface with a series of long, narrow, parallel wells, or grooves, on it. The wells are of either fixed or of varying depth, but of constant width. A sound ray impinging on this surface finds itself interacting with reflections from the bottoms of the wells. The phase (or time relationship) of the reflections from the wells varies with the depth of the wells; hence the arriving sound ray must interact with well reflections delayed varying amounts. Dr. Schroeder related the well depths to various mathematical sequences [45] which result in sound coming from a given direction to be scattered throughout 180 degrees. In this way mysterious mathematical names have come to identify different types of these diffusers: quadratic residue sequence, primitive root sequence, maximum length sequence, etc. They all scatter

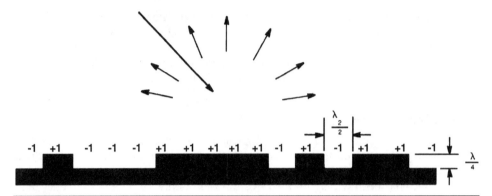

FIGURE 20.10   A diffusing surface based on a maximum length sequence. The grooves of quarter wavelength depth offer a reflection coefficient of –1, the high spots +1. Good diffusion a half octave above and below the design frequency is realized.

sound better than what was available in the past, but some are more efficient for specific tasks than others.

Figure 20.10 shows a specific type of diffraction grating sound diffuser, the kind Dr. Schroeder first tried. He needed different reflection coefficients of + and −1. A groove a quarter wavelength deep gives a reflection coefficient of −1 while no groove at all gives a reflection coefficient of + 1. This emphasizes that a design frequency must first be selected. Let us take 1000 Hz as our design frequency. The wavelength of a 1000 Hz tone is 13.56 inches, a half wavelength is 6.68 inches, and a quarter wavelength is 3.39 inches. As shown in Fig. 20.10 the well depth must then be 3.39 inches and the unit well width 6.78 inches. It can be made of wood, metal, or plastic or any other substance which is a good reflector of sound. The question now is, "Will it diffuse only 1000 Hz sound?" This maximum length sequence diffuser will work well over about one octave, a half octave above and a half octave below 1000 Hz. The diffused energy is confined to a hemidisk at right angles to the face of the diffuser.

The more complex grating diffusers perform much better. A quadratic residue diffuser is shown in Fig. 20.11. Here the sequence length is 17 and the relative well depths are as indicated and shown graphically. Two sequences are shown in the figure. Thin metal separators maintain the identity of each well. The width of the wells is about a half wavelength at the shortest wavelength to be scattered effectively. The maximum well depth is determined by the longest wavelength to be diffused. Although more difficult to construct, quadratic residue diffusers scatter sound effectively over most of the audible band.

Grating Diffuser Application

In the past the designer and builder of a studio, control room, or listening room had only reflection and absorption to work with; diffusion was largely out of reach. That has now changed. Standard diffusing units are now commercially available Such units are being installed in recording studios and control rooms as well as concert halls, churches, and home listening rooms.

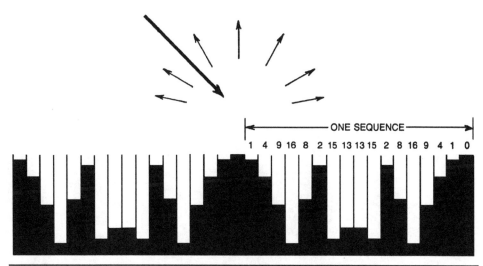

FIGURE 20.11 A diffusing surface based on a quadratic residue sequence. Good diffusion over most of the audible band can be achieved with this type of diffuser.

Diffraction grating sound diffusers are also being applied in small budget recording facilities. They are not a cure-all for the acoustical problems of small rooms but they do add a new tool to the designer's toolbox.

Are the principles elucidated in earlier chapters of this book outdated by the coming of effective diffusing elements? Not at all, but the prospects of better sound from small rooms are much improved.[46-53]

NOTE: *If this explanation is not technical enough, you can download Dr. Manfred Schroeder's original* J. Acoust. Soc. Am. *April 1979 paper titled "Binaural Dissimilarity and Optimum ceilings for Concert Halls" or get the full deal by purchasing* Acoustic Absorbers and Diffusers, Theory, Design and Application *by Trevor J. Cox and Peter D'Antonio.*

New Remedies to Common Acoustic Problems

An Introduction to Today's Premanufactured Acoustic Panels and Systems

Ionce wrote:

> In this modern, money-crazed world of fetch-it icons and buy-me buttons, you have to be more vigilant than ever about performance claims from overzealous advertising or media "whiz-kids" that are simply too good to be true. As an example, take these actual quotations from an editorial blurb about an acoustical correction panel system that ended up becoming "a complete solution to great sound" with which "users can convert *any* ordinary room into a *high performance* recording/audio work-space" (my italics). I know this did not come from the manufacturer because foam, or any other material for that matter cannot make up for serious isolation or dimensional design flaws. If anyone were to make claims like these in front of a gathering of knowledgeable acousticians such as those in the Acoustical Society of America, they might as well wear a pink electrocuted-hair wig, a large red stick-on rubber nose, and a pair of size 28 shoes because they are going to be rolling in the aisles anyway.

I am not about to eat my words because sound transmission loss and solutions to gross acoustic anomalies cannot be provided in cookie cutter "fix-it-all" prepackaged boxes. However the fact is that today you can do quality audio production work in the worst room possible, a perfect cube, if you have the $$$.

Time for a Little Perspective

Thirty years ago, folks who wanted to get involved with owning a recording studio were right-up-front, looking at both the cost of the hardware along with the building of the studio, and this could easily add up to a half million dollars or more. The mixing console alone could run $100,000, and the check you signed to purchase it would barely be out of your hand when, according to almost every possible buyer, the "used" value of that piece of equipment would be $50,000 at most! It was the same story with all the tape recorders and the outboard gear you just *had* to have. How could they justify these expenditures? Simply as the *rental* cost of doing business. Look, if you bought an SSL console for $250,000 because it would increase your bookings by 50%, and this meant an additional $2000 a month (or $24000 per year), and you ran it for five years before selling it for $100.000, the cost of your *renting* that board would only be (250,000 − 120000 + 100,000 = $30,000) or $6,000 per year. Fat chance! But it was all considered part of the cost of doing business.

In the past, the investment recording studio owners put into their facilities for noise reduction, acoustic treatment, HVAC, and furnishings was often at least the same as they had to put up for their electronic recording equipment, and it was not unusual for the major studios to lay down a million dollars for the brick and mortar side of the installation. Today laptop recorders eliminate a bunch of the zeros from the equipment side of the price tag. In fact this hardware comparatively costs beans, so there are a zillion "recording studio owners" out there. Problem is, they are also trying to match the cost of the equipment with the cost of building their room(s)! "Beans" for soundproofing, transmission loss and acoustics may not make any difference when you are overdubbing an electronic keyboard, but your "client base" is going to be severely limited.

Thing is, acoustics becomes even *more* critical in *small, one-room* facilities where the "bedroom" recordist must utilize a too-small-sized room as both a critical listening environment *and* as a space conducive to aiding the propagation of acoustic instruments. This *will* be a tough go to say the least, but take it from me, thirty years ago it was insurmountable. Today it may cost you, but it is now a "can-do" situation with most any sized room.

With today's economic conditions it is often just not possible to build a new brick and mortar (sheet rock and 2 × 4) world-class recording studio. However adjusting an existing room to achieve professional sound quality is now more feasible than ever. Unfortunately in most cases we are not talking about a studio and a control room, but combining the control room and studio into a one room "facility." The acoustical demands placed on recording as opposed to mixing differ widely. Critical listening requires that there be little to no "acoustic liveness" between the speakers and the listening position, while recording requires an environment that is somewhat reverberant so as to be conducive to the propagation of sound from music instruments (except when tracking electronic instruments where no acoustics are involved).

Fact is, small size means larger acoustic problems, yet in most cases both critical listening during mixdown as well as live recording while monitoring with headphones is being attempted in less than adequately sized, acoustically untreated rooms.

Luckily things have changed to the point where you can actually forget about modifying the dimensions of your room to correct modal problems because it is now possible to acoustically treat small rooms by simply adding premanufactured acoustic panels and systems to control unwanted reflections or remove excessive resonance. You may not be surprised by this if you grew up with it, but believe me, forty years ago it was unheard of.

Oh, there were companies manufacturing acoustic absorption panels back then, but as soon as you said the words "recording studio" all conversation ended just about instantaneously. These folks were interested in auditoriums, airports, schools, government buildings (think OSHA) and factory floors where dozens upon dozens of panels would be utilized. Not some "studio-thing" that *might* be interested in purchasing a half dozen at most! And *nobody* was building floor and ceiling vibration mounts designed specifically for recording studios, let alone diffusers. What changed? *You* did!

Back then there were *at most* a dozen viable working recording studios in an average metropolitan area, and a few hundred or so individual recordists around. Today due to the quality of "laptop" recording technology there are millions of recording studios and a kazillion recordists. That's a huge market and it seems everybody and his brother wants a share. And market demand is *high*!

Just about the only acoustical advantage a small room has is its short reverb time, (moderately sized rooms seldom have reverberation times higher than 0.5 seconds.) On the other hand, it is difficult to achieve quality sound in small rooms simply because the dimensions of the walls, floor and ceiling are very close to the half wavelengths of the low frequencies being reproduced. Here a six-foot spacing between two walls will produce a fundamental mode of 94 Hz, and its harmonics will add in mud at 188 and 280 Hz. This sets up numerous acoustic modes that cause multiple reflections and standing waves which not only distort the sound, but cause peaks and dips in level throughout the space. As already stated, it can be bad enough that some locations will not be fit for microphone placement.

Smaller rooms are all commonly plagued by the same problems, low-frequency build up (boomyness) due to multiple modes in the same frequency area, flutter echoes caused by parallel walls and mid- to high-frequency beaming.

In the past, several methods were employed to correct these deficiencies, such as building bass traps and mounting foam or fiberglass high frequency absorption panels on the walls. Traditional bass traps helped but required the building of a large diaphragmatic panel in front of an air space lined with fiberglass that was often the size of a closet. This equated to a big loss in the room's cubic footage. Thankfully mineral wool could be used instead of fiberglass so that this space could by used as a tape machine room, tape library, or mic storage room. I've even thrown a small guitar combo amp in them when recording basics to reduce track bleed.

In the past, the only alternative was to do the math to calculate the likelihood of problematic modes and then change the room's dimensions accordingly by building out or splaying walls and deadening surface areas. This was a rather costly affair (see wall construction in Part II) and again resulted in a loss of room size. Granted some of the larger bass-trap-sized areas could be used for storage, the areas between could be splayed, and preexisting walls could be filled with barrier material to help reduce sound transmission to and from neighbors. I have even run HVAC ducts and electric mains power through them and once built an equipment rack in one. But the fact remains that it is still a loss of much needed space.

It does not cause as much of a decrease in room size, and it is less expensive to use nothing but absorbent materials to "suck up" extra reflections, but believe me, the end result is a room that is not conducive to critical listening or playing music instruments.

Any way you look at it, you are dealing with limitations that would not be a factor if you had a recording space larger than 4000 cubic feet. Luckily, today home studio recordists can forget about readjusting their room's dimensions to correct modes and instead just use premanufactured acoustical treatments to control them.

Let's Take a Look at Some of These "Treatments"

Okay, right off the bat, I must make something perfectly clear. *Any* prices listed are subject to change. They were included only to give you an *idea* of the cost. Furthermore, the following "overviews" of acoustical products are not meant to be taken as reviews.

1. I no longer review professional audio products.

2. You cannot actually pass judgment on *any* acoustical product because what might be worthless in one situation can excel in others.

3. By giving you examples of actual acoustic products and going over the method of seeking out specifications, laying out the pros and cons, and determining where they would be of value, I will be showing you how to do it for yourself.

Always remember that while there are many "This is *not* the way to do it" in Acoustics! there are very few "This is the *only* way to do it." As an example let's review some reflection theory.

Axial modes reflect from two surfaces, tangent modes three, and oblique modes six. Tangent modes have half the power of axial modes, and oblique modes are down one quarter of that.

So it is obviously very important to first and foremost go after the axial modes. Which axial modes? The monitor in the mirror on the side wall test will give you the number one cause of early reflection phase distortion, comb filtering, and flutter echo. Next comes the ceiling and the floor. Any path that goes from the speaker to a boundary and then directly to your ears *must* be dealt with. The best correction is diffusion because it does not rob any of your high end. Next come sectional wall splaying panels because they also retain the high-frequency content, redirect the reflection around the listening position, and may even add a bit of low-frequency absorption. However if you cannot afford the former and the latter is beyond your DIY capabilities, then you must at least put up some absorptive material to attenuate those early reflections from your side walls.

I put a good deal of time into the following section; space was limited so I had to choose products that helped make specific points, such as how important it is for you to dig deeper and deeper into the manufacturer's literature as well as call them with any remaining questions that need clarification. Believe me, these folks have *no* problem with potential customers requesting information!

RPG Products

We'll start with RPG Inc. because the whole field of commercially available diffusion panels began with their work back in the 1980s and they've been at the forefront ever since. Most people think that other manufacturers "steal" ideas from RPG. That's not really true because RPG gave it all away by writing a book on the subject telling everyone exactly how it is accomplished. You see they do not *need* your business because their panels are made from furniture grade materials and are used in corporate headquarters, high power meeting rooms, the lecture halls of prestigious institutions of higher learning, and other places such as world class recording studios where money is not an object. Yet they *continually* try to help the small studio owner by researching newer materials and methods of manufacturing.

Here's a quick run through of a couple of RPG's products to give you an idea of their product line. But *do* go online to rpginc.com for the full catalog.

The RPG Binary Amplitude Diffsorber

RPG ceiling clouds reduce axial reflections from the ceiling, and they are perfect right above the console. With RPG's Binary Amplitude Diffsorber™ (BAD™) you have a flat, fairly thin panel doing the job of absorbing low frequencies while simultaneously diffusing the highs. *No* construction budget, *no* studio down time, *no* loss of cubic footage and *no* itch. They *are* bad. In a nutshell: Sound freely passes through the fabric facing. Then it impinges on a ridged panel with mathematically calculated perforations that are designed to let the bass through to the absorbent material behind it while diffusing high frequencies outward. Like I said you *have* to go online to rpginc.com. You'll get a mega acoustic education free of charge. Now let's take a more in-depth look at some of their acoustic products.

The RPG Quadratic Residue Diffusor (QRD) Family

The QRD 734 is an early RPG diffuser dating all the way back to the 1980s, which is famous for its even sound diffusion (see Fig. 21.1). It resulted from RPG's never ending research into *reflection phase grating*, which provides sound diffusion from a series of divided reflective wells of equal width and differing depths. The depth of the reflective wells of the QRD is based on mathematical number theory sequences. Furthermore, depending on its orientation, it will disperse sound in either the vertical or horizontal plane.

However because the QRD is made of wood, it can be bit heavy when mounting multiple units on a wall, and it can be difficult to ceiling mount. So RPG simply came out with a molded, cost-effective copy of the QRD. These are called Formedffusors (Fig. 21.2) and because they are molded from thin Kydex, they are very light weight. In addition to being much easier to wall mount, they are sized just under 2' × 2' (1'11.625") square) so they can be dropped right into hung ceiling frames. They offer the same broadband dispersion as the original QRD, and depending on their orientation, these panels also disperse sound in either the horizontal or vertical plane uniformly for *all* angles of incidence. Due to its thin thermoformed body, it can function as an active panel absorber and thus provides some low- to mid-frequency diaphragmatic absorption as well (see Fig. 21.3). They can be painted any color you want, if you are not partial to white, and are fairly inexpensive to boot. A low-cost, lightweight, broadband diffuser that also furnishes a bit of low-frequency control? Sounds like just the thing the average small studio is in need of. But my favorite way to use them is to insert them right into sheetrock walls for virtually zero loss of room space. Now wait a minute, before you start smashing holes in your walls, you need to make sure they are mounted to studs that are on 24" centers *not* 16". This can be accomplished by using an inexpensive stud finder from your local hardware store. With both wood and metal studs, the depth is

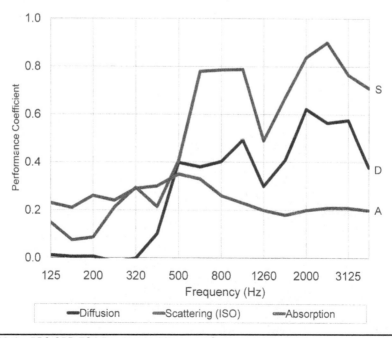

FIGURE 21.1   RPG QRD 734 Random Incidence performance data graph.

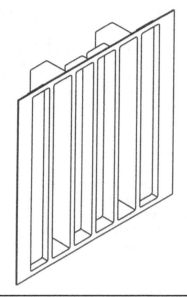

FIGURE 21.2   RPG Formedffusor cutsheet isometric drawing.

generally three and a half inches. The depth of a Formedffusor panel is 3.875″. Therefore if you use 2″ × 2″ (1.5″ square) wood furring strips to frame around the opening, you will end up with five inches of space, which is just enough room to mount a one-inch-thick fiberglass board on the rear wall to help with low-frequency absorption and leave one quarter of an inch of air space so nothing comes in contact with the back of the panel. Sheet rock saws make this a breeze and outside of that one and a half inch there is *no* real loss of space.

RPG also added the Diviewsor™, a clear Plexiglas version of the QRD, to its line, which can be hung in front of the control room window, providing much needed diffu-

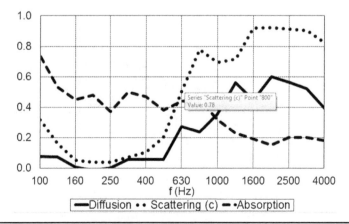

FIGURE 21.3   RPG Formedffusor Random Incidence coefficients graph.

sion to that area while still allowing visual contact. Then there's the original "Two Birds" panel. The Abffusor is just like the QRD except it is an absorption phase grating panel. Here the combination of diffusion and absorption helps end flutter echo problems. This panel absorbs sound in two ways. The first is the reduction of sound energy by converting it to heat by friction in the fibers of its porous cloth covering. The second is from "viscous losses due to high particle velocity flows across the well dividers induced by the pressure differences in adjacent wells." This results in added low-frequency absorption, so now you get the same even diffusion of sound *with* some added absorption. Abffusors mounted on the side walls minimize flutter echoes by diffusing the sound while providing some broadband absorption. Two birds? No actually three, because they not only look good, (they come in a wide variety of custom fabrics), they also match the QRD so that the two can be used together with eye appealing results. The combination of diffusion and absorption can mean an end of flutter echoes, while not overly reducing the room's reverberation time. The graph (Fig. 21.4) shows the absorption capabilities of the Abffusor with both the direct on-the-wall A mounting and the E 405 mounting, which has a rear cavity depth of 16 inches. Here the increase in low-frequency absorption when mounted with the rear air cavity as opposed to when attached directly to a wall is obvious. I've also listed the inter-octave frequency absorption coefficients in Table 21.1 to give a better indication of the Abffusor's absorption capabilities. As you can see they are impressive.

But RPG continued their research and found that the periodic repetition of the exact same shape, as when QRD panels are lined up side by side to diffuse longer walls, caused lobbing "in specific diffraction directions," which translates to less than perfect dispersion. So then they came out with a new "modulated optimized" diffuser called the Modffusor™. This panel is different from the QRD in that it is asymmetrically shaped, and when flipped over 180 degrees, becomes a mirror image of itself. Each of these two orientations

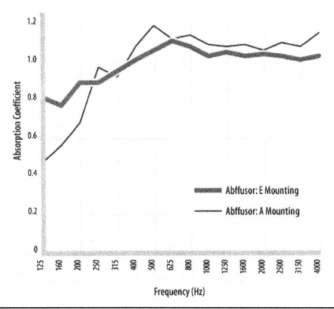

FIGURE 21.4 RPG Abffusor Absorption Coefficients A and E mount graph.

Frequency in Hz	E Mounting	A Mounting
125	0.82	0.45
160	0.82	0.48
200	0.78	0.57
250	0.90	0.69
315	0.90	0.98
400	0.96	0.93
500	1.07	1.20
625	1.12	1.13
800	1.09	1.15
1000	1.04	1.10
1250	1.06	1.09
1600	1.04	1.10
2000	1.05	1.07
2500	1.04	1.11
3150	1.02	1.09
4000	1.04	1.16

TABLE 21.1   Abffusor Inter-Octave Absorption Coefficients

can be given a binary designation of either "0" or "1" making it easier for them to be arranged in an optimal sequence for different array sizes. One example given is 0, 0, 1, 0, 1, 1 (Fig. 21.5). The result is better low-frequency response and reduced lobbing, and not only is there greater diffusion and scattering, but both are more even with frequency from a panel that ended up being 16% thinner. Talk about turning the caldron! That's quite an improvement. Check out the specifications for these panels yourself (Fig. 21.6).

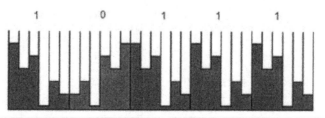

FIGURE 21.5   Modffusor Type "0" and "1" unit drawing.

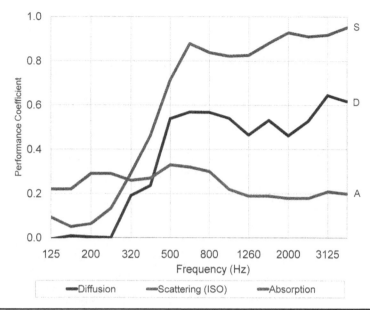

Figure 21.6   RPG Modffusor Random Incidence performance data graph.

RPG Diffractal Diffusers

"Fractals" or fractionally smaller-sized diffusers, are nested in the wells of reflection phase grating diffusers to produce a *fractal diffuser*.

Dimensional nested sound diffusers consist of a group of high-mid-frequency quadratic residue diffusers nested into the wells of a low-mid-frequency reflection phase grating diffuser, which in turn are nested in the bottom of low-frequency wells (Fig. 21.7). Each nested diffuser has a specific frequency range and a wide area of coverage. This grouping of different-sized diffusers provides uniform scattering over an extended bandwidth. The performance data graph (Fig. 21.8) provides a clear picture of the overlapping mid- and high-frequency bands. Thus, dimensional nested sound diffusers provide a broader diffusion spectrum from a single diffuser. Straight out, RPG has gone beyond what I thought was possible in terms of increasing diffuser bandwidth. Their line of diffractal panels includes the DFR-72MH, which measures 2' × 4' and has a 400 Hz to 17 kHz bandwidth, their DFR-82LM, which measures 3' × 11' and has a 100 Hz to 5 kHz bandwidth, *but* their DFR-83LMH, which measures 3' × 16', has a bandwidth that ranges from 100 Hz to 17 kHz. While these things are designed for concert halls not studios, I'm still speechless!

The FlutterFree Family of Diffusers

We all know that parallel walls can cause nasty flutter echoes, and that the old school way of adding copious amounts of absorptive material is unacceptable due to the resultant loss in high-frequency content. RPG's "Flutter-Free" diffusers wipe out flutter echoes without high frequency absorption, thus your room is more natural sounding (see Figs. 21.9 and 21.10).

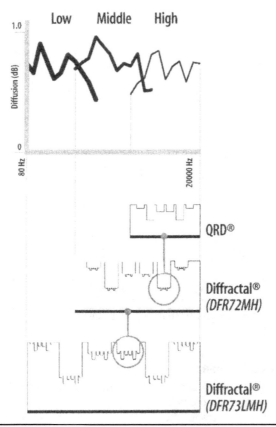

FIGURE 21.7 RPG Diffractal layout drawing.

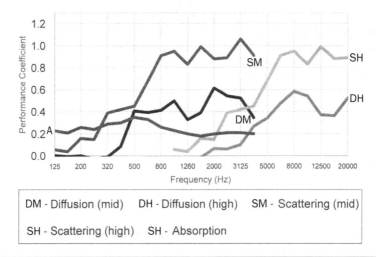

DM - Diffusion (mid)   DH - Diffusion (high)   SM - Scattering (mid)

SH - Scattering (high)   SH - Absorption

FIGURE 21.8 RPG Diffractal Random Incidence performance data graph.

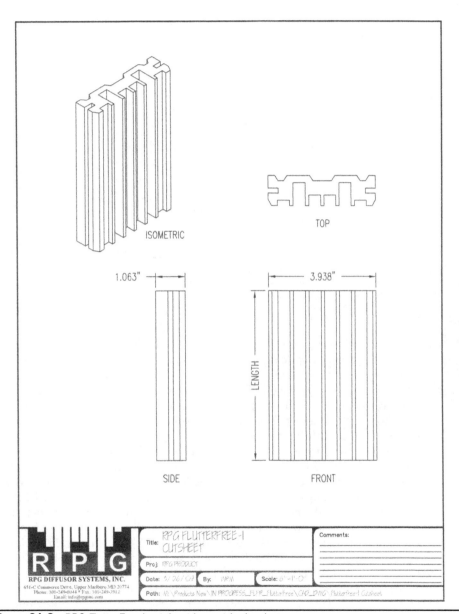

FIGURE 21.9   RPG FlutterFree-I cutsheet isometric drawing.

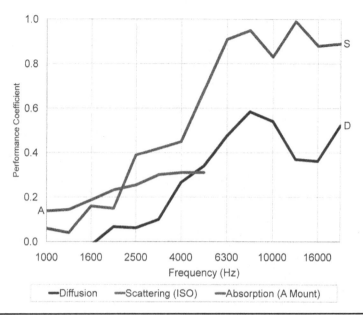

FIGURE 21.10 RPG FlutterFree Random Incidence performance data graph.

These panels seem very simple, and the name itself sums up their function. They're wood planks, 4 inches wide × 1 inch thick by either 4 or 8 feet long. Mounted on a wall, they do exactly what it states they do, eliminate flutter echoes. So why would anyone mess around with it? Well again RPG found that when placed repetitively side by side lobbing can occur, which ends up providing less than ideal diffusion. Furthermore, due to the short depth of the wells, diffusion was effective only at higher frequencies. First, just as with the Modffusor, they made the layout of the facing less symmetrical by putting the deep wells to one size and the shallow to the other. Then on the sides they added an "L" shaped undercut with the bottom of the "L" facing into the plank. This not only lengthened the depth of those wells, but when placed side by side, the two ends formed a large upside down "T" shaped well (see Figs. 21.11 and 21.12).

This broadened the bandwidth of the diffusion, and by flipping panels a half turn (180 degrees) they become a mirror image of each other. Next came a binary coding where one orientation is a "0" and the other is a "1," which aids in achieving their proper arrangement. This ended the lobbing problem, and the new FlutterFree became FlutterFree-T. Was RPG finished? You kidding?

They then started to mount them separated by $\frac{1}{16}$ of an inch so that they made up the slats of a slotted Helmholtz resonator shown in Fig. 21.13. This worked out nicely, so they started to mill slots across the back side of the planks, which made rectangular shaped holes at the bottom of the deepest wells for the standard Helmholtz effect. Check out the specifications and graphs in Fig. 21.14. If that isn't impressive enough, the panels can also be supplied in the materials and finishes shown in the table on page 328.

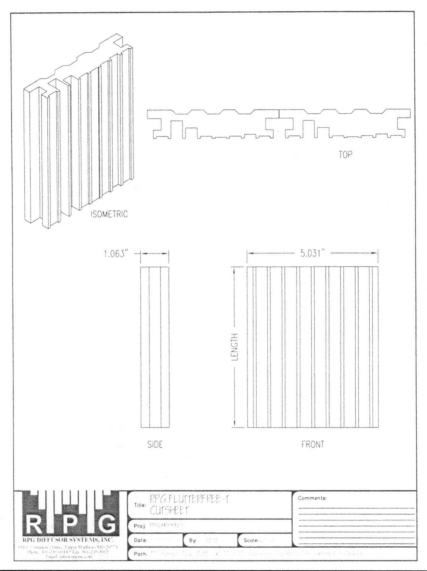

FIGURE 21.11   RPG FlutterFree-T cutsheet isometric drawing. The two side-by-side top views plainly show the undercut "T."

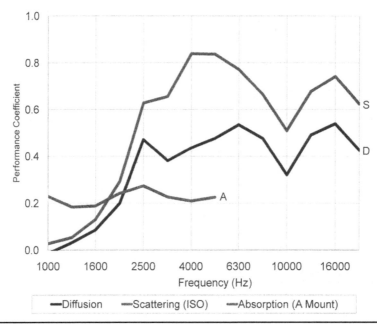

FIGURE **21.12**   RPG FlutterFree Random Incidence performance data graph.

Materials		Finishes	
AC	American Cherry	L	Lacquered
HM	Honduran Mahogany	P	Painted
P	Poplar	R	Resin-Filled
RO	Red Oak	S	Stained Only
WA	White Ash	T	Stained and Lacquered
WB	White Birch	U	Unfinished
WM	White Maple		
XX	Custom Wood Species		

The cut sheet drawings provided show the original FlutterFree-I (Fig. 21.9), the FlutterFree-T (Fig. 21.11), and the FlutterFree-I Slotted (Fig. 21.13). The specifications and graphs in Fig. 21.14 show the improvements made in both diffusion and absorption by the changes. I also included more detailed drawings of the three mounting methods utilized (see Fig. 21.15). In the "A" mount, there is one inch of air space behind the panel. In the "C" mount, that inch of air space is filled with a one-inch-thick panel of 6-lb.-per-foot density fiberglass board. In the "D" mount, a one-inch-thick, 6-lb.-per-foot density fiberglass board sits up against the back of the panel and behind that is 3½ inches of air space. When matched with the graphs you can readily see the effect the mounting has on absorption. Bottom line? FlutterFree looks great, wipes out flutter

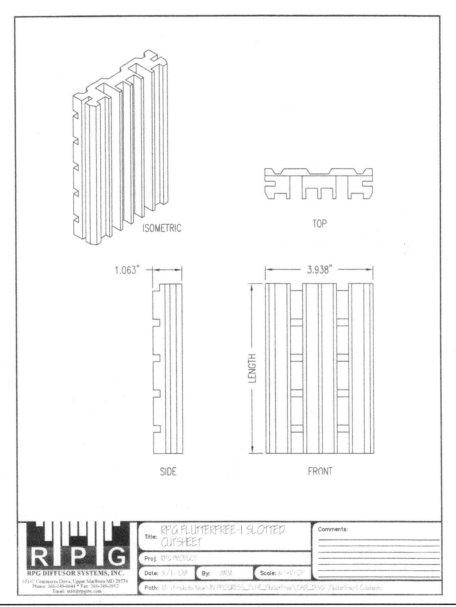

ISOMETRIC

TOP

1.063"

3.938"

LENGTH

SIDE

FRONT

RPG DIFFUSOR SYSTEMS, INC.
651-C Commerce Drive, Upper Marlboro MD 20774
Phone: 301-249-0044 * Fax: 301-249-3912
Email: info@rpginc.com

Title: RPG FLUTTERFREE-I SLOTTED CUTSHEET

Proj: RPG PRODUCT

Date: 9/1/09   By: WRM   Scale: 6"=1'-0"

Path: M:\Products New\IN PROGRESS_FLTR_FlutterFree\CAD_DWG\Flutterfree-I Cutsheet

Comments:

Figure 21.13   RPG FlutterFree-I Slotted, cutsheet isometric drawing.

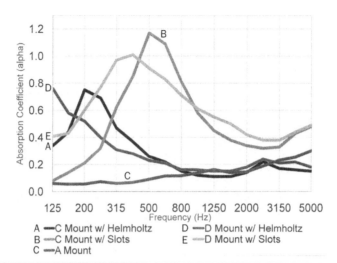

A ▬C Mount w/ Helmholtz   D ▬D Mount w/ Helmholtz
B ▬C Mount w/ Slots   E ▬D Mount w/ Slots
C ▬A Mount

FIGURE 21.14 RPG FlutterFree Slotted, Random Incidence performance data graph.

without high frequency absorption, and if used in a Helmholtz mounting, can also reduce bass buildup significantly to get you *"Three Birds!"*

RPG'S Abflector™

Someone told me that the Abflector absorbed low frequencies and at the same time reflected high frequencies around the listening area. I thought that this way to help stop early first order reflections from causing flutter echoes and phase anomalies was ingenious (see Fig. 21.16).

Splaying walls helps to cure flutter between two parallel walls. But it can be a very expensive undertaking except during initial construction, not to mention studio downtime and the cleaning up of debris. Furthermore, because the amount of splaying often depends more on the available space that can be given up as opposed to tried and true mathematical formulas, the end result will not be known until the project is finished.

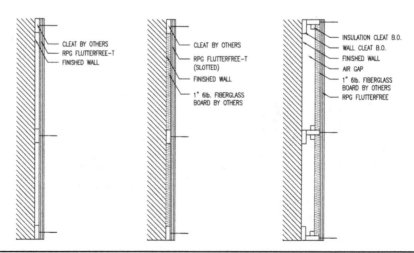

CLEAT BY OTHERS
RPG FLUTTERFREE-T
FINISHED WALL

CLEAT BY OTHERS
RPG FLUTTERFREE-T
(SLOTTED)
FINISHED WALL
1" 6lb. FIBERGLASS
BOARD BY OTHERS

INSULATION CLEAT B.O.
WALL CLEAT B.O.
FINISHED WALL
AIR GAP
1" 6lb. FIBERGLASS
BOARD BY OTHERS
RPG FLUTTERFREE

FIGURE 21.15 Details showing FlutterFree-T with A, C, and D mountings.

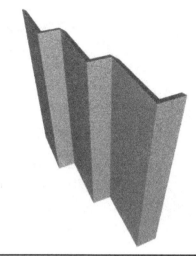

FIGURE 21.16   RPG Abflector.

Premanufactured panels come with specifications, dispersion charts, absorption coefficients, and easy-to-use mounting fixtures. No real downtime, with a known result before purchase! I have splayed whole walls and even partial sections of a wall to get rid of flutter echoes, but a portable, tunable array of sectional splayed panels is so simple and so promising that I am surprised nobody thought of it prior to this. Then I received the specifications (Fig. 21-17). It was not only disappointing, but it also left me scratching my head over RPG's thinking. Look at the absorption coefficients in the following table:

Hz	Absorption Coefficient
125	0.46
160	0.55
200	0.68
250	0.69
315	0.94
400	1.09
500	1.13
630	1.20
800	1.18
1000	1.15
1250	1.16
1600	1.18
2000	1.16
2500	1.18
3150	1.18
4000	1.17

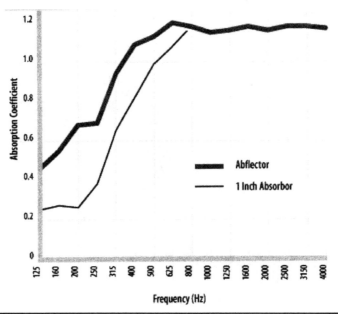

FIGURE 21.17 RPG Abflector Absorption Coefficient graph.

This panel absorbs *some* bass, but at 400 Hz and above, it is a perfect "open window" absorber. I understand that it is nearly impossible to diffuse or reflect bass. The wavelengths are so large that they just go around obstructions. However, active panel absorbers can eliminate a great deal of bass. Fairly thin, cloth-covered fiberboard panels do not eat much bass but they are *very* effective at swallowing up your high end.

So why bother to bring it up?

Because it is still a great idea and I think you should steal it. Start with Fig. 21.18, but instead of using the fiberboard as a panel, mount it flat on the wall behind a plywood

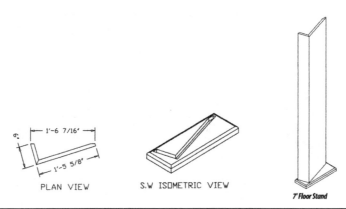

FIGURE 21.18 RPG Abflector top, base, and 7' full-length plain view.

panel. Go back to the design jig for active plywood absorber panels in Part I. Start with a somewhat wide bandwidth like "C" at 100 Hz. The spacing required is only 8 inches, which will give a two-foot-wide panel a good deflection angle. You can first check it out with cardboard, but you'll want to use at least ¼-inch plywood for stability. Hinge mount the rear side to your wall with at least four hinges, and copy the mounting bass that RPG uses if it is going to be floor mounted. Once you have the angle of your deflector figured out, you can add a length of wood between the top front corner and the wall to keep its positioning repeatable. If you rip a 4' × 8' sheet of plywood across at the midpoint (four feet up), and stack the two remaining 4' × 4' pieces and rip them at the two-foot mid-point, you end up with four 2' × 4' panels. Pick the best-looking side of each one and varnish it.

Lose a bit of bass muddiness, deflect first order high-frequency wall reflections around the listening position, and add some nice looking wood panels to your studio, for very little cost. Could be a *three birder*!

The RPG Modex Corner

Low-frequency modes used to require the addition of large amounts of absorption material. The result was often a greater decrease in the high-frequency content than in the low. Additionally "dead" rooms, while they can sound okay to a microphone, are not what you would consider to be an acoustical environment favorable to the propagation of sound from a music instrument. The only reason it worked at all was because all the musicians were wearing headphones providing them with a mix of artificial reverberation added to their instruments.

Today with the use of premade units like the Modex Corner from RPG, the bottom end can literally be sucked out while the high end is left alone.

Talk about intelligent designs, the average home studio is plagued by low-frequency resonance (they sound boomy) due to the fact that their small size is on the same scale as the low-frequency half wavelengths that are part of the music content. The majority of these rooms have 8-foot ceilings. The equation used to find the initial mode of resonance or *fundamental* frequency is the speed of sound (1130 feet per second) divided by twice the height, length or width of the room. For a ceiling height of 8 feet, it works out to 1130 divided by 16 or 70.625 Hz. The Modex bass trap from RPG exhibits near-perfect absorption (Sabine's "open window") at its peak. Where is that peak centered? Right around 71 Hz. If you think that this is one of those "happy" acoustical accidents, think again. It took many years of research and development to accomplish this feat, and they flat-out nailed it!

How does it work?

Rule number one of noise reduction is to start at either the source or the most problematic locations. All modes end in the corners of the room and low frequencies tend to build up there, so it is obviously beneficial to locate bass traps in corners. The overall shape of the Modex Corner is that of a wedge or triangle because it was designed to fit into corners where there is a lot of bass resonance.

This unit has an internal "membrane" panel that vibrates sympathetically with a resonant frequency peaking at 71 Hz. This vibration moves the air behind it through an absorber. All ducks in a row? I've heard what these units accomplish, and for their size, they provide roughly the same amount of absorption as the old quarter-wavelength-sized bass traps that we used to build into back wall soffits, which required a whole lot more real estate. How much more? It was not unusual for a diaphragmatic

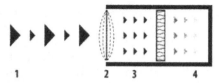

Sound (1) strikes the membrane (2) which sympathetically vibrates, converting the sound pressure to air (3) motion. The air loses velocity (4) as it moves through the internal absorber and air cavity. Now you absorb more bass in less space!

FIGURE 21.19   RPG Modex Corner drawing.

wall made of ⅛" plywood to be erected a couple of feet out from the back wall, and some of those walls were 8 feet high and 18 feet wide! So you are talking about a lot of lost space.

Yes, you could use the room for an amp room, but amps tend to heat up in fiberglass-lined rooms. It could also be used for storage, if what you are storing does not need to be retrieved from the "itchy room" too often. Let me get this across as succinctly as I can. No real loss of space, no construction headaches, no fiberglass to deal with, and they accomplish exactly what is needed! Yeah, I'm a fan.

The drawing (Fig. 21.19) shows the membrane in action, and the specifications (Fig. 21.20) mostly speak for themselves. But the impedance tube test results (Fig. 21.21) are a different story. Hey, don't get me wrong; it was a surprise to me too. Oh, I know all

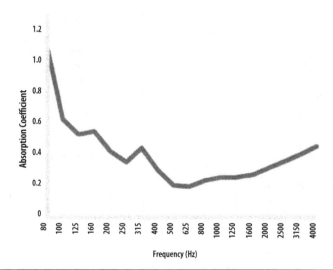

FIGURE 21.20   RPG Modex Corner Absorption Coefficient graph.

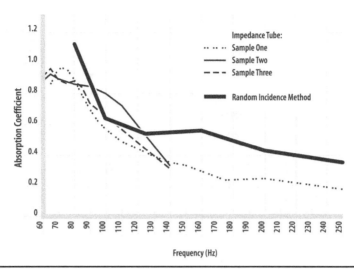

FIGURE 21.21 RPG Modex Impedance Tube Absorption Coefficient graph.

about impedance tube testing wherein they take a sample of a material, place it in a tube-like channel, fire different frequencies of sound at it, and measure the result with a movable microphone. But I must confess that before this, I had never heard of a two-foot square impedance tube! This test must have cost RPG a pile.

Now let's take a look at what some other companies have to offer the small studio owner in the way of acoustic honesty.

Acoustic Sciences

TubeTraps™

These are cylindrical devices with cloth covers. Under those covers is a woven screen followed by a limp mass membrane, which wraps half way around the tube. It was designed to reflect mid and high frequencies. This is followed by a *fiber seal* inside of which is a medium-density glass fiber tube that is hollow except for fiberglass "chamber" separating bulkheads. The top and bottom are *sealed* (see Figs. 21.22 and 21.23).

In short, low frequencies pass right through all the outer material and enter the center hollow fiberglass cylinder where they are attenuated. So are we talking about an absorber? Partially. TubeTraps *are* designed to reduce bass buildup in small rooms. They are cylindrical (ASC provides models that are half and quarter cylinders as well) so they cannot be laid flat against a surface; therefore, test results are given only in Sabins. Furthermore, I cannot determine the total surface area of these tubes so I cannot even try to calculate absorption coefficients. But ASC provides a single graph that show the amount of Sabins per frequency of their 9", 11", 16" (as well as super 16"),

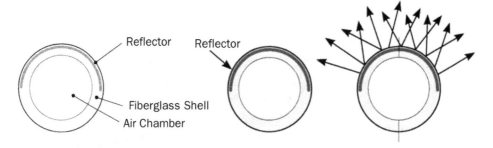

FIGURE 21.22 Acoustic Sciences simplified TubeTrap drawings. The half of the tube with the reflective face diffuses high frequencies and absorbs bass, while the other side absorbs high and low frequencies.

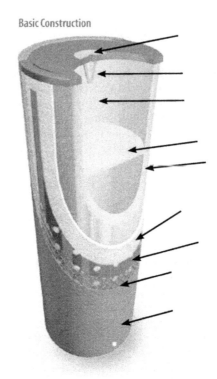

FIGURE 21.23 Acoustic Sciences TubeTrap basic construction drawing. From the top down its make up consists of: a threaded insert, seal cap, air chamber, bulkheads, medium density fiberglass shell, the fiber seal, perforated diffuser sheet, exoskeleton woven screen and finally its outside fabric cover.

and 20" diameter, four-foot-high TubeTraps (see Fig. 21.24). However unlike common absorption devices, TubeTraps can also be tuned to reflect midrange frequencies (above 400 Hz).

Picture this—the half of the cylinder that has the reflective layer in front of the fiberglass absorption cylinder has virtually no effect on low frequencies, which pass right through it. However due to the nature of this barrier *and* the fact that we are talking about a cylindrical surface, higher frequencies are not only reflected, but are also diffused into the room from this side of the unit.

When placed in a corner with the non-reflective side facing inward, it becomes a super bass removal tool, while at the same time on the opposite side it presents only a small amount of absorption and a high degree of reflective dispersion to higher frequencies. Win-win. How so?

The fact is that there is nothing new about high-frequency, absorptive, cloth-covered, glass fiber panels. They have been around forever. Additionally bass traps have been successfully employed along with resonant chambers to reduce low frequencies for just about the same amount of time. So what's the big deal here?

Well those old bass traps might need to consume a three-plus-foot depth of your control room to accomplish what a couple of TubeTraps will at less than twelve inches each! Still not impressed?

While absorption as a rule is almost always better at removing high-frequency content from your signal and leaving you with a resulting sound that is dull, unclear, inaccurate and/or muddy, ASC's TubeTraps take away the mud and actually improve the clarity of the signal by dispersing the higher frequencies. That's a big help in preserving more than just speech intelligibility! Try a broader sound stage, clearer depth, more defined localization, and a much more honest presentation of the audio program. No lie, the facts back it up. Pushing it? Think so? Try this on for size.

By using a wall of these devices (see the "attack wall" in Fig. 21.25,) in front of and wrapped around to the side of the listening position with your monitor speakers soffit mounted within them, you can get dependable honest playback monitoring in rooms with the worst dimensions ever, even an eight foot cube! You *will* have to deaden the

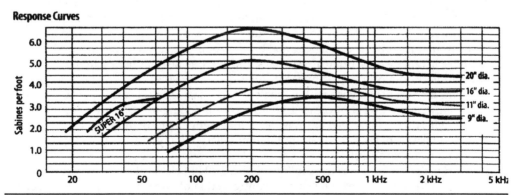

FIGURE 21.24   Acoustic Sciences Sabins per frequency graph.

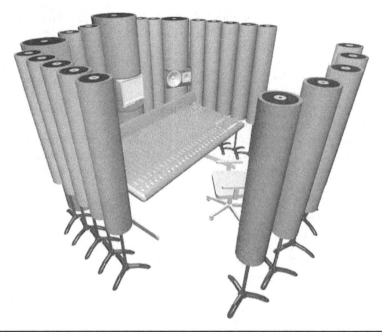

Figure 21.25 Acoustic Sciences TubeTrap attack wall layout.

back wall, add a throw rug, and at least, float a cloud above the listening area, but you will have accomplished what was once considered "impossible."

However, it can be a bit claustrophobic sitting up close to those near-field speakers and facing into a wall of grey cloth. Certainly better than being locked inside a set of headphones, but wouldn't it be nice to back up, lean back, stretch out your legs, and take in the esthetically pleasing soundscape you've just finished creating?

This can be done, but we are going to need some additional premanufactured products to help us out. You see, I go along with the whole concept of a dead front end designed to eliminate unwanted early reflections, but I do not want to feel like I am stuck in a specific listening position while mixing, *and* have nowhere to put clients. However, because of the dual-sided nature of these columns, you can use them across the front to absorb low frequencies and bring a couple more out to the sides to diffuse the highs. In other words, dump excess bass, and at the same time, honest up your listening position. Add a couple of diffusers farther out to the sides, and sit yourself *and* your clients anywhere you wish within the now more sizable sweet-sound pocket.

Furthermore, when it comes time to record anything acoustically (in other words through a microphone) you are going to want to move some of those TubeTraps into the recording area of your room around the musician. What you end up with is premade, instant, variable acoustics.

Have a 1:00 PM voice over, followed by a 2:30 PM classical violin overdub?

At 12:45 spin a horseshoe shaped configuration of TubeTraps so that the reflective sides are all facing out. Position the talent inside and track the voice overdub as dead as a door knob so that post can add whatever amount of reverb they need to match picture. At 2:15 PM spin the tubes around so that the high-frequency reflective sides are facing in, and capture all the glorious overtones from the violin played within this more reflective side. After a short catch-your-breath break, haul the tubes back around the console for a 4:00 PM mix. *Critically important* to the "hauling" part is that internal "fiber seal," which means these columns *do not* expel fiber and they do not weight a lot. But they are *not cheap*! So how can you talk about them in a book about *budget* studios?

1. TubeTraps have been around for over a quarter of a century.
2. TubeTraps do not significantly deteriorate over time.
3. There are a whole lot of TubeTraps out there.
4. Secondhand TubeTraps are fairly easy to come by for half price.

They even show up on Ebay, but here's a big hint: wealthy audiophile/home theater aficionados purchase these things by the dozens. You can sometimes find them being offered for sale in high-end consumer audio chat rooms and cyber clubs for greatly reduced prices. Grab them up if you can, you will not be sorry, *and* you will get back just about all of your investment when you resell them. In my book, purchasing *used*, well-made, premanufactured, acoustic panels is one of the shrewdest financial moves in all of professional audio. So jump online, and after you download reams of info on the Acoustic Sciences TubeTraps at www.acousticsciences.com or call them at 1-800-272-8823, then *immediately* start seeking out preowned TubeTraps up for sale.

Acoustics First

This company provides an impressive line of acoustic supplies, but I particularly want you to know about their economically priced diffusers.

They actually manufacture eight different types, including four of what are called Art Diffusors. Their model F comes in (CLASS 1A) thermoplastic ($150). It is 2-foot square but is only two inches deep so it has a limited bandwidth of 1 kHz to 1600 Hz. This is fine for spreading high frequencies around, but if you need to spread low-mids, you will need a bit more depth. I'm also not going to discuss their wood model W, which is 9½ inches deep and is made of beautiful hardwoods like poplar, maple, red oak, mahogany, cherry, or walnut because, while they *are* economical, they cannot be termed "budget."

That leaves us with their models C and E. The two-foot square (actually 23⅝ inch-square) by 4½-inches-deep model C, which can be placed in hung ceilings, is also thermoplastic and also priced at $150. The four and a half-inch depth gives it a bandwidth of 250 Hz to 1600 Hz. Our last choice, model E, is sized smaller at 15″ × 15″, but it is 9 inches deep, which drops the bandwidth all the way down to 125 Hz. These run $66 each, but a box of two costs $110 instead of $132. That's only $55 each! Even though this was all covered in one of the manufacturer's publications titled "TecSpecs," something bothered me about those high-frequency specifications. I had received a lot of data from Acoustics First, and I went through everything. I found what I believed to be a printing error, and I needed it cleared up. In the "TecSpecs" publication, the upper bandwidth specification for models C and E Art Diffusors is listed as 1.6 kHz (1600 Hz). In the

"Technical Product Overview" publication under "Architectural Specifications," the top end of the bandwidth is listed as 16 kHz. I believed the latter to be correct because, while it is difficult to diffuse low frequencies, high frequencies, even though they tend to beam, are easily diffused by most any irregular surface. Furthermore, unlike low-frequency bandwidth limitations, the size of the panel is less important with highs because the bandwidth of 1600 Hz is around 1.7 inches and at 16,000 Hz, one tenth of that. But I make it a point to check this kind of thing out long before mistakes are published, especially mistakes that could detract from the value of someone's product, lower confidence in their product line, and possibly negatively impact opinion of the company as a whole.

The folks at Acoustics First got back to me right quick with the answer—"a typo in the TecSpec publication." They thanked me for pointing it out. Said they would take care of it in the next printing and then informed me that one of the reasons these panels do so well at high frequencies is because they angled the tops of the blocks instead of leaving them parallel to the wall, which gave me a nice tip to pass on to all you DIYers out there. Take a look at Table 21.2.

While we are not too concerned about absorption, any time you can pick up some low-frequency absorption to go along with high-frequency diffusion, it's a plus. But something troubled me about these specs. The E 405 or suspended ceiling mount with 16″ of space behind it usually yields better low-frequency absorption than the A mount, which is mounted directly on the surface. But let's stick to diffusion for a moment. The choice between these two panels can be approached on several different levels: the price is big, the area of coverage is not unimportant, but the difference in diffusion is slight.

But these do not seem to make any sense as far as diffusion specifications are concerned. It's almost as if we were talking about absorption coefficients. That can't be proved out either way because of the fact that Model E's absorption is not yet available. Furthermore, we have a 4½-inch-deep panel with a bandwidth down to 250 Hz and a diffusion rating of 0.71 at 125 Hz, while the 9 inch deep panel with a bandwidth down to 125 Hz has a diffusion rating of 0.68 at 125 Hz! Are we doomed here? Not really because there are other clues. Luckily we are also provided with polar pattern diffusion charts for both the horizontal and vertical planes in Figs. 21.26 and 21.27. In the horizontal plane, we find that below 1 kHz the model E has better diffusion and above this frequency, the diffusion is superior for the model C. This goes right along with the difference in their

		125 Hz	250 Hz	500 Hz	1000 Hz	2000 Hz	4000 Hz	NRC	Diff Coef.
Model F	TYPE A	0.05	0.60	0.07	0.09	0.07	0.13	**0.20**	-
	E400	0.20	0.10	0.06	0.05	0.06	0.14	**0.05**	-
	Diffusion	0.79	0.74	0.66	0.67	0.69	0.67	-	**0.70**
Model C	TYPE A	0.32	0.20	0.10	0.29	0.20	0.16	**0.20**	-
	E400	0.20	0.12	0.12	0.31	0.23	0.22	**0.20**	-
	Diffusion	0.71	0.71	0.72	0.75	0.72	0.71	-	**0.77**
Model E	TYPE A	Absorption not available at time of printing.							
	Diffusion	0.68	0.69	0.70	0.73	0.75	0.72	-	**0.77**
Model W	TYPE A	Absorption not available at time of printing.							
	Diffusion	0.69	0.69	0.7	0.75	0.74	0.72	-	**0.74**

TABLE 21.2 Acoustics First Art Diffusor Performance (Absorption/Diffusion) Table

C   E

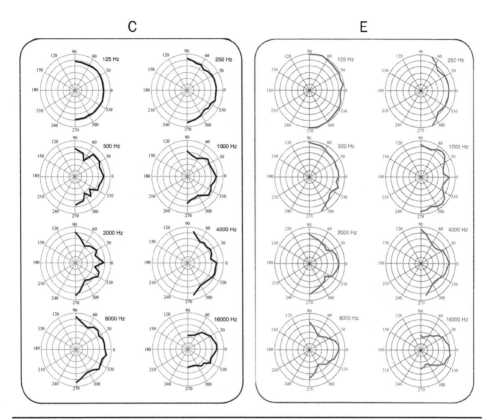

Figure 21.26 Comparison of Acoustics First Art Diffusor Models C and E horizontal response plots.

depths or thickness. On the vertical side, E is slightly better at diffusing 125 Hz, 500 Hz, and obviously better at 16 kHz, while model C is slightly better at diffusing 250 Hz and 8 kHz but obviously better at 1 k, 2 k and 4 kHz, which kind of goes along with what you would expect from the difference in their…, but *wait a minute*!!!!

Stop the presses! A closer look at the specifications straightens everything out because we find that the polar response testing was at 0 degrees incidence using *four* of each panel. This comes to a total area of 48 × 48 inches for model C and 30 × 30 inches for model E. If this is the explanation, then the hands-down choice might be the Model E with a 9″ depth for roughly 40% less in cost!

Not so fast, this is acoustics we are talking about, and just as with logarithms, 1 + 1 does not necessarily equal 2.

Now as you know size has to do with the area that needs to be covered which may be *very* important after you have performed the "see the monitor in the mirror" game on your side walls, especially if the mirror reflective area is in the range of 8′ × 8′ on both side walls. But that's all about mixing, and I'm a huge fan of quality acoustics for recording as well as neutral mixing.

Where am I going here?

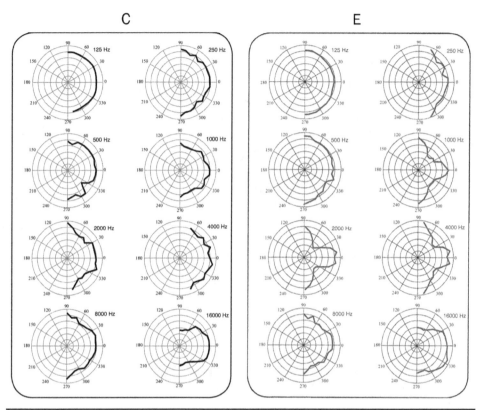

FIGURE 21-27 Comparison of Acoustics First Art Diffusor Models C and E vertical response plots.

Okay here are the numbers: E is 15″ × 15″ = 225 square inches. (I'm going to call 23⅝ inches 24 for ease of mental duress). C is 24″ × 24″ = 576 square inches. That's a *big* difference, but so is the difference in cost between $55 and $150. Best take this to the end.

6 E panels at $330.00 cover 1350 square inches or a little over 9 square feet.

4 C panels at $600.00 cover 2304 square inches or 16 square feet.

12 E panels at $660.00 cover 2700 square inches or a little over 18 square feet.

8 C panels at $1200.00 cover 4608 square inches or 32 square feet.

What does it all boil down to?

If you have a large sidewall reflective area, you may be better off going with the larger 32-square-foot coverage of the model C panels for $1200.00. But if your reflection of the monitors in the mirror area is closer to 18 square feet, you will be better off with the model E panels. Why?

Simply because it gives you 12 instead of 8 diffuser panels to spread around the area of the room where you record acoustic instruments. So you'll have three panels for each of the surfaces (back wall, two side walls, and ceiling). And should that fit your budget, you'll have more versatile acoustics for almost half the cost. Mounting should not be a problem, as the model E panels only weigh *one* pound each (the model C just four)! So four squares of Velcro at the corners of each panel should do nicely.

It all comes down to that sidewall mirror test, or if you have the capabilities, spectrum analysis and reverberation time tests, along *with* the sidewall mirror tests.

On the other hand, if you have a hung ceiling, those two-foot-square model C panels will fit the important need of dispersing reflections from the ceiling area above the console *very* nicely.

This all just goes to show that if you want to think like an acoustician, you have to twist your mind around every possibility just as they have to—meaning leave no stone unturned including, durability, cost, specifications, and overall performance along with fulfilling as many requirements as possible (multiple birds) and *dig, dig, dig* through every specification you can find. Or you can just go on line to www.acousticsfirst.com, or give them a call at 1-800-765-2900.

That's what I did, and they sent me copies of the original 2003 lab test results of absorption for the Model C Art Diffusor with both A and E 405 mounts. So now I can tell you that there *is* an increase in low-frequency absorption when these panels are placed in a suspended ceiling, but it occurs *below* 125 Hz. See the following table.

Frequency in Hz	50	63	80	100	125
A Mount Absorption	0.11	0.08	0.06	0.19	0.32
E 405 Mount Absorption	0.40	0.47	0.41	0.33	0.20

Acoustics First published only the six-octave absorption coefficient frequencies. But below 125 Hz you get better absorption from the E 405 mount, as shown in the above table.

Acoustics First also makes thermoformed plastic polycylindrical diffusers (see Fig. 21.28). This translates to light weight so they are easy to mount. Sizing is 2 × 2, 2 × 4, 3 × 3, 3 × 4, 3 × 5, 3 × 6, 4 × 4, 4 × 6, and 4 × 8. They come with "L" brackets for wall

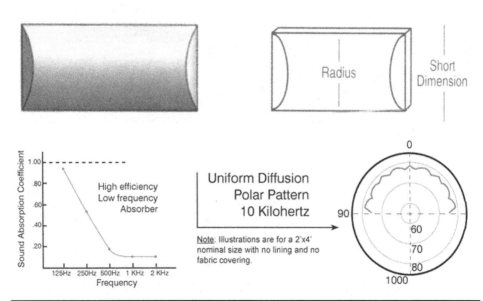

FIGURE 21.28 Acoustics First Double Duty Polycylindrical Bass Trap/Diffuser with absorption graph and diffusion polar pattern (at 10 kHz) graph.

mounting, and there is also a size that matches standard suspended ceilings. But I'm thinking of the listening area behind the mixing console. These diffusers are by no means inexpensive, but today's budget mixing area will not require a four-foot-by-eight-foot, $590 polycylindrical diffuser. Instead, a four-foot length should be adequate, and the two-foot wide model is only $180. Acoustics First adds fiberglass to the largest models to "prevent vibration at low frequencies." I think that's great for large concert hall type walls, but I'm thinking of backing it with a 2′ × 4′ piece of wire screening followed by one or two pieces of two-inch-thick UltraTouch and hanging it from the ceiling over the console with heavy gauge fishing line. Absorbs lows, scatters highs, and looks good. Three birds, and it will only cost about $200 plus shipping and taxes.

Now *that's* budget!

UltraTouch?

Acoustical Surfaces

This company has a product line that is huge. They handle just about *every* aspect of acoustics: noise reduction, transmission loss, absorption, diffusion, test equipment, consultation and much more. But I was excited by the prospects of their B.A.P. (bonded acoustical pad). This is made up of recycled cotton and denim products. Which translates to *no itch*! You could sleep on this stuff and one of their ads even shows kids hugging it!

So I asked for and received a 6″ square sample of the all-cotton (white) B.A.P. a few years back and the first thing I did was what I always tell everybody to do with absorptive material, "Yell at it." Even though I have been doing this for well over FORTY years, I was still surprised by the lack of reflection back into my face. The "sixes" and "twos" I yelled into it also told me that this material was effective at both high (SSS) and lows (woo) frequencies. But you do not have to take my word for it, Table 21.3 gives the absorption coefficients for their R-13 and R-19 "UltraTouch" material.

WOW, ratings like these usually require a material thickness of 8 inches or more, and even at that, these results are not common. Density? I do not know, but I can tell you the weight.

R-13 (3.5 inches thick) is around 0.425 pounds per square foot.

R-19 (5.5 inches thick) is around 0.585 pounds per square foot.

R-21 (? inches thick) is around 0.63 pounds per square foot.

R-30 (? inches thick) is around 0.90 pounds per square foot.

	Absorption Coefficients @ Octave-Band Frequencies					
	125	250	500	1000	2000	4 kHz
R-13 (3 inch) A mounted	0.95	1.30	1.19	1.08	1.02	1.00
R-19 (5 inch) A mounted	0.97	1.37	1.23	1.05	1.00	1.01

TABLE 21.3   Acoustical Surfaces ("Acoustic Performance") UltraTouch R-13 and R-19 Absorption Coefficients @ Octave-Band Frequencies Table

R-Value	Length	Width	Thickness	Sq. Ft/Bundle	Weight
R-13	94 inches	16 inches	3.5 inches	104	39 lbs.
R-13	94 inches	24 inches	3.5 inches	126	39 lbs.
R-19	94 inches	16 inches	5.5 inches	52	30 lbs.
R-19	94 inches	24 inches	5.5 inches	63	37 lbs.

TABLE 21.4   Acoustic Surfaces UltraTouch, Available Sizes

I could not find the weight per cubic foot of the 4-inch Bass Buster but the 5.5-inch Bass Buster is 1.2 lb./cf, and they both have absorption coefficient specifications (at the six-octave spacings) that are identical to the 5-inch-thick R 13.

Getting this information together took hours of work and then things started to get *real* complicated. I put a couple of days into this. Why bother? Because the stuff promised to be a thin, perfect, (open window) absorber!

In one way it is pretty straight ahead; heavier and thicker material makes for more absorption. Common sense, right? The stuff is *really* good at what it does right? So why is sussing it all out like pulling teeth?

Granted I'm getting *old*, but even with over forty years of experience specifying acoustical materials under my belt, I could not crack this nut! The gates of the promised land slammed shut in my face at *every* turn, and I was doing hurdles through multiple web sites *and* using the manufacturer's data sheets.

Here's another BB in the gear box of understanding. There is a *third* type of this material. It is also called UltraTouch, *but* it comes in what they call a "Handybag." This got my attention because the price was $10 for a 48 inch × 16 inch × 2 inch thick sheet. That's only $2 per square foot but there are *no* SPECS outside of thermal. *Bingo*, we are talking about insulation with more than likely the density of cotton candy. *and* it comes with this warning: "Attention. Do not install insulation adjacent to any areas where there is incomplete work that involves flame or heat-producing sources. The same practice shall apply to electric heat coils or any similar heat-producing device. Do not place insulation in air spaces surrounding metal chimneys, fireplaces or flues."

I found that they also make a 2-inch-thick, 3 lb./cf panel that has an egg-crate (convoluted) surface. It is called the "echo eliminator," and its absorption coefficients are:

125	250	500	1000	2000	4000 Hz
0.35	0.94	1.32	1.22	1.06	1.03

Here is some of the information I pulled from the manufacturer's web site. I found out that the R-Value is not only a product designation, but it also refers to the material's thermal properties wherein the higher the number the greater the insulation. Table 21.4 is an "Available Sizes" table I found on a page titled "Technical Data,"

Obviously something is wrong with one of the R-13 weight specs because you cannot have 22 additional square feet without added weight. While this is of little concern, it does point to the likelihood of other typographical errors.

Table 21.5 gave me:

Product	R-Value	Length	Width	Pieces/Bundle	Sq. Ft./Bundle	Bundle Weight/lbs.
UTR1316	R-13	94 inches	16.25	8	84.88	35.6
UTR1324	R-13	94 inches	24.25	8	126.36	54
UTR1916	R-19	94 inches	16.25	5	53.04	31
UTR1924	R-19	94 inches	24.25	5	79.15	46.5
UTR2116	R-21	94 inches	16.25	5	53.04	35.5
UTR2124	R-21	94 inches	24.25	5	79.15	52.5
UTR3016	R-30	48 inches	16.25	5	27.10	24.5
UTR3024	R-30	48 inches	24.25	5	40.40	36.5

TABLE 21.5   Acoustic Surfaces UltraTouch, More Available Sizes

As for absorption coefficients, Table 21.6 is what I was able to download from the manufacturer.

I also downloaded two absorption coefficient lab tests—one for R-13 at 4-inch thickness and another for R-19 at 5½-inch thickness. As you can see, the numbers are mighty impressive. But something troubled me about the graphs (see Figures 21.29 and 21.30). 200 Hz on the 4-inch graph is labeled as 0.14 when it is actually 1.14. Furthermore the graph level *is* correct, but it is just labeled wrong. The same thing happens with the 5½-inch absorption graph at 160 Hz. Here the graph is labeled 0.08 when it is actually 1.08. Again the graph is right, but it is labeled incorrectly. These are the two points where there is the most significant difference between the two thicknesses. Again, the labeling shows the 4-inch with the correct absorption coefficient of 0.93 at 160 Hz and the 5½-inch thickness with *only* 0.08 instead of 1.08. At 200 Hz, the

	Mount	125	250	500	1000	2000	4000 Hz	NRC
1 inch 3 lb./cf	A	0.08	0.31	0.79	1.01	1.00	0.99	0.80
2 inch 3 lb./cf	A	0.17	0.60	1.16	1.21	1.14	1.22	1.05
Bass Buster								
5.5 inch 1.2 lb./cf	A	0.97	1.37	1.23	1.05	1.00	1.01	1.15
UltraTouch								
R-13 (3 inch)	A	0.95	1.30	1.19	1.08	1.02	1.00	1.15
R-19 (5 inch)	A	0.97	1.37	1.23	1.05	1.00	1.01	1.15
B.A.P. Bass Buster 4 inch w/1.5 lb. density	A	0.97	1.37	1.23	1.05	1.00	1.01	1.15

TABLE 21.6   Acoustic Surfaces B.A.P. Composite Panel: Sound Absorption/Noise Reduction Ratings

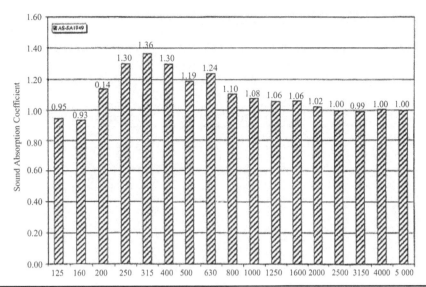

Figure 21.29   Acoustic Surfaces: 4-inch UltraTouch $\frac{1}{3}$-octave-band, center-frequency Absorption Coefficient graph.

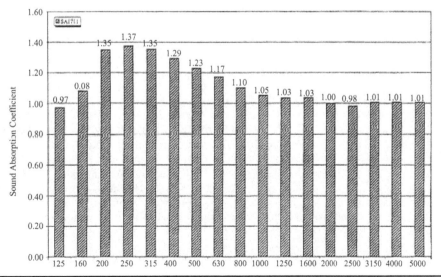

Figure 21.30   Acoustic Surfaces: 5$\frac{1}{2}$-inch UltraTouch $\frac{1}{3}$-octave-band, center-frequency Absorption Coefficient graph.

4-inch graph is labeled as *only* 0.14 instead of 1.14, while the 5½- inch graph is correctly labeled as 1.35. A simple typographical error, so what troubled me? The laboratory's "Technical Director" *signed* both of these reports! And now these misprints run throughout *all* of Acoustic Surface's promotional material and their spec sheets! Here are the lab report listings for the A mounted ⅓-octave-band center frequencies:

Hz	4 Inch Thick R-13	5½ Inch Thick R-19
125	0,95	0.97
160*	0.93	1.08 *
200*	1.14*	1.35
250	1.30	1.37
315	1.36	1.35
400	1.30	1,29
500	1.19	1.23
630	1.24	1.17
800	1.10	1.10
1000	1.08	1.05
1250	1.06	1.03
1600	1.06	1.03
2000	1.02	1.00
2500	1.00	0.98
3150	0.99	1.01
4000	1.00	1.01
5000	1.00	1.01

Transmission Loss

A ton of transmission loss data can be found at www.acousticalsurfaces.com including two lab reports that not only give Transmission Loss data in ⅓-octave and octave bands, but also STC and Indoor/Outdoor Transmission Class Ratings for a:

> Typical Residential Wall 96 × 96 × 4¾ inch plain/single, plain/single stud partition (single stud with a single gypsum board on each side) having ⅝ inch US Gypsum sheetrock gypsum panels, FireCode (Type X) core directly mounted to 2 × 4 wood studs on 16-inch centers with 3-inch average depth R-13 UltraTouch between the studs.

and an

> Asymmetrical Staggered Metal Stud Wall Assembly w/Two Layers ⅝" Firecode Gypsum (Source side), One Layer ⅝" FireCode Gypsum on RC-1 Channels (Receive Side) with the stud cavities filled with UltraTouch R-19.

I could have included the specs but I think it best for you to go online and see *the* most detailed description of an asymmetrical, staggered stud wall I have ever come across. I mean these lab guys get down to the length of the screws used. Yet I was *still* left in the dark as far as thickness goes because they *left out* that information leaving it as simply "R-19 UltraTouch."

There is also a B.A.P. Composite Noise Barrier and Sound Absorber with "⅛-inch 1 lb./sf. Mass Loaded Vinyl noise barrier septum bonded between two layers of Ultra-Touch "acoustical pad absorber and decoupler." It can be used above hung ceilings or applied directly to ceiling and wall surfaces, wrapped around HVAC ductwork and piping. You can order it with "laminated foil or other washable facings for easy cleaning, radiant heat, and moisture barriers. Its overall thickness and weight are 1⅜" and 3 lb./cf.

Its specs in sound absorption/noise reduction are:

Mount	250 Hz	500 Hz	1 kHz	2 kHz	4 kHz	NRC
A	17/0.31	21/0.79	28/1.01	33/1.00	42/1.99	27/0.80

TABLE 21.7   Acoustic Surfaces B.A.P. Noise Barrier Composite Panel: Sound Absorption/ Noise Reduction Ratings

Wait a minute, they *reversed* the two ratings, *These* specs are for noise reduction/ sound absorption! In addition, while those are pretty impressive ratings, you have to wonder why 125 Hz was omitted. If you are interested primarily in blocking low frequencies, it definitely necessitates a call to the manufacturer.

$ Cost?

I can only tell you a little about the cost of all of this material because there is no pricing on the acousticsurfaces.com web site, but they have a site called asistorefront.com that does list 4-inch Bass Buster. It comes in 2' × 4" panels (32 sq. ft.), four panels per box, which sells for $224 (or $1.75 per square foot) plus shipping. 128 square feet of material with octave ratings of: 0.97, 1.37, 1.23, 1.05, 1.00, and 1.01, should take care of just about any noise problem in a small studio.

They had Quiet-Duct Wrap™, which comes 4' × 15' with a lightweight foil facing on what looked to be about an inch of UltraTouch. At $90 it's only 67 cents a square foot, but there were no TL or absorption specs! They had the UltraTouch "Multipurpose" roll which is 16" × 48" of 2-inch-thick R-8 for $10 a roll or about $1.85 per square foot. But the store had *no* specifications on it or *any* information on R-19 UltraTouch! At this point I turned to retail houses!

The Home Depot catalog claims that they sell UltraTouch by the piece (or "roll"). It is 16 inches wide by 48 inches long. It is not *"Handybag,"* and a roll retails for $6 (six rolls for $36). But *this* UltraTouch is only two inches thick (multipurpose?), and there are no absorption ratings for it *anywhere*! A trip to a local Home Depot proved fruitless. The folks on staff had never heard of it but told me that Lowes carries a lot more insulation and they "should" have it. No dice at Lowes either, but I told everyone all about the stuff at both places, and several in each said "I'm gonna check *that* stuff out online!"

Time for a call to Acoustical Surfaces. I talked to a salesman who told me:

UltraTouch R-13 runs about 80 cents a square foot.

UltraTouch R-19 runs about 93 cents a square foot.

UltraTouch R-21 runs about 99 cents a square foot.

UltraTouch R-30 runs about $1.28 a square foot.

Then I tracked down an UltraTouch distributor. They sell building materials to contractors, and they told me:

R-13 is 4 inches thick, and in 24 inch widths you get 126 square feet for $96 (76 cents a sq. ft.),

R-19 is 6 inches thick, and in 24 inch widths you get 79.15 square feet for $90.67 ($1.15 a sq, ft.),

He did not know about R-21, but R-30 is 8 inches thick, and in 24-inch thicknesses, you get 40 square feet for $68.80 ($1.70 a sq. ft.).

Time for phone call number two to Acoustical Surfaces. First off there were a couple of price changes: the R-21 is $1.20, the R-30 is $1.80 a square foot, and the R-13 is 70 *cents* a square foot. As for sizing, R-13 is 3 inches thick; the R-19 *and* 21 are *both* 5.5 inches thick. Now wait a minute, I've got this stuff in 2, 3, 3.5, 4, 5, 5.5, 6, and 8 inch thicknesses!

My next call to Acoustical Surfaces finally clears a few things up.

R-13 comes 3.5 inches thick and "fluffs out to" 4 inches thick.

R-19 comes 5.5 inches thick and "fluffs out to" 6 inches thick.

R-21 comes 5.5 inches thick and "fluffs out to" 6 inches thick.

R-30 is 8 inches thick.

I still do not know difference between R-19 and R-21 in terms of absorption, but it makes no difference for budget recording studios because it all boils down to this. Although the absorption coefficients for the 3-inch R-13 are not as good as those for the 4- and 5.5-inch Bass Buster, they are not significantly lower. However the UTR1324 UltraTouch (R-13 in 24 inch widths) comes in bundles of eight 94-inch × 24.25-inch sheets. That means 126 or almost 128 square feet per bundle. At a cost of (depending on whom you talk to) of $0.70 or $0.76 per square foot, and the price of a bundle is under $90 or $96 plus shipping. The bundle weight is 54 pounds.

One more time, that's less than $100 plus shipping for almost 128 square feet of material with the absorption coefficients shown in Fig. 21.31 and Table 21.8 :

That is enough to completely fill the space between the two-foot on-center studs of an 8- by 17-foot wall. Add some horizontal slats and turn your whole back wall into a Helmholtz resonator. If you choose a slat material that either reflects or diffuses high frequencies, your high end will remain, while all the excess bass is soaked up. Can you say "Two Birds"? How about "Three"?

This stuff is not just environmentally friendly, it has 85% recycled content, uses natural fibers, contains no carcinogens or formaldehyde, is mold and mildew resistant, is light weight, and is it *ever* fire retardant. I once saw a demonstration where a propane

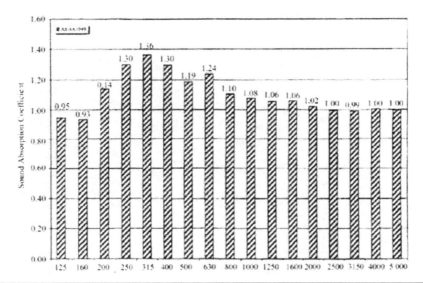

Figure 21.31   Acoustic Surfaces R-13 UltraTouch: ⅓-octave-band, center-frequency Absorption Coefficient graph.

torch was used to melt a piece of copper lying on top of some UltraTouch. Even though the copper was liquefied, the UltraTouch was merely blackened. They achieved this fire *stopping* capability by adding boron (think borax) into the material. Borax is also used as an insect deterrent so UltraTouch gained some serendipitous insect repellent along with the ability to completely *stop* fire.

I also need to introduce you to another product from these folks because Ultra-Touch, when combined with it, provides what might be the *perfect* solution for small room acoustic anomalies. This second product manufactured by Acoustical Surfaces, which should be of interest to small room owners, is called Woodcube™ (see Fig. 21.32).

It comes in 4-inch-deep open-backed squares (think tic-tac-toe with sides). The "I modular" version can be dropped into hung ceiling frames that are ¹⁵⁄₁₆ or ⁹⁄₁₆ wide, meaning they are just under 2′ × 2′ (36 four inch squares each). These folks even provide "matching adhesive-backed veneer strips" that cover the metal T bars for an all-wood look! But the II monolithic is a "solid wood suspension system" that accepts *pre-assembled 2′ × 4′ panels* (72 four inch squares)! Somebody pinch me because this is a small studio owner's dream come true.

Overboard? I don't think so. Just look at the facts: UltraTouch sucks up *every* frequency. Evenly spaced four-inch square louvers diffuse the high end very well. Low frequencies have wide wavelengths so they will be more likely to impinge on the R1324

UltraTouch Sound Absorption – ASTM C423							
Frequency	125Hz	250Hz	500Hz	1KHz	2KHz	4KHz	NRC
3″	0.95	1.30	1.19	1.08	1.02	1.00	1.15
5″	0.97	1.37	1.23	1.05	1.00	1.01	1.15

Table 21.8   UltraTouch Sound Absorption

Figure 21.32 Acoustic Surfaces' Woodcube II Monolithic Open Cell Wood Ceiling Lay-In Louver.

UltraTouch behind the louver and be absorbed. High frequencies beam at multiple angles so they will more than likely impinge on the sides of the wood squares and be dispersed back into the room in many directions.

Need more high frequency reflection? Plug up some 4″ × 4″ squares with reflective material (wood, ceramic tiles, or composite flooring) to block off more of the Ultra-Touch. Furthermore, because this combination can be assembled in 2′ × 4′ sections, you can add them into the room one at a time until the bass buildup is *gone*, while the rest of the spectrum is more *evenly* dispersed throughout the room. Could be a dream come true. Because you want to mount the UltraTouch behind the panel, your best bet is to add 1″ × 8″ wood planks to the outside of the panel. 1″ × 8″ wood is actually ¾″ × 7″ so you end up with an extra 3 inches all around the back. Add the UltraTouch with some screening behind it to hold it in place, and you're done. Make the two side pieces longer for a stand mounted Go-Bo, add a piece of carpet to the back for dual absorption (absorb bass and reflect highs at the front, and absorb bass and highs at the back). Add 1″ × 8″ shelf in between the side planks for stability and another at the bottom for attaching wheels. Taken to the max you end up with a *great* looking, super versatile, easily varied acoustic panel.

Cost? Again it is under $100 for what can be cut into *sixteen* 2′ × 4′ sheets of R1324 UltraTouch. The Woodcube II Monolithic 2′ × 4′ panels run around $20 a square foot ($120 per panel). Here's where it gets a little tricky. If you *could* purchase the R1324 UltraTouch in single 2′ × 4′ sheets, it would run you about $6.00, so we are talking about $126 plus shipping per panel. But they do not sell UltraTouch by the sheet. Don't figure on purchasing enough Woodcube to use up all of the R-13 bundle because $96.00 for the 16 sheets of UltraTouch plus 16 Woodcube panels × $120 comes to $2016 plus shipping. But you could team up with three other studio owners, each getting four of these panels (which could easily correct the acoustic response of most small rooms) for around $500 each plus shipping. A very fair price for a very good panel. Still not quite "budget" enough?

Okay, so *now* we bring in wood patio flooring tiles. *What?*

These are squares made up of wood slats with a space in between each slat for water to drain through (think Helmholtz). How Helmholtz? That depends on

the slat width in inches (w).

the width of the slots between the slats in inches (r).

the air space behind the slats or the depth in inches (D).

the "effective" depth of the slot in inches, which is equal to 1.2 × the thickness of the slat (d).

You simply multiply 2160 × the square root of (r) divided by [(d × D) + (w + r)], and you'll have the resonant frequency in Hertz.

Don't feel like playing Master Mathematician? Then unless you are building the resonator from scratch, just "Forget About It."

1. You know that high frequencies will bounce off of the slats.

2. You know that low frequencies will go around or through the slats.

3. Finally you know that just about everything that gets past the slats to the UltraTouch will be absorbed.

Reduction in bass, reflection of highs, *and* nice looking wood panels on your studio walls for a very low price—Three Birds. Okay, *Three Birds* are nice, but how low a price are we talking?

I am writing this in the late spring when everyone in the suburbs is going patio *crazy*. High demand equals high cost, right? Well, at this moment I am looking at an ad for eucalyptus wood flooring tiles that measure 18 inches by 18 inches each. The price is $11 for three or $3.99 for one.

Let's get wild and go for a full 6-foot-high by 9-foot-wide panel or 24 individual squares, which we can spread out for better absorption. The cost of the wood tiles would run you $88. Wait until the fall, and you may get them for $66, and it may be even less expensive in the winter!

Who advertised these eucalyptus wood floor tiles? Christmas Tree shops no less. That means that just about *everybody* carries them!

So let's go crazy and try and use up the full UltraTouch R1324 bundle by purchasing 56 of these tiles for $209, meaning we'll end up with enough squares to cover an area 6 feet high by 21 feet long for under $300 plus shipping!

But you may think your best bet is to design a slatted Helmholtz resonator to deal with the particular low-frequency problem of your room and then purchase the lumber to build it. If you have, or know someone with a table saw, it will be a cinch. But after a few hours of hand sawing slats, $209 will not look that expensive anymore.

NOTE: *As long as I was already researching this offbeat area of budget acoustics, I figured I'd dive in a little further to save you some time. For example, the slatted cedar flooring made for baths and saunas looks great, but it is meant to be stood upon, so the manufacturers had to concern themselves with lawsuits. This translates into panels that are way too heavy to mount on walls.*

GIK Acoustics

It is best to deal with every acoustical product as a separate entity, that is, using the ratings per frequency to tell you which item within a manufacturer's line will best suit the particular situation you are dealing with. If you have a 4 kHz ring, it's obviously not going to be their bass trap. GIK Acoustics had Riverbank Test Labs perform J mount testing of panels with open and caped perimeters, when both wall and corner mounted. Okay your best bet here is to go to *http://gikacoustics.com/products.html* then click on the 244 Bass Trap in the Absorption Products list. Scroll down and click on the red "Absorption Report" just above "GIK Acoustics Greensafe." This brings you to the GIK "Test Results" page. Ignore the graph and the lists of numbers for now. Instead scroll down to the red PDF file markers midway down the page. Click on the Corner Mounting PDF under "Test results of the GIK Monster Bass Traps." This will get you the actual lab report giving the absorption test results in Sabins per frequency for the GIK Monster Bass Trap when corner mounted.

Download and print these two pages. Go back to the "Test Results" page and click on the "Recommended Wall Mounting" PDF. This gets you the actual two page lab report giving the results of the absorption tests in Sabins with the Monster Bass Traps positioned in GIK's "recommended" wall mounting. Download and print these two pages. Now go back to the GIK "Test Results" page and download and print that page as well. This gives you five pages of data. Don't bother trying the "Click to view detailed PDF Report" because it will not open. Right off the bat you'll notice that there are no actual lab reports available for the 244 Bass Trap, and while I believe the "recommended" corner and wall mounting Sabin ratings are from actual laboratory tests, labs do not give J mount ratings in absorption coefficients. Furthermore, due to the 244's being furred out by 40 mm (leaving 1½ inches of air space behind it), it cannot be tested flat against the wall as in the standard A mount. What is being called an A mount is actually a D-20 mount. However, we have to play through here because if you're going to nitpick every mount deviation, you'd have to state that the three side views of the RPG FlutterFree mountings are not actually "standard" A, C-20 or D-20 mountings either.

The standard ASTM A mount specifications call for tests to be performed with units laid against the surface. While this provides ratings that can be readily compared with those of other products, it does not relate very well with how they should actually be mounted in recording studios. It is obvious that panels with open sides are better at absorbing sound than those that are capped. But there is also the question of placement. Do they work better on the wall or in the corners, and at what frequencies?

Because of these J mount tests, we now have a window into panel placement that pretty much gives an answer that is plain as day. Below are the ratings for the GIK 244 Bass Trap panel 2' × 4' × 4" (with 1.5" furring for a total depth of 5.5") w/capped or sealed perimeters.

Hertz	125	250	500	1000	2000	4000
Wall Mount	0.89	1.84	1.63	1.07	0.8	0.68
Corner Mount	1.64	1.75	1.65	1.08	0.71	0.57

Pretty close to the same except at 125 Hz.

Below are the ratings for the GIK Monster Bass Trap panel 2' × 4' × 6" (with 1.5" furring for a total depth of 7.5") w/open perimeters.

Hertz	125	250	500	1000	2000	4000
Wall Mount	2.06	1.99	1.92	1.36	1.04	0.96
Corner Mount	2.0	2.06	1.87	1.16	0.83	0.78

Pretty close to the same except at the high end, but something does not seem to make sense here!

The 244 Bass Trap does a little better when it is mounted in the corner, and the Monster Bass Trap does a little better when mounted on the wall. But the difference is slight for the Monster Bass Trap, as the wall mounting is only +0.06, −0.07, −0.05, +0.20, +0.21, and +0.18 that of the corner mount. But a closer look at the lab report tells us the "recommended" wall mount was of "8 units with 4″ spacers placed underneath each unit. Units are evenly spaced." Go back to the GIK Test Results page online and look at the photos. The Monster panels are not only spaced four inches up, but also are not placed fully flat against the wall. Absorption when spaced out from the surface is going to be about the same as corner mounting until you go below 125 Hz.

However the biggest difference in all of this is the doubling of the absorption at 100 Hz of the 244 Bass Trap when mounted in the corner as opposed to the wall mount. Thanks to GIK Acoustics' going through the effort (and expense) of having their panels tested using ASM J "recommended" mountings at both corner *and* wall mountings, we are given a perfect picture of what happens when the sides, top, and bottom of an absorptive panel are left open and exposed to sound.

Thinking about the *facts*, spells it all out. Mounted on the wall, the wood framed sides of the 244 are not absorbing. Now if you take that exact same panel and move it into a corner where *both* its front and back are exposed to *more* bass, the absorption of low frequencies doubles. A very succinct demonstration. Luckily, GIK Acoustics also had the lab check specifications not just for the six-octave frequency bands, but also at full ⅓-octave center frequency bands. Therefore, we are able to see that the increase in absorption of the corner mounted 244 is substantially better below 200 Hz, but the absorption ratings of the recommended wall and corner mountings begin to match at 200 Hz and continue to be very close (within one Sabin) right on out to 5000 Hz.

Now let's take a look at the Monster panel data. With its sides, top, and bottom open, it does better absorbing frequencies at 400 Hz and above when it is spaced four inches off the floor and a couple of inches out the from the surface than when it is corner mounted. The open-sided Monster bass panel absorption seems a little more complicated, until you lay it all out in the Sabins per unit lab results. The recommended corner mount was measured using eight panels: "7 units 45 degree angle on wall spaced 8″ apart. Placement of units: South and West walls 2 units each, North wall three units. Long dimension parallel with wall/floor joint. One unit in NW corner" (vertically?). Refer to Table 21.9 to find the absorption coefficients for these panels.

Facts are facts, so open-sided absorbers can be much more effective at dealing with frequencies below 125 Hz when they are mounted in the corner. From 125 to 800 Hz, it's six of one, half dozen of the other, with no substantial difference in absorption with either mounting method. But from 800 Hz all the way out to 5 kHz, the wall mounting consistently provided a full Sabin-per-unit increase in absorption over the corner mounting. I just wished GIK had their Monster Bass Trap tested in the standard *flat* against the wall mounting with the results given in Sabins. Then you would *really* have been able to see the difference between the standard flat against the wall and "recommended"

Hz	Recommended Corner	Recommended Wall	Difference with Wall Mounting
	All in Sabins per Unit		
50	5.06	1,46	−3.6
63	10.17	3.25	−6.92
80	23.97	4.52	−19.45
100	18.69	7.57	−10.94
125	16.03	16.46	+0.43
160	14.83	14.11	−0.72
200	15.76	16.21	+0.45
250	16.48	15.92	−0.56
315	16.70	16.67	−0.03
400	16.82	17.47	+0.65
500	14.92	15.32	+0.4
630	13.31	14.05	+0.74
800	11.34	12.51	+1.17
1000	9.25	10.84	+1.59
1250	8.05	9.78	+1.73
1600	7.14	8.78	+1.64
2000	7.14	8.32	+1.18
2500	6.63	7.83	+1.20
3150	6.22	7.63	+1.41
4000	6.21	7,67	+1.46
5000	5.81	7.65	+1.84

TABLE **21.9** Absorption Coefficients GIK Published for the Monster (Open Edge) Panels

mountings. However, you should still take advantage of this research. By adding 2 x 2 (1.5 x 1.5) inch furring across the back of your panels at the top and bottom, you'll also end up with broader band absorption, and they will be easier to both mount and re-position.

We can also use this information as a clue to the best placement for these panels, per frequency. As far as the closed edged 244 is concerned: below-160-Hz corner mounting provides superior absorption, but at 200 Hz and above, wall mounting is more favorable. The same goes for the open-sided Monster panel as far as providing less attenuation with corner placement, but again here the crossover frequency starts at 800 Hz.

Folks (meaning other manufacturers) are going to be bent all out of shape about the use of J mounting or "manufacturer specified" mounting test methods because these methods, while they may be very illustrative, can end up as an apple and orange comparison with products tested using standard mountings.

My point of view is a little different; I am just looking to inform, so I see this as an invaluable instructive tool. Furthermore, GIK Acoustics *does* provide the absorption coefficients per frequency for the 244 (closed edge) Bass Trap when mounted in a standard mount, so it can be compared directly with other panels. However, you have no way to compare the Monster trap to anything tested with normal laboratory mounting methods. See the Absorption Coefficients for the 244 Bass Trap in Table 21.10

Looking over GIK's specifications for their 244 Bass Trap we see that below 125 Hz their "recommended" *corner* mount provides more absorption than the standard A

Frequency	A Mount	Recommended Wall Mount	Laboratory's Sabins per Unit	Corner Mounted	Laboratory's Sabins per Unit
50	0.47	0.26	2.0679	0.58	4.65183
63	0.45	0.23	1.82965	0.78	6.26138
80	0.66	0.46	3.67901	2.02	16.14634
100	0.89	0.63	5.02655	1.85	14.77995
125	0.93	0.89	7.1542	1.64	13.13826
160	0.84	1.02	8.13273	1.48	11.81404
200	1.05	1.61	12.94586	1.40	11.16875
250	1.00	1.84	14.74053	1.75	14.0313
315	1.09	1.79	14.33658	1.67	13.36324
400	1.17	1.79	14.32796	1.78	14.27246
500	1.12	1.63	13.05388	1.65	13.21372
630	1.03	1.40	11.19675	1.49	11.93043
800	0.84	1.23	9.87555	1.30	10.40749
1000	0.69	1.07	8.55999	1.08	8.63589
1250	0.58	0.96	7.7059	0.92	7.32633
1600	0.51	0.85	6.82576	0.79	6.28432
2000	0.46	0.80	6.39408	0.71	5.60028
2500	0.43	0.73	5.84027	0.64	5.14079
3150	0.41	0.68	5.45497	0.58	4.63396
4000	0.4	0.68	5.40179	0.57	4.56003
5000	0.41	0.66	5.01443	0.53	4.22132

TABLE **21.10**   The Absorption Coefficients GIK Published for Their Model 244 (Sealed Edge) Panels.

mount, but in this same frequency range, the standard A mount performs better than their "recommended " *wall* mount. For frequencies at and above 200 Hz, their "recommended" wall mount provides more absorption than the standard A mount and their "recommended" *corner* mount.

It is incumbent on us to look at some of these test results as GIK states them in "absorption coefficients" with these two equations in mind:

1. Total Sabins = total area in square feet multiplied by the absorption coefficient. The result of that, divided by the number of panels gives you Sabins per panel.

2. The absorption coefficient = total number of Sabins divided by the total area in square feet, or the Sabins per panel divided by the total area of the panel in square feet.

Six panels of 244 (48 square feet) placed in the "recommended" wall mounting render an "absorption coefficient" of 0.89 at 125 Hz.

7.12 Sabins per panel divided by 8 square feet = an absorption coefficient of 0.89.

0.89 × 48 square feet = 42.72 total Sabins. That divided by 6, (the number of panels) = 7.12 Sabins per panel.

The manufacturer's recommended "bridging the corners" J mounting method rendered an absorption coefficient of 1.64 at 125 Hz.

13.13826 Sabins per panel divided by 8 square feet = an absorption coefficient of 1.6422825.

1.64 × 48 square feet = 78.72 total Sabins divided by six panels = 13.12 Sabins per unit.

Try a few out for yourself, and it will be obvious, due to this little exercise, that GIK Acoustics is *not* "cooking" the numbers, but all the J mounting tests are certainly skewed. This becomes more apparent with the data on their Monster Bass Trap.

First off, here they are using 8 panels instead of six for a total of 64 square feet. But because the square footage is included in the equation for absorption, there is no gain specification-wise in using more panels. Again the test results are given in Sabins per unit and absorption coefficients, but the lab reports give only absorption in Sabins per unit and total Sabins. So let's check out those absorption coefficients.

At 80 Hz, the "recommended" *wall*-mounting method renders 4.52446 Sabins per unit and an absorption coefficient of 0.57.

4.52446 divided by 8 square feet = an absorption coefficient of 0.5655575.

At 125 Hz this mounting method renders 16.45525 Sabins per unit and an absorption coefficient of 2.06.

16.45525 divided by 8 square feet = an absorption coefficient of 2.0569062.

From 125 Hz out to 2 kHz, it acts as a perfect absorber. At 80 Hz the manufacturer's "recommended" *corner* mount rendered 23.97018 Sabins per unit and an absorption coefficient of 3.0.

23.97018 divided by 8 square feet = an absorption coefficient of 2.9962725.

At 125 Hz this mounting method rendered 16.03005 Sabins per unit and an absorption coefficient of 2.0.

16.45525 divided by 8 square feet = an absorption coefficient of 2.0569062.

From 160 to 1250 Hz, it acts as a perfect absorber.

Again, it is apparent that GIK Acoustics is *not* cooking the numbers, but the test results are certainly skewed. How skewed? Here's my analogy.

We are both plumbing fixture manufacturers. The standard method of measurement for the flow rate of faucets is to measure the amount of cold water that passes through the faucet in a given amount of time. You have that measurement taken for your faucet, and the lab comes up with a result of one gallon per minute.

I have my faucet tested and say: "But most folks run hot *and* cold water at the same time, so let's test my faucet that way." The lab report on my faucet with this "J" method renders one and a half gallons per minute. How are you going to feel? That is why other manufacturers of acoustic panels are crying "Foul!"

However, I honestly believe that GIK is actually *not* trying to pull the wool over anyone's eyes. If that were the case, they would have recommended a mounting method that placed their panels perpendicular to the walls (at a 90-degree angle) for twice the absorption as the standard "A" mounting method. Furthermore, they do not try to hide the fact that the numbers provided cannot be used to compare their products to any others by publishing statements on their web sight like: "those J-mount measurements are not comparable with standard A and E-mount measurement standards expressed as absorption coefficients."

Furthermore, I found *all* of the above information on *their* web sight! So as *always*, it is up to the individual to do the research and try to find out exactly what the facts are. Note: I *do not* want this to come off as a criticism of GIK Acoustics because:

1. The mounting methods used are actual mountings normally used in recording studios.

2. The company is totally frank about your not being able to compare their test results to any others.

3. These products are *very* good.

How good? This manufacturer offers a full 100% refund on anything you are not totally satisfied with. Yes, you do have to pay for the return shipping, but these folks are located in Atlanta Georgia. There have always been a whole lot of recording studios close by in the Southeast region of the United States. So any products deemed inadequate would be sent back right quick, and I am not aware of UPS handling a stream of acoustic panels with Atlanta, Georgia as the destination. Acoustic manufacturers have a high enough mortality rate without offering 100% money back warranties on less than adequate products. The 2' x 4' 244 Bass Traps sell two per box for about $140, and the Monster Bass Trap runs around $120. That's certainly budget pricing!

Do go back on line and check out other products from this company. You will be as impressed as I was with their full line of acoustic products.

Yet with crazy graphs showing absorption coefficients of 3.0 at 80 Hz and statements like—"with new certified lab tests, the GIK Monster has a rating of 3.00 absorption coefficient at 80 Hz. There is no other product on the market that even comes close to this number. Most products hit only half that number or don't post any numbers at all."—they're just kicking a hornet's nest.

It's too bad, because these U.S.A. made panels *are* actually very good at what they do. Furthermore, GIK absorbers use ECOSE™, naturally occurring and/or recycled raw materials that are bonded using bio-based technology and *no* formaldehyde. It is "Greenguard," indoor-air-quality certified for use in schools and meets the Collaborative for High Performance Schools green building rating program especially designed for K-12 schools' low-emission materials criteria. It does what it's supposed to do at a budget price and is safe to boot! What more could you ask for?

Have a look at the graphs in Figs. 21.33 and 21.34. These show the Monster Bass Trap graphic specifications in Sabins per frequency as given by the laboratory for the recommended corner and wall mountings. Personally I'd have been proud to publish only the lab specs and happy to avoid all the flack.

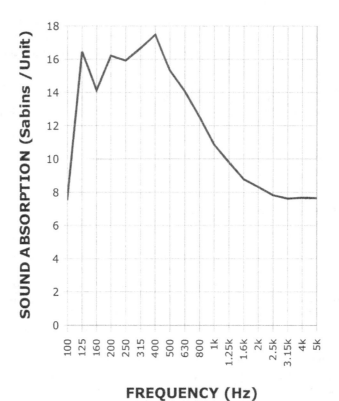

FIGURE 21.33 GIK Acoustic Model 244 Bass Trap laboratory Sound Absorption Sabins per frequency graph. (Riverbank Acoustical Laboratories)

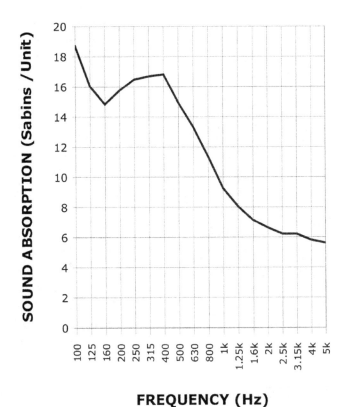

FIGURE 21.34 GIK Acoustic Monster Bass Trap laboratory Sound Absorption Sabins per frequency graph. (Riverbank Acoustical Laboratories)

Compare those graphs with Fig. 21.35, the "Absorption Coefficient" graph from GIK Acoustics. My guess is that either a design or an applications engineer came up with the idea of using those particular "J" mount tests with an eye to improving the design or its implementation, but it was probably someone in sales that came up with the above quote. I think it best you remember that faucet analogy *every* time you check specifications.

Auralex Acoustics

These folks have been around for many years and are continually adding new developments to their catalog. Today their product line is about as extensive and diverse as it gets. They provide everything from "U" shaped rubber boots used to float walls to wood sound diffuser arrays. They also offer isolation mounts for speakers as well as glass fiber panels. But acoustic foam is their bread and butter, and nobody can touch them in terms

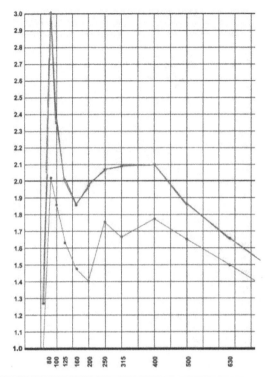

FIGURE 21.35   GIK Acoustics "Absorption Coefficient" graph for its Monster (upper trace) and 244 (lower trace) bass traps.

of variety when it comes to shapes and colors. It would take close to a whole book to describe every product they offer in detail, and that's not what I intend to do.

However one of their products needs to be brought to the attention of everyone attempting to accurately record and monitor sound in smaller rooms. You know all about the most common problem here. Bass build up. If you add enough absorptive material to sufficiently drop the low-frequency content you end up losing a lot of your top end as well.

Bass traps of old helped to tame the bass while leaving the top end alone, but at a fairly large price in terms of lost real estate. This is something that is unacceptable in smaller rooms. Auralex has a product they now call the "Venus Bass Trap." When I first ran across it, it was called "Paramount" but the two are identical. What is it? A series of very deep wedges cut into 12-inch-thick foam (see Fig. 21.36).

What does it do? It is the proverbial open window! Check these absorption coefficients:

100	125	160	200	250	315	400	500	630	800	1k	1.25k	1.6k	2k	2.5k	3.15k	4k	5kHz
1.19	1.63	1.30	1.31	1.34	1.32	1.36	1.29	1.25	1.25	1.26	1.20	1.22	1.25	1.21	1.18	1.20	1.24

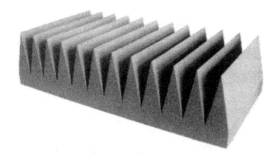

FIGURE 21.36   Auralex Acoustics Venus Bass Trap.

This stuff sucks up just about everything, but that's generally not what anyone wants—except in *one* place, behind the diaphragmatic panel of a bass trap.

Check out the advantages. It allows for the reduction in depth to just a little over one foot. Storing something in any remaining space is not completely ruled out, as is the case when the absorber is lined with glass fiber. Near perfect broadband absorption. While it is not dead even at every frequency, it comes very close. Only low frequencies will pass through a diaphragmatic panel so the high end is unaffected while the bass is sucked out of the room.

Don't want to give up a little more than a foot of space across your whole back wall? These come in 24 x 48 sizing, and two of them run around $350.00. Stacking them comes to 2' x 8', perfect for filling a large across-the-wall polycylindrical absorber! The framing is the most important part of the project, and I would suggest 2 x 4s for strength. Okay, your diameter is a little over 24 inches, which gives you a radius of about a foot. The full circumference of the *circle* we are dealing with is 81.5 inches, but we are only building a half circle, so ripping a $\frac{3}{16}$" thick 4' x 8' plywood panel down to 41" should give us the right-sized panel to wet down and attach to our frame. This process will be greatly aided by making two half-circle end caps and some center dividers of the correct diameter to help shape the curve and hold the polycylindrical panel together. Total size: slightly over 2' wide x 8' high x 1' deep, and light enough for two people to move easily.

Lay it down facing out across the back or front wall/floor junction while mixing. Straddle a wall corner behind a musician for no bass boom and very good high frequency diffusion. Then once you decide that you like the difference they make in the overall sound of your room, you can buy another pair.

On the other hand you may be better off with multiple 2' wide x 4' tall x 1' deep polycylindrical absorbers for even more variable acoustic flexibility.

Note $\frac{3}{16}$" plywood has some absorption on its own, so it may be best to varnish it for increased reflection at high frequencies as well as to lessen the chance of clients getting slivers! Plywood is pretty tough, but musicians moving equipment can be formidable! I have wall papered denim cloth onto $\frac{1}{8}$" panels followed by varnishing (denim is also absorbent). This toughens it all the way up to rehearsal room standards. If the thought of blue denim does not fit with your decor, denim also comes in black. A lightweight shallow polycylindrical diaphragmatic absorber that works wonderfully all because Auralex came up with the answer to the proverbial acoustic open window. Check out additional specifications on-line at auralexinfo@auralex.com. Then request a catalog and look over their full product line. They may very well have a solution to other problems you are wrestling with. Here are some examples of what you'll find in Table 21.11

	125	250	500	1000	2000	4000 Hz	NRC
1" Studiofoam Wedge	0.10	0.13	0.30	0.68	0.94	1.00	0.50
2" Studiofoam Wedge	0.11	0.30	0.91	1.05	0.99	1.00	0.80
3" Studiofoam Wedge	0.23	0.49	1.06	1.04	0.96	1.05	0.90
4" Studiofoam Wedge	0.31	0.85	1.25	1.14	1.06	1.09	1.10
2" Studiofoam Pyramids	0.13	0.27	0.62	0.92	1.02	1.02	0.70
4" Studiofoam Pyramids	0.27	0.50	1.01	1.13	1.11	1.12	0.95
2" Studiofoam Metro	0.13	0.23	0.68	0.93	0.91	0.89	0.70
2" Sonomatt	0.13	0.27	0.62	0.92	1.02	1.02	0.70
2" Studiofoam Wedgies	0.15	0.21	0.70	0.99	1.05	1.05	0.75
2" DST-114/244	0.16	0.29	0.57	0.75	0.90	1.00	0.65
Max-Wall Panels	0.81	1.02	1.06	1.05	1.02	1.02	1.05
Venus Bass Traps	1.63	1.34	1.29	1.26	1.25	1.20	1.30
LENRD* Bass Traps	1.24	1.28	1.45	1.39	1.27	1.31	1.35
Sunburst Males	0.8	1.23	1.14	1.07	1.05	1.08	1.10
Sunburst Females	0.65	1.02	1.00	1.08	1.05	1.08	1.05

*(Low-End Node Reduction Device)

TABLE 21.11   Some ASTM C423 Test Results of Auralex Foam Products

Further Information

More than likely you will be continuing your search for acoustic products online—the land of *too much* information. Stick to one product at a time and research it *thoroughly*. Dig out every technical specification you can: test results, graphs, mounting methods, size, weight (both mounting and shipping), warranty, return policy, surface material toughness, and color options. Sometimes the digging required is fairly shallow. Take Fig. 21.37.

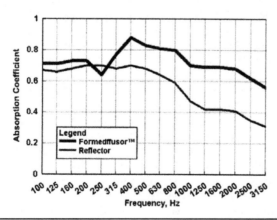

Diffusion

The Formedffusor™ is based on the QRD® reflection phase grating introduced by RPG® in 1983. It offers broad bandwidth wide angle diffusion. The graph illustrates the average diffusion coefficient (1 is ideal) for all angles of incidence. Compared to a flat reflecting panel, the QRD® maintains uniform diffusivity as a function of frequency above the diffraction limit.

FIGURE 21.37   RPG Formedffusor Diffusion Coefficient graph.

This old RPG Formedffusor diffusion coefficient graph is mislabeled "Absorption Coef-ficient." How old? It's from before the ISO diffusion standard upgrade that "normalized" or corrected the testing procedures to show-low-frequency diffusion and scattering more accurately. Big difference? Compare this graph to the one in Fig. 21.3. RPG is generally pretty meticulous about their literature, so if *they* can have typographical errors, *anyone* can. *And* as we've seen, these can often unfavorably misrepresent the manufacturer's own product. Personally, I think it is important to know that I'm going to pick up some low-frequency absorption (0.41) at 80 cycles free of charge when I drop-ceiling mount an Acoustics First Model C ArtDiffusor and that the diffusion extends to 16 kHz not 1600, that four-inch-thick R-13 UltraTouch has an absorption coefficient of 1.14 instead of 0.14 at 200 cycles, and that at 5½-inch thickness its absorption is 1.08 at 160 cycles as opposed to 0.08. So to me the digging is well worth the effort.

But in the case of RPG's older Formedffusor graph, the paragraph *right next to* it plainly states: "The graph illustrates the average diffusion coefficient (1 is ideal) for all angles of incidence." That's not what I'd call strenuous research!

In general this is what you want to see offered by an acoustic panel manufacturer:

- Green products as in bonded material that was recycled from cotton, newspaper, or wood fiber

- 100% recyclable products (No problem getting rid of it, once you're done with it.)

- Class A fire retardant (The biggest problem I have with putting egg crates on your wall is not that they do not work, but because it's a huge fire hazard.)

- Free online room analysis (Let them do the math.)

- Full refund policy

- Test result documents from independent labs

- Complete list of standard specifications (lots of useful information and it shows they have nothing to hide)

- Low cost ($$ the bottom line).

Obviously you are not going to get *all* of this, so you have to prioritize. Take a com-pany that uses recycled materials and provides some specs. Their wall panels' absorp-tion coefficients are 0.34, 0.96, 1.31, 1.24, 1.05, and 1.02. But there are no specs provided for their bass traps. Pricing varies from $4 to $15 per square foot with no apparent indi-cation as to what causes the price to increase. *Some* of their products are class A fire rated, and they do have a fairly good return policy. Is this a company to avoid?

Certainly not, but I doubt if there is a long waiting list to purchase their products. Hey, these days it is not going to be difficult to find a half-dozen companies providing exactly what you need to straighten out (that is, "Honest-Up") the acoustic side of your monitoring system. Now that you know what is important, what to look for, and where to find it, the only thing left is the easiest and, at the same time, the hardest part of your decision-making process: deciding how much you can afford to spend. Luckily there are also a bunch of companies selling kits and a world of "Do It Yourself" information out there.

Forty-five years ago, when I was in your shoes, outside of *DB The Sound Engineering Magazine* there was *zip*!

Start out by trying to determine the applicability of acoustic products as remedies to specific problems using just the graphs and tables supplied by their manufacturers.

Right now you may not need something, but when you do, that product *will* pop right back into your mind.

Now go back and look at the three Sabins per frequency graphs in Figs. 21.24, 21.33, and 21.34.

All kinds of conclusions can be derived from looking at these three graphs, *but* they cannot be compared with each other. Figs. 21.33 and 21.34 show the results in Sabins of adding the single function GIK Bass Trap to a room, and in this function, it does an outstanding job of eliminating bass build up. On the other hand, the ASC TubeTrap does a righteous job of trimming excessive low frequencies while at the same time scattering frequencies above 400 Hz. In other words, the two cannot be compared to each other because they were designed to meet different requirements.

Ignoring the TubeTrap because it does not attenuate low frequencies as well as the Monster Bass Trap negates its usefulness in evening out the overall response of a room. Passing on the Monster because it offers low-end attenuation that isn't even with the high-frequency attenuation negates the value of its bass reducing function.

You must become what can almost be termed an acoustic detective in these situations and unfortunately you are pretty much on your own. I just hope I've given you enough insight to navigate your way through what can often seem to be an insurmountable quagmire. I don't know if this will be helpful or add to the confusion, but here's a lab report where the test results were given in both absorption coefficients *and* Sabins for a 48-square-foot specimen under test when laid directly against the surface. The lab also included the % of uncertainty.

⅓-Octave Center Frequency (Hz)	Absorption Coefficient	Total Absorption in Sabins	% of Uncertainty with 95% Confidence Limit with Specimen
125	1.63	78.21	2.87
250	1.34	64.44	1.98
500	1.29	61.86	1.32
1000	1.26	60.27	0.67
2000	1.25	59.84	0.62
4000	1.20	57.36	0.42

Remember: The Absorption Coefficient = Total Sabins divided by the square feet. (here 48), and Total Sabins = square feet (48) × the Absorption Coefficient. For instance, at 1000 Hz, 60.27 divided by 48 square feet = 1.255625, and 48 × 1.26 = 60.48

That's close enough for me, and it shows that with a level playing field (meaning using the same mounting method), absorption as rated in Sabins and absorption coefficients *can* be meaningfully, honestly and easily compared. Unfortunately, asking every manufacturer to spring for the cost of dual testing may be a bit out of the question.

In addition to these difficulties, somewhere along the line (maybe due to the proliferation of "electronic musicians" out there) the art of recording acoustic instruments started to be overlooked. For you, that must change because you are going to need as much knowledge as you can gather to pull off capturing all the nuances of acoustic instruments played in a less than ideal setting (such as a bedroom). The first step is to understand how music instruments produce and propagate sound.

PART IV

Music Instruments

CHAPTER 22

How Music Instruments Produce and Propagate Sound

"Wait a minute, I thought this book was about acoustics?"

It's really all about recording in a less than an ideal environment, and how to improve that situation as much as is economically possible. Part of that equation is often the requirement of going in closer with your microphone placement on music instruments to help reduce acoustic anomalies that might conflict with the correct capturing of that instrument's timbre. In-tight mic placement is often essential to getting the best possible sound from an acoustic instrument. Let's face facts, if an electronic keyboard player stops by your apartment at 3:00 PM to lay down some tracks, you plug him into your laptop, hit record, and let him rip. Now, for the other side of the coin.

You walk into an 8000 cubic-foot wood-paneled room covered with intricate decorative reliefs. Dead center sits a freshly tuned Steinway 12-foot grand. You and your *three* assistants carefully remove the piano's lid and move it off to the side of the room. You then hang your absolutely mint condition 1962 Neumann U-67 tube microphone set to omnidirectional pickup pattern six feet above the now wide open piano. You then walk into the control room as the master pianist, wearing cans (headphones), begins to "warm up" by playing the most difficult passages of the piece to be recorded.

In the control room you pay very close attention to the tempo of the piece *and* how it fits in with the reverberation time of the room. This piece has lots of big fat swells that change to quiet "airy" passages. You reach over to under the right hand side of the console and move a toggle switch lever away from yourself.

As the ceiling rises, the swells linger a bit longer. At a certain point when the lingering notes just start to conflict with what is currently being played, you reverse the upward motion of the ceiling by pulling the toggle switch toward yourself. The playing style of the pianist opens up noticeably, and the overdub is completed in a single take.

Pipe dream? No, I've *been there*, and when they tore down the old RCA recording complex in NYC, I literally shed a tear because I knew that the hydraulically lifted ceiling that made recording a string section a joy beyond belief, would *never* be duplicated because of the prohibitive space and cost requirements.

In less than ideal environments, it is often best to go in close with your microphone placement. Where you place the microphone(s) depends on several factors including the instrument, how it propagates sound, and the timbre you want to capture. And *that* is what this section is *all* about.

I wrote this with the intention of saving the reader a lot of time and trouble. I tried to write it so that it could be easily understood; therefore, I avoided the common practice of proving out points by making use of mathematical equations at the end of every paragraph. At times the text may come across like a mass of jumbled facts, but some repetition was required because some points apply to many aspects of this subject.

I wanted to provide enough factual information so that the reader would be able to understand everything that I was trying to relate about capturing the sound of music instruments. While I did not come close to completing this job, I've reduced the amount of work that's left for you to do, mightily. You will find that assimilating all of the available information out there can be a tough go, but it is very much worth the effort.

For example, take the longitudinal and transverse forces of a string in motion. You'll find that strings transfer sound energy to the top plates of guitars and violins via sound waves through the air along with vibrations through the bridge. Furthermore, the force from the vibration is so much stronger than the sound through air that the only locations where the through-air force is noticeable are those that are furthest away from the bridge. One of these areas ends up being where the top plate meets the neck. Since lower frequencies travel more efficiently by vibration, it also means that the sound at the neck area will have more high-end content. Because of this, I knew enough to point a mic at the bottom of the neck on an acoustic guitar when I wanted to pick up more high-frequency "string sound" from the instrument so that it would cut through the din when it came time to mix. 'Nough said? But we really need to start this discussion back at the beginning.

Okay it's 10,000 BC, and you're a caveman wandering around looking for some food. The wind picks up and blows across the open end of a hollow log producing a low-frequency "*hum*". It also loosens a ripe coconut, which falls from the tree hits a solid log for a "*bink*" and then bounces to the hollow log, "*bunk*." You walk over, pick up the coconut, and before you break it open, you hit the same two logs right after the wind excites the column of air inside that hollow log again for the same "*hum*"–"*bink*"–"*bunk*". You then make it a "*hum*"–"*bunk*"–"*bink*" or two, and a star is born. Soon you find that it's easier to use a couple of sticks. A few centuries later, felt mallets become the rage only to be replaced by super-ball-ended percussion sticks.

Humor aside, a lot went on in producing those three simple notes. But before we move on to a detailed breakdown of how tube, block, membrane, string, and reed instruments produce sound, there are a few facts to you need to understand. As an audio recordist, you should find this information of great interest.

Some Basic Facts about Sound Production by Music Instruments

Resonance

Resonance causes an increase in both the level and duration of a tone. Blocks and tubes have resonant frequencies that are determined by their size and, in the case of tubes, the amount of air within them. Resonators are used to increase the level and to improve the tone of an instrument. For example, if a partially enclosed body of air like a tube is placed near a ringing bell, it will increase the level and duration of the bells tone. Yet the bell's fundamental frequency may be a little more obvious when it is rung without the increase in harmonic content caused by the tuned resonator. Adding water to a bowl increases its mass and lowers the pitch of the tone. Yet when air is blown across the top of a more enclosed cavity, such a bottle, exciting the air column to vibrate, adding more water will raise the pitch. This is because the total volume of air has been decreased.

Note Duration

After the "attack" or beginning of a note, the fundamental frequency continues to sound while additional harmonics build up around it. First there is a rise in the level of the fundamental; then this periodic pulsation is joined by harmonics, which may either augment or mask it. We readily perceive this change and appreciate it as opposed to a pure steady state tone, especially when there's vibrato or a wobble-like variation of less than 10 Hz within the frequency content. But our brain is very quick at picking out variations that are too regular to be natural. We find natural vibrato more pleasing precisely because it is not as exactly timed as those that are simulated electronically. The note's full sound as it progresses through these changes is termed its "envelope," and the envelope of a waveform is critical to the overall sound. As an example, if a pure tone like a sine wave is made to vary in level to match the envelope shape of a triangle wave's slow rise and fall, the note will sound like an organ or a violin. If the tone is made to match a sawtooth wave with a sharp rise and a slow decay, the note will sound more like a plucked string.

All tones are made up of separate components called "partials." The combination of partials is not only different from instrument to instrument, but from note to note, even on the same instrument. Try using a harmonizer to lower a note one octave, and compare it to the same unprocessed lower note. If they do not sound alike, it may not be the fault of the processor.

Attack

Hearing the beginning of a note is essential to our brain's being able to recognize a sound and its source. Try making a recording of a horn section and editing out the beginning of a sustained note. When you listen to only the part after the edit, you will have a hard time discerning the individual horns within the section. However with the attack in place, you will easily be able to pick out each individual instrument, even when all the horns start to play at the exact same time.

Because the attack of an acoustic piano is so unique, editing out the beginning of a played chord will cause this fact to really jump out at you. Yet it is not just the waveform, level, timbre, attack, or harmonics that make the source of a sound recognizable, but all of these components and more. That is why, even though conventional violins, violas, cellos, and double bass instruments have frequency responses that overlap, it is easy to pick out each instrument, while listening to a string quartet, because of their distinct individual tonal qualities.

Wolf Tones

There are notes which, when played on violins and other instruments, can jump up in level to the point where the note will sound a full octave higher than intended. Here you'll find that the fundamental frequency of the note being played matches the resonant frequency of the instrument. A string may overly excite the body, which in turn acts upon the string, causing an effect much like feedback. The fundamental then decays leaving the higher harmonics sounding, which makes the pitch of the note seem to increase. There are all kinds of theories about this phenomenon, but luckily outside of a slight mic repositioning, it is not of too great a concern to the recordist. On the other hand, tones that come across lower in level, such as the transitional notes across the register or "octave key" of woodwind instruments, may require the addition of a second mic.

Impedance

Impedance must also be considered. As with electronic circuits where more resistance equates to lower voltage, impedance to airflow affects the output level of woodwind instruments.

There is little to no resistance when blowing across edge tone instruments. They are said to have very low "wind" resistance, but reed instruments can present a great deal of resistance to the flow of air.

Checking out the player's breathing tells the whole story. Someone playing a long section of music on a recorder or a flute may have to take in a breath before it is completed. A reed player may also have to take a breath, but they often must first exhale unspent air that has been held back by the resistance of the reed.

This little fact may seem insignificant, but to the audio recordist who has the job of micing or editing a performance, it can mean the difference between breathing sounds all over your tracks and a nightmare editing session, or just another job well done. Once while following along with a set of charts, I noticed a short silent passage in the middle of a long clarinet solo. When the piece reached that point I muted the clarinet mic and then dropped it back in just before the solo was to resume. Not big deal, but the producer and musicians were blown away by the fact that I knew beforehand exactly when the clarinetist was going to expel a loud, half lung-full of air during the overdub.

Sound Radiation

The sound radiation pattern of an instrument changes with frequency. Sound is generally radiated in all directions at lower frequencies, but as the frequency increases, the pattern narrows down more and more with different frequencies output at differing levels from different locations. Furthermore, just as with microphones, no two instruments (even if they have consecutive serial numbers) will sound exactly alike. So there is absolutely no way to positively layout a definitive directional characteristic for all violins or any other music instrument.

Yet musicians need to know that the intonation of the instrument that's being radiated out into the room is correct, and this can be confirmed only if they can hear the "room's" response. Remember a musician is very close to the instrument and thus needs to hear more of the immediate or nearby reflections from the room than does the audience. While the structure around the stage is very important to the musician, out in the general seating area the room's reverberation will have more of an effect on the makeup of the sound that reaches the listener's ears. This "room response" is also important to the audio recordist. For example, in a small room with a ceiling height of 8 feet, the lowest resonance will be about 70 Hz, and the next resonance will be at 140 and then 210 Hz. The added vibration from the use of a floor peg may produce peaks in the low-frequency content of an instrument in this room. Therefore in this case, placing cello and basses in a more open space and on risers may make better sense for those wishing to do a good job of recording these instruments.

For music instrument radiation patterns, I believe the best bet for the audio recordist is to go directly to Meyer's work (circa 1972) because he attempted to lay it all out pictorially, showing the patterns from acoustic instruments per frequency. Again, this will still be just a starting point as nothing that has to do with acoustic instruments can be definitive. Furthermore the frequencies at which the testing was done generally do not pertain to music notes, keys or even octaves. But you will find that a cello emits lows (150 Hz. and under) mostly from its top plate on right hand side, while for the low-mids (to around 300 Hz.) the radiation is greater from the back plate. At 350 to 500 Hz, the pattern is more from the top plate, but it also sweeps around the sides angling to the rear, looking much like a cardioid microphone's polar pattern. From 400 to 800 Hz, the radiation pattern narrows down a bit with the sound projected more from the top plate. At 800 to 1 kHz, the right-hand side of the instrument does most of the radiating. From 1 kHz to 1250 Hz, the radiation comes more from the upper part of the top plate, and the highs (2 kHz to 5 kHz) are projected upward and downward straight off of the top plate. This means that if you want to beef up the low-mids and reduce the high-frequency content of a cello, you'll want to mic it from the rear or go in low at the front. Spending the time going through all of this and noting the test results for each instrument is a lot of work, but I think it was very much worth my effort.

Sound Transmission

A vibrating body will transmit energy to other body parts and out to the air. The method of transferring the energy to the air is crucial to instruments like strings and percussion whereas with woodwinds, no interface is required because their sound is delivered directly into the air.

Grand pianos are certainly loud, but this just proves the point of heavier is better at sustaining sound. A grand's strings are more like cables in size and are strung at great tension to a massive metal harp attached to a huge soundboard.

Gongs also illustrate why large solid bodies can produce loud sounds. They can handle the heavy hit of solid wood mallets or even logs, which cause them to vibrate wildly, and they transmit this energy very effectively into the air. This is not the case for a metal "I" beam. Oh it can vibrate with the best of them, but it cannot transmit that vibration very efficiently. This is because the beam just isn't as flexible as a gong and it doesn't have a large enough surface area for transmitting sound well. So its vibrating sound fades quickly especially at lower frequencies.

Radiators and Resonators

These help if the initial sound carrier can drive them and they have a large surface area, but in order to be easily driven, they must also be light and flexible. Yet heavy solid resonators sustain vibrations longer than light ones. This also applies to ones that vibrate wildly like metal plates and ones that snap back to their original positions very quickly like metal reeds and tines.

On their own, strings are not very efficient at radiating sound. This is because while strings vibrate wildly they transmit sound poorly. Thus they require the use of light, thin soundboards as found on violins and guitars or the drum-like heads of banjos. Strings are more like the prongs of a tuning fork than the handle. The prongs visibly move when they vibrate (high voltage/weak current) while the handle doesn't seem to move at all (low voltage). But the handle has what could be called a high current because when it is placed against a large radiator (such as a tabletop) it can easily excite it. This will not be the case with the prongs since their weaker current is easily damped. All you have to do is touch one to confirm this fact for yourself.

However if you touch a vibrating string away from its mid-section (where it is vibrating wildly) and more toward the areas where it is fixed (at the nut or bridge), the vibrating string will have less movement (voltage) but it will have enough current to easily vibrate the bridge and an extra light soundboard. In fact, the first time you pick up a "Flamenco" guitar, you'll be astounded at how lightweight it is. These instruments are specifically designed to render louder more percussive sounds. Conversely electric guitars have heavy bodies in order to facilitate a longer sustain. What they lose in amplitude is, as you know, made up for with electricity.

Resonators, soundboards, and air columns are used to help amplify and project sound when they are excited by the vibrating energy from strings, tines, wood blocks, etc. By using a soundboard, the sound energy is spread over a larger surface area that is able to move the surrounding air more efficiently. Air columns within chambers, such as the tubes under vibes or drum shells, are excited by the sound, amplify it, and pass it out to the surrounding air. The sound boxes of guitars and violins utilize both types of sound resonators. To function properly, some air columns must have a resonant frequency that matches the sound source.

An effective soundboard must have the right weight so that it can be easily driven by the source and yet be large enough to move a lot of air. It must be somewhat rigid so as to help spread the vibrations but also remain flexible. How do you achieve rigidity without adding too much weight? One example is the use of struts or bracing to thin soundboards. The curves of arched-top violins and the thin drumhead radiators of banjos also achieve this result.

When it comes to low-frequency propagation, size counts. Once more it's the old "greater than one half the lowest frequency's wavelength" rule. But adhering strictly to this rule would have an upright player hugging a bass the size of a grand piano. Luckily for contrabass players, those "F" hole openings in their sound boxes pump enough air to provide a healthy amount of low frequency.

One of the most remarkable resonators is the system used on Dobro or National steel guitars. It utilizes a unique approach to increase sound radiation. Under the decorative metal plate inset into the guitars body is a thin aluminum cone-shaped diaphragmatic resonator that is connected directly to the bridge. The strings vibrate the bridge, which has no trouble exciting the lightweight aluminum cone, which in turn vibrates wildly projecting the sound out into the surrounding air. The sound of these instruments is also enhanced by the normal low-frequency boost of the guitar's sound box. This instrument was designed in the days before electric amplifiers, and today they still do a mighty good job of amplifying the sound. But my favorite radiator has to be the drumhead combined with the chamber and tone ring of the banjo.

I'll never forget the first time I *really* heard a banjo being played. It was January 7th, 1967, and the "Charles River Valley Boys" were playing at the Club 47 in Cambridge Massachusetts. I had just finished explaining to the bouncer/bartender that I was indeed under-aged (not quite 16), but that I was only there to hear the band and had no interest in alcohol but could afford the price of a cup of coffee, when the banjo player entered the room playing a solo. The way that thing just filled the room with sound had me awe-struck.

I got more when Lester Flatt and Earl Scruggs and the Foggy Mountain Boys rolled in the next week. Hey for that matter, The Otis Rush Blues Band was there the following week after which came Richie Havens, the Pennywhistlers and Sandy Bull. In between was the Paul Butterfield Blues Band at Jordan Hall, not to mention blues and jazz shows at "Paul's Mall" and the "Jazz Workshop" where I had similar non-drinking "I just gotta hear this band" deals worked out. At the time things were very much smoking for this young recordist, and my ears were the size of an elephant's.

Some Specific Sound Producing Mechanisms

Tines, Rods and Tongues

These are fixed at one end, and when struck, plucked, or bowed, they emit a tone. But they usually need some sort of resonator to amplify their sounds. Examples are music box comb shaped tines, hollow box drums with slots cut in their tops ("tongue drums"), and thumb pianos (or African "kalimbas"). With these instruments, the size of the vibrating part is relatively small so by rigidly affixing them to a sound radiator (board or hollow box) their weak sounds can be amplified enough to be easily heard. I once met someone in the Caribbean who made kalimbas, and he explained the tuning method to me.

Bigger or heavier tines mean lower frequencies. Therefore filing down the vibrating (plucked) end raises the pitch—the more rigid the mounting, the higher the frequency. So the maker could adjust the mounting screws to change the frequency by loosening (lower) or tightening (higher). If he filed down the tine at the area next to where it is attached, it would lower the frequency, and as already stated, he could file down the plucked end of the tine to increase the frequency. Adding solder will either lower (when added to the plucked end) or raise the pitch (if placed where the tines are attached). Therefore, a heavier "played" (or vibrating) end adds to the sound's fundamental. This maker was able to add overtones to his large Jamaican bass kalimbas to increase their ability to radiate the sound while still producing very low frequencies.

Jaw harps fit into this category, but they add in the ability to vary tones by the use of different tongue and lip positioning, along with changing the size of the mouth cavity as well as the forcing (blowing and sucking) of air across the vibrating tine. A player who uses a lot of air movement in their playing style requires the use of a more distant mic positioning so it's always best to give these instruments a good one-ear-listen up close before setting up your mics.

Tongue drums are usually hollow logs with single slots, "U" shaped slits, or "H" shaped (dual tone) cut outs in them. Today many are made in the shape of a wooden box or simply lengths of bamboo with these slots cut into them. In all cases, the sound produced depends both on the body's size and the thickness of the tongues. Woodblocks, "boos," and especially, temple blocks render that wonderful "ka-tonk" sound that I just loved to delay (repeat) and modulate to get waterfalls of beats.

Sound Transmission of Acoustic Guitars

The guitar must be looked at as a complete system. The head stock with its machine tuners, neck from the nut, fingerboard, and frets all the way to the foot where the neck attaches to the body and the body's sides or ribs, top and back plates, bracing (whether cross or fan shaped), sound hole, bridge, and its tail piece, all combine to produce a tone when a string is plucked. The hollow body section of acoustic instruments is really a series of vibrating interconnected parts.

The strings excite the top plate, which projects the higher frequencies. They also vibrate the bridge/which moves the top plate/which moves the ribs/which moves the back plate/which projects the mid-range frequencies. When the strings move the bridge, they also excite the air cavity, which pumps sound out of the sound hole, which helps project low frequencies.

Although the moving string has little real energy (or current), it can easily vibrate the bridge, which excites the top plate. Because these two parts transfer the energy to the sides, braces, air cavity, and back plate, sound is radiated in all directions from the guitar body in addition to the sound hole. This all makes for an efficient system. Individual parts may be more effective at radiating particular frequency ranges (air cavity and sound hole, the lows, the back plate the low-mids, the lower section of the top plate the high-mids and the upper half of the top plate the highs) but all together they make up the instrument's uniform coherent sound. Noteworthy, is the fact that the decorative inlay "purfling" work around the outside perimeter of the acoustic guitar body acts as a hinge and lets the top plate vibrate more freely.

In a nutshell:

The string's motion both vibrates the bridge and excites the upper section of the top plate at the area just under the neck, which emits the high-mid tones.

The bridge in turn vibrates the top plate causing it to both radiate midrange content and transfer the vibrations to the braces, sides, and air cavity.

The air cavity emits low frequencies from the sound hole and together with the sides and braces transfers the vibration to the back plate, which radiates low-mids.

Physicists spend a lot of time studying this process, and the results are there for everyone to utilize.

However test performed in anechoic chambers often point to bipolar or quadripolar radiation patterns, but in real listening situations (where the wall surfaces are reflective) the sound appears to be more omnidirectional, even if the frequency content varies slightly from location to location. This variation is the reason physicists prefer to use anechoic chambers and close mic positions for their tests. Yet they do take the normal listening environment into consideration when they attempt to judge the quality of an instrument. I've seen comments stating that a faster decay rate is preferable in large halls while longer decays may be more suitable to small rooms that are used for chamber ensembles. Some physicists even get involved in the debate over appropriate instrument body bracing methods, but I've yet to see any indication as to how all their testing relates to different music tempos.

Holographs are generally used to depict vibration modes or the excitement of parts of the guitar body. However the body is often submerged in sand during the test and the excitation is not via string movement but by a continuously rotating vibration generator that is brought into contact with the bridge. The sound produced is seldom related to an actual note, and as you can imagine, there's a lot more vibration when the instrument is strummed or played normally as opposed to when it is buried in sand.

Utilizing test results of one model from a single maker can provide you only with generalities because even with the same guitar a difference in excitation frequency of as little as 2 Hz can change the radiation pattern. Yet once you understand how these instruments produce sound, you begin to glean a lot of information from those holograms. While the resulting knowledge is not definitive, it does point you in the right direction.

As always it depends on what you're after. For instance 150 Hz is radiated from the lower sections of both the back and front plates, but because it is also emitted from the sound hole, a mic can be positioned a little higher up in front. Should you want mucho bottom, go right for the sound hole; if mostly low-mids with some lows mixed in, the center section of the top plate; and for pure low-mids, the lower half of the back plate ends up being a better place to position a vibration pickup or a ceramic (electronic) transducer. However test results should only be used as clues or starting points because nothing is definitive when it comes to guitars, violins or any other acoustic instrument.

Making Use of Physicists' Research Experiments

Tests resulting from using a single mic in an anechoic chamber can be unpredictable when placement is left up to the individual performing the test. The best approach often uses several microphones, rectifies their outputs, and then sums the results. Even when a frequency response curve is recorded in an reverb chamber, many microphone positions are often used and the results averaged. Again, music instruments are usually not "played" but are "driven" by attaching rotating or oscillating pressure directly to the bridge. Physicists utilize this method to eliminate the problems of inconsistencies in playing level and the musician's body "shadowing" or blocking sound propagation paths.

Whenever you delve into the acoustics of music instruments, you very quickly find yourself swamped by mathematics. The equations physicists use to prove out certain points, such as the vibration excitation properties of music instruments, makes the math used for room acoustics seem as if it were preschool level.

I did not work out all of these equations when I was researching this subject area, since (even at 17 years old) I was much too busy to spend half a day attempting to find out what the cryptic equation components referred to (see Figure 4.1), or to fill reams of paper with numbers just to find out the exact elasticity coefficient of a single vibrating string. After a couple of hundred pages of physics with copious mathematics covering vibration, you eventually get to transverse and longitudinal wave equations for tubes, bars, plates, and membranes. Next come various time and frequency analyses of strings when plucked, bowed. or otherwise "driven" or set in motion. This is followed by the propagation of sound waves in air. Believe me, you can easily spend a day and a half sussing out the math for the transverse (directional change) and the longitudinal (change in length) that make up the forces of a string in motion.

Checking out how sound reacts in the presence of a hollow cavity is a major area of research, and physicists tend to back up their findings with arduous equations. Yet for the recordist, spending a little time messing around with an inexpensive jaw-harp is more fun and more rapidly discloses some of the basic facts. Plucking the metal reed

$$(x, t) = \sum_{n=1}^{\infty} a_n \sin \frac{n\pi x}{L} \sin n\omega t.$$

Figure 4.1

that sits between what looks like a tuning fork produces a fundamental and a host of overtones. Holding the instrument against your teeth you can accentuate individual tones by adjusting the size of your mouth cavity.

Physicists often study the individual parts of music instruments to detect vibration modes. When their interest is in only a single section of an instrument, every other vibrating or sound producing part of the instrument will be silenced in order to obtain the most reliable test results. For instance, when testing vibration modes on a top plate, the vibrations at the back plate and sides are of no interest. Due to this fact, they will often bury a guitar lying down in sand right up to the top plate to eliminate other vibration interference. The sound hole's lowest resonance depends on the air cavity size and the sound hole diameter (Helmholtz). The top plate and any other part of the guitar body not buried will also affect this resonance.

The results of these tests are quite often given as shaded-in areas showing the largest resonance per frequency. Caution—while the holographic pictures of sound waves of specific frequencies moving across a top plate are somewhat revealing and useful for determining vibrations and pickup mounting positions, they should not be blindly accepted as good locations for microphone positioning. You have to read the small print. Again these folks are often studying only top plates, not the cavities or even the influences of the back plate. It is common for the body to be buried up to the top plate in sand in order to eliminate the back plate's influence. Hey, I've seen some of these tests done where the air cavity was filled with sand to eliminate any sound from that portion of the guitar as well! While these test methods are used to provide the most exacting picture of top plate vibrations, they can vary widely from the results of the same instrument when it is played conventionally.

Though this information is very helpful to physicists and designers, two problems obviously limit their application for the recording engineer. The results only suggest the instrument's frequency range, and the actual vibrating frequencies applied to the bridge are seldom related to music notes. Furthermore, vibration modes do not necessarily equate to sound propagation, and again, the response of an acoustical guitar buried in sand will bear little resemblance to the sound when played normally. Measuring the sound pressure from multiple points surrounding an instrument may be more beneficial, but considering the number of positions needed (let alone the number of frequencies to be tested) makes plotting a complete diagram close to impossible. Even if you had the time, budget, and number of assistants to pull it off, these experiments and tests seldom use methods that duplicate the real playing conditions in which audio recordists normally find themselves.

I once saw a physicist's descriptive layout of the radiation pattern of a clarinet, which had *all* of the high frequencies (above 2 to 4 kHz) emerging from the bell. It was immediately obvious to me that the testing was performed in an anechoic chamber using a close micing technique.

While it is true that a great deal of the high frequency content emerges from the bell of woodwind instruments, this is generally in the form of overtones and the sounds that come directly from the reed. In this experiment, the mid-range dispersion was correct in that it was depicted as emanating from the horn's tone holes from just below the midpoint of the upper section (clarinets are separated into two halves) to almost the end of the instrument. No radiation patterns were given for the lowest notes, which in fact radiate from bottommost side ports (the bell acts as a additional length to the tube to produce the lowest note), nor for the highest notes, which not only appear at the bell but also at the top of the horn, just below the mouthpiece.

This physicist should have spent a little time discussing the experiment with a few musicians, as they would have known about things like over-blowing to achieve higher notes and that the lowest note emanates from the bell.

While musicians and instrument makers often provide the recordist with opinions that are purely subjective as far as the playability and sound quality of an instrument, the objective findings from scientific experiments are often based on testing procedures that may have narrowly defined parameters and are thus incomplete in their conclusions. For example, scientific prediction of the fundamental pitch of a flute can be off by as much as a semitone simply due to the internal irregularity of the instrument, and it's rarely mentioned that varying the amount of lip overhang at the mouth hole allows the player to change the instruments tonality and tuning on the fly.

It can seem that everything these scientists expound is out of whack. After spending many hours plowing through a thirty or so page dissertation on the way clarinets produce sound, the author concluded by stating that it did not matter what material was used for the inside surface of the horn's tube because the dimensions were more significant. This statement made me angry. You see bass clarinets are made of wood, hard rubber, plastic, and metal. I've played them all. Yeah, bodies made from hard rubber can sound very much like wood-bodied instruments, and even some plastic material bodies may come close, but metal-bodied bass clarinets are far raspier sounding. The resonant frequency of the metal (think of the sound of a sax) body along with its cylindrical shape negates some of the dark mellow sound this instrument is known for. Hey, don't take my word for it, check the pricing. The reason instruments made of wood can bring prices that are thousands of dollars higher than their counterparts is due to the sound, not ease of playing.

Yet that physicist was not out and out wrong. He was analyzing the horn as far as its dimensions were concerned. This guy measured every sound-hole diameter, position and even rim height. Moving a port by an eighth inch will radically affect the pitch. So from his point of view, the subtle difference made by the use of different body materials (which does not in fact alter the pitch) was insignificant. Yet for the musician, it makes all the difference in the world. As an audio recordist, I prefer dealing with a wood horn because its sound is less harsh and grating. In fact if someone were to make a tenor sax with a wood body, I'd buy one. So as you can see, I'm a bit opinionated too.

Tubes and Pipes

In effect, that prehistoric hollow log whose air cavity was stimulated by wind can be thought of as a tubular instrument. Exciting the column of air within a tube produces a resonant frequency that is contingent on the size of the air column within the tube. In fact, the length of the air column will be one half the wavelength of the resonant frequency. Wavelength = the speed of sound (about 1130 feet per second) divided by the frequency. So should you decide to make a giant panpipe out of logs, you now have the formula.

When the air inside a tube vibrates, it also produces harmonics. If the tube is cone shaped (or conical like trumpets, saxes, and bones), the even harmonics will be accentuated, and if cylindrical in shape (like flutes, and clarinets), the odd harmonics will predominate.

Tubes

Pulling a cork from or blowing across a bottle opening or tube will cause a note to sound. Here the vibration is in the form of a wave traveling up and down through the air in the tube. These are called "longitudinal" vibrations or oscillations. Many African percussion instruments are based on this principle. Try slapping your palm against tubes of different lengths and see what you can come up with.

The task is simply to set up a vibration in the air column within the instrument's body. The simplest example is blowing across the top of an open-ended tube (as with panpipes). The pressure of the air flow is adjusted to match the resonant frequency of the tube, and the time it takes for this matchup to occur adds significantly to the overall timbre or tonal quality of the instrument. With increasing pressure, the pitch will at some point jump up an octave. Here the number of pulses will be doubled in the tube, with one pulse reaching the end of its travel at a point when another pulse just begins to make its way down the air column.

A tube's length can be "tuned" to match the resonant frequency of a block positioned near its opening. In the case of vibes, marimbas, and xylophones, each note has its own resonator. The resonant frequency of a tube with both ends opened is twice as high as that of the same sized tube with one end closed. This is due to the fact that at the end of its travel, a vibrating airwave is reflected back up the tube. Only one half of the number of pulses need to be fed into a tube with one end closed in order for it to match the resonant frequency as opposed to when both of the ends are open. Therefore, closing one end causes the frequency to drop an octave, even though the tube is the same length. You can prove this by blowing across the aperture of a tube. Then stop up the open end with a cork. If done correctly, the frequency will drop by an octave.

When you open both ends, things get more complicated, as now the sound is projected from both the lip opening and the end of the tube. Additionally these two sounds will be in phase for odd harmonics but out of phase when even harmonics are produced. Vibraphones add a circular disc within the tube that almost completely cuts off the length of the air column. A rod passing through all of the tubes holds these discs in position. When the rod is spun by a motor, all the disks rotate, opening and closing off the airflow, which produces a vibrato effect.

Tubular Instruments

Resonant vibrations are generated within tubular instrument by several methods. A stream of air can be made to flow through a slot and past a sharp edged lip or a pointed wedge. This causes the air to rotate rapidly, setting up a vortex (or swirling motion). This vortex causes vibrations in the air column in the tube. This is how organ pipes and recorders produce sound. In other instruments, vibrating closures, such as a reeds (saxes, harmonicas, bagpipes, and clarinets) or a pair of lips (trumpets and bones) are used to interrupt the flow of the air stream. In these cases after stopping the airflow, they go back to their original position, only to repeat the cycle again and again. The resonant frequency of an organ pipe is determined by its size, while with recorders and flutes it is based on both the size and the number of holes or ports that are closed off, with all closed providing the lowest frequency.

The Brass and Woodwind Instrument Family

Brass and woodwind instruments are divided into those that have edge-tone actuators (recorders and flutes), wind-caps with reeds that are not pressed against by the player's lips (bagpipes), mouthpieces with reeds (saxes and clarinets), and those where the lips alone create vibrations (bugles, trumpets, and trombones).

Musicians, manufacturers, and physicists seldom agree on the properties that make up a good sounding instrument.

Edge-Tone Instruments

Because the transition of the alternating wave of air is very smooth from edge-tone instruments, they almost have what could be called a sine wave or pure tone output. Their tone holes work the same as other woodwinds; however, half covering the ports and over blowing are also available to those playing these instruments. When the pressure of the air column across the wedge is increased, the note sounded may jump two (one octave) or three times higher in pitch. Students trying to master the recorder have the most trouble with accidental over blowing, which causes squeaky high note jumps. What's happening here is that additional harmonics are being added to the fundamental. These added harmonics are called "overtones." The first overtone is twice the fundamental frequency, and the second overtone would be three times the fundamental frequency. As any inexperienced recorder player will tell you, keeping these overtones under control can be quite a chore.

Flutes, unlike recorders, are cylindrical with the standard orchestral models running about 26 inches long. The end closest to the aperture sometimes has a movable cork stopper that can be repositioned for tuning. The embouchure plate itself is raised above the body. There are all kinds of designs and theories as well as controversy on this subject, but I can attest to wooden end pieces producing a mellower tone. The pitch of an orchestral flute ranges about three octaves above middle C, while the piccolo flute is an octave higher in pitch. It can be helpful to know the names of the parts that make up an instrument such as a flute:

The *body* is considered the middle part of the instrument, which contains the entire key section. The *crown* is the end cap of the *head joint,* which is the section that contains the embouchure plate and hole. The *foot joint* is the bottom section of the flute. Using a "C" foot joint makes it possible for the flute to range down to C, while a "B" foot joint extends the range down to B.

The thickness of the flute's metal cylinder walls has a lot to do with its tone. Heavier flutes sound darker than brighter light-walled flutes. The difference? As little as 2/1000 (two one-thousandths) of an inch! Don't think there'll be much of a sound difference between one head joint and another? Then explain why someone would pay a thousand dollars for one made of wood. It's simply because a wooden head joint produces a warmer and mellower sound.

Ports

The pitch of a note coming from a tube is changed by opening and closing the ports along its side because this varies the length of the internal air column. Woodwind instruments also use this method of changing the pitch. The side holes also introduce the sound into the air around the instrument. In fact playing some notes with the bell section removed will not greatly affect the pitch because most of the sound will emanate through the side ports. Different notes are radiated from different areas of the instrument and in different directions. Out in the audience, this is not all that noticeable, as reflections bring the full sound of the instrument together at the listener's ears. But when recorded using highly directional microphones outdoors or in rooms with deadened acoustics, the levels of some notes will vary according to the proximity of the mic.

Pipe Organs

Two types of pipes are used for organs: edge-tone (or "flu") pipes, which can be opened and closed at their end to change their tuning, and those that incorporate reeds that have a small rod that is moved up against them to change the tuning. This wedge or reed is used to trigger vibrations in the pipe's inner air column.

The variables of these pipes can seem endless. Diameter, length, the pressure of the jet stream, the size of the opening the air passes through, the size of the wedge, and its angle and exact location are just a few of the factors that determine the tone of flu pipes. The length of these pipes range from around 18 feet for a 31 Hz pipe to 2 ½ inches for a 15 kHz pipe. The C two octaves lower than middle C requires an 8-foot pipe, which has a frequency of 70 Hz and a 16-foot wavelength.

Usually reed pipes are angled or bent to face into the audience. Unlike the "swell" section of the organ, which uses shutters to control the level, these are used strictly for solos. On the other hand, the lowest pitched keyboard (called the "choir") is not as loud as the other sections and is used when backing up vocalists.

These instruments are the equivalent of a full orchestra and are controlled by up to three keyboards as well as a pedal board. They can have several pipes sounding the same note, or a "rank" of pipes with the same tonal sound grouped together under the control of a "stop." Pipes can be linked together according to harmonic content, octave spacing, or a smaller interval in pitch by "couplers." There are also drawbars, buttons, and presets to deal with. It surprises me that the organist does not require a copilot to handle the thing. In fact some of these settings can now be microprocessor controlled and thus preprogramed to allow the musician to "step through" the changes required to play a particularly difficult piece. Fact, every pipe on every organ is different so there can be no set micing method. However watch out for the phase cancellations that can occur with close microphone positions. If the acoustics of the room are good, just back off on the mic location and enjoy the performance!

Reeds and Brass

A stream of air fed into a tube can be periodically cut off by a reed, which is a tongue-like slip of wood or synthetic material like those used in saxes and clarinets. The reed acts like a valve that opens and closes allowing jets of air to be fired into the tube by repeatedly cutting off the airflow. This causes vibrations in the air column within the tube. The sounded frequency depends on the instrument's overall size and the number of ports that are closed. When all are closed the instrument will produce its lowest tone and when all are open the highest. But different amounts of pressure fed into the tube can produce additional harmonics and also excite a greater amount of energy. Top players use this phenomenon to produce tones that are theoretically below or above the capabilities of the horn. Most wind players strive to take advantage of this ability to play "off the horn."

No matter what the wind instrument, lip control is a dominant factor in correct intonation. The better players with good ears are able to "pull" a note into tune regardless of horn condition, temperature, humidity, or even fingering position.

When the wave of air shoots down the tube some of it will be projected out of the first open port. The rest of that initial wave will expand within the tube and some of this expanding wave comes back up toward the reed. If the musician adjusts their playing (mouth and lip formation, air pressure, etc.) the next jet of air will be released at the same time as the first jet's expansion returns back up to the reed, and a sustained note of a matching frequency will be emitted.

The partial vibrations that occur along with the fundamental of the note that is being played have a great deal to do with the makeup of the distinct sound of an instrument, but this will also vary with pitch. Reeds are not linear. They do not vibrate in nice smooth sine waves. They have more of a square wave shape to one side of their movement simply because they are positioned against a hard surface, which limits their movement in one direction.

It is difficult to predict the sound of any instrument especially ones that have tubes that are not completely smooth on the inside due to sound-hole ridges. In addition, with woodwinds the ports are always partially covered to some degree by the key pads.

With bassoons, the output of the higher harmonics can far outweigh the fundamental. Yet we still perceive the "missing fundamental" because our brains are capable of filling in what should be there in accordance with the higher tones we hear.

Bagpipes

The reeds in bagpipes do not come in contact with the player's lips but are enclosed in the pipes, which are "capped" off. Hence the name "wind cap" instruments. The result is that they do not produce a continuous pure tone but actually a chain of sharp pulses. This gives the instrument its "edgy" sound.

There are two main types of wind caps used in bagpipes. The "chanters" utilize double reeds and are used to play the melody while the "drones" use a single reed and are used for a single frequency background accompaniment.

Accordions

These instruments utilize metal reeds, which are bent up and rest over a slot. When air is forced across the reed it vibrates. The frequency or pitch of this vibration depends on the thickness of the reed and its length. The reeds are mounted in two boxes (or "cases") that are attached to either end of a set of bellows. Note selection is via buttons on the left hand or "bass" box and piano-like keys on the right hand "treble" box.

These buttons and keys open valves that direct airflow across specific reeds. Inside each box, a soundboard is used to amplify the sound of the reeds and broaden the timbre of the instrument. This is also aided by the air chamber within the boxes.

In addition to the buttons and keys, slide stops are often used to change the tone, and in some cases, they can add a note an octave below the one played or cause a single keystroke to trigger an added semitone. The player can use this to produce tremolo or a beating effect that occurs between the two tones.

The openings on each box that emit the sounds usually have an adjustable port. These are easy to locate as a decorative grill usually covers them, but they always project the sound to the audience through the forward facing side of the cases and to the player from the top panels.

Older "single action" accordions vibrate a single reed for each key. With the introduction of "piano" accordions, each button and key can cause the air stream to set a second reed in motion. Here one reed is used when the bellows are squeezed, and the other when pulled open. Thus every note is available at all times. To cut down on air losses, the internal slots are automatically blocked by leather strips when their keys are not pressed.

Mouthpiece Reed Instruments

The double-reed oboe has a conical shape (which provides more even harmonics). The alto oboe is larger and has an added bell, which has more of a Helmholtz resonator shape than the common horn bell. The tenor oboe is the bassoon. This instrument is so large that the port positions are not within easy hand reach. Therefore a sidewall is added to the instrument making it thick enough so that the ports can be reamed at an angle, putting the furthest openings within the player's reach along with the closer openings.

Clarinets have a cylindrical shape and utilize a single reed attached to a mouthpiece. Lip pressure is used to control intonation, and their additional ports and key configurations make the fingering a little more complicated than for flutes, saxes, or oboes. A clarinet's pitch is an octave lower than the same sized oboe because of their cylindrical shape.

Three registers make up the full range of a clarinet. The lowest matches the fundamental associated with the size of the tube and is the easiest to play. The two upper registers produce notes that are increasingly more complex as far as harmonic content, and are more difficult to play. Try hitting the F above the staff on a bass clarinet some day, and you'll understand what I mean. Yet over blowing an Eb bass clarinet for the G above the staff is possible with a stiff reed and a lot of practice. Science cannot factor in human determination.

Trumpets and Other Cone-Shaped Instruments

Conical instruments are smaller at the input end so the air wave increases in size as it moves down the tube, but because it is recompressed on its way back up, the harmonics of conical tubes can act much like those in cylindrical tubes. Yet, clarinets, which have cylindrical tubes and an end closed by a reed, produce odd harmonics when overblown, while oboes, which have conically shaped tubes and an end closed by a double reed, produce both odd and even harmonics when overblown. Once a bell is added to a tube, the sound waves in the air column take longer to develop, and during this time the timbre changes. This changing tone is one of the main factors that make up the distinct sound of an instrument. Yet the timbre of individual instruments can also be altered by the player's lip formation and stance, the level at which they are played, as well and the acoustics of the environment in which they are played.

The Trumpet Family

Horns evolved over the centuries to what we now consider to be standard configurations. The mouthpiece itself is very important in that it must provide a comfortable seat for the player's lips. It also has its own resonant frequency. If you slap the opening, the tone you hear will be that frequency.

The low end is increased by the shape of the bell, which also increases the radiation of high frequencies, and this affects the harmonic content of the horn's overall sound. A horn's cutoff frequency refers to the point below which much of the energy is reflected back up the tube, while the frequencies above this point are more readily projected.

I had once heard that French horn players kept one hand inside the bell while playing to achieve different pitches. Upon questioning a French horn player, I found that this was the case in the distant (pre-valve) past, but today the practice is continued only because it allows the musician to vary the cutoff frequency and therefore change the horn's overall sound quality or timbre while playing.

Without valves or slides, horns are only able to produce a single harmonic sequence of eight or nine notes, and older so called "natural" horns were capable of little more than the type of melody that is typically obtained from bugles.

In order to play in different keys, short sections called "crooks" were added to the horn to effectively increase the length of the instrument. Side holes had been added previous to this, but the real breakthrough came with the addition of the sliding crook as is found on the trombone. Permanent crooks were then included on trumpets and their lengths added into the horn body by the using pistons or valves. This developed into the current three-valve system, which lowers the pitch of the horn by a half, whole or a step and a half.

The Human Voice

The vibration excitation mechanism of the voice consists of two thin flexible sheets of tissue through which air is forced, very much like when stretching the neck of a deflating a balloon. Our vocal chords' ability to vary pitch is not only used for singing but for expressive speech as well. The resonant cavity we use is unique in that it consists of the throat, nasal passages, mouth and all their movable parts. Pitch is adjusted not only by varying the tension of the vocal chords but also by the size and shape of these cavities. Think of how many tones you can get from a jaw harp simply by changing the size and shape of your mouth.

The waveform expelled by the vocal chords ends up being a series of pulses with sharp peaks, which include many harmonics. At the same time, the air turbulence that's expelled along with the sound causes full bandwidth (white) background noise. The sound leaves the vocal chords, goes through the larynx, then out the sinuses and nostrils along with the mouth and lips. All these parts can be changed along with the tongue for a vast variety of tones. Yet every person has his or her own characteristic sound. Can you not recognize the voice of most of the people you know? These tones can vary from the bite of anger to the cooing of a mother to a child—all performed automatically without any effort. So what?

The training a professional vocalist must master includes the placing of all the muscles that control the opening and closing of the vocal cavities into exact tensioning. The conscious control of these muscles is not something that is normally possible. In addition, illness, stuffiness, or allergy problems throw the whole process off. So the next time you're recording a vocalist who is having a difficult time, cut them some slack with a kind word, a glass of water, a comfortable seat, and maybe mention that *you'd* like to take a short break right about now.

AN UNIMPORTANT NOTE: *I always thought that helium made voices rise in pitch because it affected the vibration of the vocal chords. The fact is that the vocal chord frequency is not changed at all. I found that the rise in pitch occurs because sound passes through helium three times faster than through air. Therefore when our voice cavities are filled with helium the sound comes out three times faster.*

Most recordists place the voice in the same category as brass instruments as far as its radiation pattern is concerned. This holds true to some extent in that you'll often be more concerned with breath pops and sibilant mouth sounds than with trying to pick up the "full frequency content" of a voice. Omnidirectional mics will reduce proximity effects or the increased bass that comes with in-close mic positionings, while a cardioid mic placed off axis will cut down on some of the highs and pops. There's also a world of processors devoted solely to taming vocal problems, such as de-essers, filters, and compressor/limiters that can be utilized.

Yet while high frequencies are projected very directionally out of the mouth, sound also radiates from the nose, chest, and head. During my ambitious experimental days, I had a single ear headphone setup with a vibration pickup attached to the off-ear support base. While I had originally designed it for use with horn players, it picked up some nice low end from vocalists as well. In both cases, the sound is affected by the shape or size of the mouth cavity. Since the low end vibrates the jaw and head, few people outside of the players themselves ever hear what a tenor sax truly sounds like.

How did I know about this? I've played some woodwind instruments. Oh, not all that well, or I'd have earned my living as a musician and had more fun, instead of calibrating equipment, replacing electronic components, and jockeying faders. But it clued me in on the possibility of this very specialized recording technique.

This is one of the reasons I believe it's important for every recording engineer to make an attempt at playing a music instrument. It puts you right in the musician's shoes, aids in your following along with charts, hips you to the lingo, and opens up your comprehension of the whole music-making process.

You wouldn't want to appear to be an alien in your own field would you?

Vibrating Strings

Instruments like violins and guitars use strings as the main source of sound excitation. Just as the pitch of the air column in a woodwind instrument is varied by opening side ports to change the tube's length, when the vibrating source is a string, varying its length by fretting it changes the pitch. In fact halving a string's length (by sizing or fretting) doubles the frequency. Because of their ability to produce multiple tones, strings, unlike blocks, do not use tuned resonators to increase the level and duration of the tone. If this were the case, the musician would have to be able to continually vary the resonator's size.

Whack a tuning fork and then hold it up to your ear, and you'll only be able to hear it faintly. Hold it a foot away from your ear, and it almost disappears. However, if you hit it and then hold the base of the handle against a tabletop, it can be heard at a good distance. What's going on here?

Like a string, the tines of the tuning fork are not capable of moving much air. However, they are able to transmit a vibration through their stem to the larger table for a short time duration. The table, on the other hand, can move a good deal of air. Therefore, the table, acting as a resonator, uses the vibrations of the tuning fork more efficiently. If you place different pitched tuning forks against a tuned tube or hollow box, the level varies with the frequency. This is because the size and density of the object used to increase the resonance has an effect on the sound. Resonators often do not react quickly enough to transmit the beginning or attack of a sound, and the attack is an important part of the overall sound because it provides major clues to our brain/ear recognition system. Furthermore, whenever a vibration is transferred to a resonator, there will be some frequency loss because no vibrator/resonator interface is perfect.

Bows

Bows like other sound actuators, such as rubber tipped mallets and serrated strikers, produce vibrations via friction. Horsehair is traditionally used for bows because it has a scale-like surface, which does a great job of holding onto the rosin applied to it. This combination increases the bow's ability to momentarily "grab" and then release a string, bent saw blade, bowl, or whatever is being excited. Another example of this is the rotating wood wheel of old-time "hurdy-gurdy" instruments whose outside edge was coated with rosin to help excite the strings.

Bowing is utilized to prolong the vibration of the string. The process requires friction as provided by the rosin, which both sticks to the string and then, at a certain point, allows the string to slip a little only to catch it again. It's kind of like what you feel when causing wet glass to squeak by running your finger around its rim. This is how glasses of water can be played as an instrument. Here the amount of water in each glass determines the frequency produced. This is due to the fact that the water adds mass and dampening to the glass; therefore, the more water added, the lower the pitch. Don't take my word for it; go ahead and try it for yourself the next chance you get—that way it will "stick" and thus increase you knowledge base on the gut level. For a real eye/ear full go to Fur Elise on glass harp on YouTube and check out a cat named Robert Tiso playing a couple dozen water-filled glasses of different sizes, while laying down basic and overdub tracks simultaneously via split screen.

With glass harmoniums (I've also seen them referred to as glass "armonicas"), water also facilitates the vibratory excitation of sound. Invented by Ben Franklin no less, this instrument is made up of a series of glass disks that graduate in size and are fixed to a rotating rod. This is set in a trough of water, and it produces sound due to the friction produced when the player applies his fingers to the rotating wet glass disks. Due to the fact that they supposedly cause neurological disorders to their players, they are often replaced by a celeste. This instrument is sort of a cross between an upright piano and a glockenspiel. It has metal bars like the later, but they are struck by piano keyboard lever-actuated, felt-covered hammers. The celeste is thus more conducive to playing chords, and because of its dampers and sustain pedal, the duration of the bell-like sound it produces can be varied by the musician.

Bows are a bit more complex. The bow sticks to the string and causes it to move to the side. When the tension reaches the point where it overcomes the frictional hold of the bow, the string slips back a bit until the bow grips the string again, and the cycle repeats. Thus a bow can continually oscillate a string using this catch and release process, as when using a figure-eight bowing method.

Hollow Bodied Instruments

Lutes have been around for over four centuries. Their hollow bodies are shaped like half pears to which a neck, bridge and strings are attached. The strings are excited when plucked, and the pitch is determined by the length of the string as per the position of the finger fretting it against the neck. Most of the strings use a second string in conjunction that makes up a "course." Therefore some of these instruments have up to 24 strings and are a bear to tune. There are countless hollow body shapes, yet all are light and rigid.

Guitar bodies are usually thin and depend on their large top plate to amplify and sustain the string vibration. The construction of guitar backs and top plates is extremely complicated and includes special strutting systems to increase various modal vibrations. Opinions on this topic are *not* held lightly. The whole guitar body expands and contracts when the lowest notes are played with air being pumped in and out of the sound port, which acts like a Helmholtz resonator. This means that the diameter of the port and the size of the cavity are important contributors to the instrument's sound. The front plate is excited by all frequencies, and it is the primary area from which sound is radiated. It is easy to find studies of the vibration modes across guitar top plates, the earliest of which used a thin layer of sand placed on top of it, which would be displaced into patterns by the vibration modes.

Violins

There are hundreds of texts available on violins. Delving into this subject can be a quagmire of conflict between subjective opinion and scientific "facts." But a short synopsis of what I've learned may prove helpful. By the end of the 1600s, the violin began to replace its predecessor the *viol*, which was not able to output much in the way of level and was thereafter used only for chamber music. The treble viol had six strings, frets, and a body that was almost the size of a small guitar. With its floor peg, it looked like a mini cello.

The best violins were supposedly made between 1700 and 1750, but by the early 1800s, almost all of these had been "rebuilt" to provide the level output required for symphonic pieces. That's right, only a very few of the surviving violins made by Stradivari are actually untouched originals, and those that are do not project as much sound as those that have been modified. *What*?

Baroque violins are like their modern violin counterparts in that they had no frets, but their necks are not angled backward, their shorter fingerboards rest on a wedge-shaped piece of wood, and they have a lower bridge. All this adds up to a sound that is not capable of cutting through the din of a symphony.

The front and back plates of modern violins not only vary in their thickness but the total open area of the "F" holes is also important to the instrument's resonant frequency due again to the Helmholtz effect. This port shape is believed to help decouple the top (neck) and bottom (tail) ends of the "belly" or top plate. Yet underneath the belly, there's usually a strip of wood called the "bass bar," which recouples the belly and strengthens the body. It is positioned under the lowest string, while a "sound post" sits under the highest string. Arguments may abound over sound posts, but they have a major effect on the violin's tone and certainly increase its level and projection, especially at lower frequencies. It is so influential to the sound of a violin that the French term for it is "l'ame" or the "soul" of the instrument (see section on Bridges).

The Making of a Violin

The wood used on the belly usually has close straight grain for stiffness, while the back is made up of two sections split from the same piece of wood to achieve "acoustic symmetry." Since all wood varies in elasticity, weight, and damping factor (even when two pieces come from the same tree), violins are made and then adjusted according to the properties of the wood used. As an example, if the grain runs perpendicular to the face, it will be stiffer; while if it is run at a 45-degree angle, it will be more flexible. Man-made materials are more uniform, but they too require adjustments according to their properties. These can be very beneficial when used for individual sections of an instrument. One example of this is the plastic back plate of an Ovation Guitar. Usually you'll find a line running around the perimeter of the top plate. This is called "purfling" and consists of an inlay of hardwood, which impedes cracking at the edge and adds a decorative touch to the instrument. It also changes the stiffness of the back or belly at the edge area where it contacts the internal braces and, thus, affects the sound of the instrument. There are so many variables it becomes mind-boggling. Yet some makers are adept at tapping on wood at certain areas in special ways to excite an unfinished plate so as to help them determine its resonant modes and point them to the needed readjustments.

How about the varnish? Fact: Hollow bodied instruments generally sound better without any varnish, but because they are made of wood, they are affected by humidity. In the 1700s, concert halls did not have central air conditioning. Fill the place up with sweaty humans on a hot summer evening, and the tuning of the wooden instruments would start to change. This was unacceptable so a light varnish was used to seal the wood and prevent moisture penetration. A heavy layer of varnish would obviously dampen the sound of an instrument, but it was found that a hard varnish could increase the stiffness of wood that was a bit too flexible. So while varnish is important, there is nothing magical about it.

Good or Bad Violins?

You will find no consensus among physicists, makers, or musicians as to which instruments sound or "play" the best. The physicist may be most interested in the evenness of the vibration modes, the maker in the timbre and visual beauty, while for the musician, the "feel" or the way the instrument responds to their playing is the most important. The response of a violin is often described by musicians as its ability to "sing out" from the very moment the bow touches the string. This quality depends not just on the elastic qualities of the varnished wood body but also on the bow used as well.

Bowing

In addition to the rosin coated hairs catching or gripping the string, the bow must release it in a uniform manner. Extraneous sounds caused by the irregular release are more readily heard with players who are just beginners. Bowing should produce an almost sawtooth wave shape, where the sound rises slowly, building as the musician runs the bow across the string, with a sharp sudden end when the bows movement ends.

Most important to the recording engineer, is the level from note to note, then comes the instrument's sound and dispersion properties. While it may be true that the better musicians are capable of evening out the tone of an instrument by adjusting their playing style, doing so will almost certainly detract from their performance. Remember, aiding the musician is part of the recordist's job so knowing sources for fine-sounding rental instruments can be very helpful. However even with an ideal sounding instrument, microphone placement is still very important simply because the polar patterns of acoustic instruments vary with frequency.

Instruments such as cello and bass also radiate sound through their floor pegs, and risers have been made to increase the power of this radiation. Cello floor pegs were not originally added for any acoustical reasons, but to release musicians from having to hold the instrument off the floor by using pressure between their knees.

We've mentioned tuning forks and how their vibrations can be transmitted to a larger radiator to increase the power output. Supposedly this is also what happens with cello and basses. If you place your hand on the area where the peg is attached to the instrument you will feel that the vibrational excitation at this "endblock" is very weak. Yet, it can be felt by the player, and when these vibrations excite the structure (floor and walls) of an enclosed space, it can also add to the sound received by the audience. Ask a bass or cello player about what their downstairs neighbor thinks about their practicing at home. The musician often can't believe that there's anything substantial to complain about, since the sound power within the space they are playing is seldom affected by these vibrations. However under the floor, which acts like a giant membrane resonator, the vibration comes through loud and clear.

In a performance situation, the average sound pressure level is affected only slightly by the addition of this vibrational radiation except at the lower frequencies. While the body of the instrument does not radiate much energy at a fundamental of 64 Hz (because the wavelength is too large), a riser can increase the SPL of these frequencies by as much as 6 dB. When musicians can feel the low-frequency oscillation coming from the riser, it aids their playing. In fact, some of the top players take not only very rare and priceless instruments on tour but also their favorite risers.

Remember that a plucked string only emits a small amount of sound. However it can easily vibrate the bridge, which transfers these vibrations to the top plate, which in turn resonates the high-mid frequency content. The top plate also causes the ribs, sides, and bracing to vibrate, transmitting the sound to the back plate, which emits low-mids, as well as to the air cavity within the sound box, which pumps low frequencies out of the sound hole.

The string's "transverse" force refers to the side-to-side movement, and the "longitudinal" force refers to the change in the string's up-and-down motion. Both vary the length of the string during the vibrating cycle. Thus string forces are both parallel and perpendicular. Therefore the force a string exerts on the bridge can be both across (parallel) and downward (perpendicular).

Bridges

The main function of the bridge is to transfer sound in the form of vibrations from the string to the soundboard. To do this effectively, it must serve as a solid end point to enable most of the string's vibration energy to be transmitted through it. If you were to add dampening material between a bridge and the soundboard, obviously the level of sound transferred would decrease. When a string moves or slides across the top of the bridge, it causes a buzzing sound. Therefore, bridges have rounded tops with notches to hold the strings in place.

The design of violin bridges goes far beyond these rudimentary design concepts. Besides standing taller and having arched tops to facilitate the bowing actuation, the bridges of the violin family are often ornately carved and have two distinct contact points ("feet") with the arched soundboard. This seemingly decorative-only carving actually modifies the transfer of sound. In most cases, it is used to reduce the high-frequency content, but there are specialized shapes and many variables.

The foot of the bridge under the high string sits on top of the instrument's "sound post." This is a solid link between the front and back plates, which restricts the left foot's up and down movement so that it only pivots. On the other side, under the low string, the right foot sits on a rib called a "bass bar," which runs just under the top plate parallel to the back. It has two purposes: to stiffen the top soundboard and to spread the vibrations transferred to it by the right foot. This configuration also helps to decrease phase cancellation. The downward motion by the right foot causes a larger amount of vibration on the right side of the top plate and excites the instruments air cavity. This causes sound to be projected out of the 'F' holes, while the out-of-phase left side of the top plate remains relatively quiet. When the string motion reverses, so do the motion of the bridge, its feet, and thus the vibrations of the top plate.

Unlike violin bridges, most of those used for hammered strings (dulcimers) do not require pivoting because the motion of the strike is transferred to the soundboard more directly. Some bridges are either glued or screwed (classical/electric guitars) in place on the soundboard. This is because the strings must be anchored to them. In other cases, the strings pass over the top of the bridge and continue on to an anchor point beyond the bridge. In this case, only the downward pressure of the strings is used to hold the bridge in place. Some of these bridges can be moved fairly easily; yet they require a larger contact surface area to facilitate the transfer of vibrations. In most cases, bridges are affixed to the instrument either permanently, as when screwed on, or semiperma-nently, as when a "tail piece" and a special end "button" is used.

Tailpieces for the violin family of instruments are generally bugle or fan shaped and some have built-in tuners for each individual string. They usually match the curve of the bridge and are attached at the end to a tail button at the bottom of the instrument. These buttons can be made of gut, nylon or steel and comprise a whole field of expertise unto themselves. The material a tailpiece is made of will determine tonal quality, with a lighter tailpiece "ringing" out more. In addition to having the same curve as the bridge, the pitch (or descending angle) of the string on the tailpiece side of the bridge should match the pitch of the string on the fingerboard side of the bridge.

A professional cellist once showed me that when the bridge and string angles are correctly set (that is the string angle being the same on both sides of the bridge), you can pluck the "G" string *behind* the bridge and the "F" one octave above the G should sound. I do not know the interrelationships between the two sides of the other strings, but I witnessed the fact that it made retuning the "G" string a cinch for this cellist.

Bridge movement is also used to tune an instrument by changing the string's effec-tive length. If flat, the bridge is moved toward the neck. If sharp, the bridge can be moved away from the neck.

Tunable Bridges

These bridges are commonly found on electric guitars, and they facilitate tuning and intonation corrections, as they allow both the length and height of the strings to be adjusted.

Middle Bridges (As Found on Zithers)

These stand higher than end bridges because they are held in place by just the string pressure pressing down on them. This accomplishes the important task of allowing the bridge to serve as a stop point for the string vibration, and increases the transfer of vibrations. Not only do you get sympathetic vibrations from the side not struck, but also by varying the mid-bridge placement, different intervals can be excited. For exam-ple at a two-thirds position, the two notes sounded will be a full octave apart.

Electric Guitars

The string excitement of a solid body guitar is so slight that very little of the string's energy is depleted. This translates to longer sustain. The use of multiple pickups helps take care of amplification and also helps with the blending of the frequency content. Pickups located near the neck produce more of the fundamental, while at the bridge will be found more upper harmonics. This is almost the opposite of acoustic guitars!

Violin Bodies

Things are little different with violin bodies due to the fact that they have arched tops and have a solid interconnecting sound post. Here you'll find a ton of information readily available. It's almost as if every physicist is required to study violin acoustics for their graduate dissertation.

The variables in the test methods used are diverse not just in terms of whether the strings are bowed or a sine-wave generator is applied to the bridge but also whether it is applied at the bass bar or sound post side. This will make a big difference in the test results. Additionally, ceramic pickups, proximity detectors, optical sensors, or laser reflections can be utilized to record the excitations. And of course, don't forget that the use of weighting scales and time averaging also factors in. All this makes utilizing the results difficult to say the least

Some Guitar Manufacturing Theories

NOTE: *These can vary from maker to maker.*

Large headstocks resonate more than smaller ones so they can cause a reduction in string resonance and sustain. Smaller headstocks also reduce the string length, which increases the strings downward pressure, which further increases the attack, and sustain. But increased downward string pressure can also be achieved by angling the headstock backward.

Headstocks should feed the strings to the nut (the slotted block of material at the top of the neck) in as straight a line as possible to reduce the chance of the string binding up against the slot edges because they will eventually move and require retuning. Arch-topped, solid-body electric guitars have a greater mass under the bridge and thus provide more sustain.

Acoustic guitar top plates made of solid wood are more flexible than laminated wood top plates and thus resonate more easily and provide better attack, sustain, and sound projection. A guitar's "scale" depends on the length of its strings as measured from the nut to the inside of the ball-end attachment. Less than 32 inches is considered "short," 32–34 inches, "medium," 34–36 inches "long," and longer than 36 inches is "extra long."

Setting a Guitar's Intonation

First you must accept the fact that no guitar will ever be perfectly in tune across the full length of the fingerboard. The frets have been mathematically positioned to provide correct octave spacing. This leaves the other intervals imperfect. A little time utilizing a strobe tuner will prove this out. Yet the discrepancies are spread out across the length of the fingerboard to make the sound acceptable.

NOTE: *Trying to tune a guitar to perfect intervals like pure thirds or fifths will leave you with an out of tune guitar.*

Since they are not familiar with every owner's intent, guitar manufacturers perform only a preliminary setup on guitars. Therefore, all newly purchased guitars require *action jobs*, whose rundown follows:

1. Set the string height for a comfortable easy-to-play feel.
2. Adjust the intonation so that it plays in tune.
3. Finish the frets, nut and bridge to eliminate all buzzes.

The steps are as follows:
   a. Remove all the strings and straighten the neck while it is not under tension.
   b. All frets are rounded, honed, and polished to a uniform height. The slots in the nut are filed for the proper string height and width to prevent the string from binding, thus making the tuning more durable.
   c. Re-string and re-straighten the neck under full tension.
   d. Then fret the bass string at the first and fourteenth frets. A slip of paper should slide between the string and the seventh fret. If the neck did not have this slight curve, the strings would buzz across the top of neighboring frets. The bridge is adjusted for the correct string length and height. These adjustments are interrelated.
   e. If there are no mechanical adjustment screws for the height, the "saddle" or the material inserted under the bridge is adjusted.

NOTE: *Another theory holds that you should always re-string an instrument one string at a time and complete its tuning before moving on to the next string.*

Buzzing caused by flat bridges and wide nut slots can be a major distraction for the recordist. Instead of suspecting every cable, connection, ground, and effect in the recording chain, a lot of time can be saved by knowing exactly what to look for on the instrument. If you hear a buzz that occurs only when the instrument is being played, check the frets, the nut, and the top of the bridge. The buzzing caused by a flat-topped bridge is used as the main sound on an instrument called a bugle or trumpet guitar. The name should clue you as to the buzzing sound it produces.

NOTE: *An outside temperature of 65 degrees can equate to a temperatures of 100 degrees inside an automobile. So you should never leave music instruments inside cars, especially guitars, or you'll end up having to file down the edges of the frets. What? Heat shrinks wood more than it does metal, so the frets may end up jutting out beyond the sides of the fingerboard. Ouch! It's an easy fix, which requires only a large flat 400-grit sharpening stone and some oil (try lemon oil). Lay the stone across several frets, and move it back and forth. By stopping often to make sure you are not grinding down the side of the fingerboard, the job will be painless.*

"Profiling" is re-rounding the top of the frets using a two-sided "safe" file (along with masking tape over the fingerboard) and wet/dry sandpaper at 300 to 400 grit.

Finally the fret is polished to a round shiny finish. Professional guitar repair people own specialized files and other tools for this purpose.

The Harp

This instrument is unique in two respects. Not only are all of its *two* sets of strings fully exposed to the player, but also their attachment to the lower soundboard is closer to perpendicular than parallel—"closer" because the angle is more in the order of 40 degrees than 90. I once read that a 90-degree attachment would have resulted in the strings transferring no first order vibrations to the soundboard. I cannot completely explain the reasoning behind this, but I know from micing it that the soundboard resonance is very important to the harp's overall sound. This lightweight soundboard is shaped like a sound box itself—long and tapered—and the strings are affixed to a heavy longitudinal brace that runs down its center. The string motion, being more perpendicular to the soundboard, makes the transfer of vibration faster. This results in a louder note, but it also lessens the string's sustain.

Some harps have rotating metal parts that when placed in contact with the strings shorten their length, which raises the pitch by a semitone. Regardless of this, the top note of a concert harp reaches only to about 2500 Hz. Yet when the stings are plucked using banjo-like finger picks, they emit a bell-like sound that's very close to the "ping" sound of a steel drum.

A good harpist can literally make this instrument sound like a Fender Strat playing a rhythm part, a Telecaster picking lead, a Gibson hollow-body electric punching out a rhythm part, or a Fender Mustang playing a walking bass line. So it's a good bet for the recordist to be set up and ready to capture everything this instrument has to offer. Because of this, I prefer to three mic it, using one from above to pick up the overall sound, which radiates from every part of this instrument including its metal harp and pole, a second at the soundboard to capture the lower or mellow sound, and a third aimed at the midsection, where the player's hands contact the strings to pick up the attack of percussively plucked string sounds.

Mechanically Plucked Strings

Harpsichords differ from pianos in several ways, the biggest being in how their strings are set in motion. While piano hammers strike the strings, the harpsichord uses goose quills (or their plastic equivalents) to raise the string until the tension overcomes the strength of the quill and the string snaps downward. The more elaborate of these instruments have double keyboards with the strings being plucked at different places to achieve different tones. Of interest to the recordist is the fact that harpsichords have no level variation regardless of how hard the musician strikes the keys. This is because the string is always moved the same distance by the quill before it is released.

Clavichords use neither quills nor hammers to excite the strings. Instead "tangents" or metal blades are used. These blades are left against the strings after the strike, which causes the string to sound more than one tone. This instrument not only provides the musician with a wide dynamic range as the tangents velocity is controlled by the hardness of the key strike, but they can use the tangents to give the tone a vibrato effect by not releasing the key completely.

More on the Harpsichord

This is a bizarre instrument to say the least. The strings on a normal full-sized (non-spinet) keyboard are affixed at the curved side, bend around metal posts that jut out of the soundboard, then run around another set of pins at the nut before attaching to the tuning pegs. The sound of a well-played harpsichord in good condition in an acoustically fine room is tough to beat, but they suffer a bit more than other acoustic instruments from age. Because old violins are supposed to have a mystical sound quality, many people think the same is true of all wooden acoustic instruments. While old guitars can rattle and buzz, and antique pianos might plink and plunk, a misbehaving harpsichord is something you *really* want to shy away from. Why? The action.

When pressed, the rear of the key pivots upward raising a ruler-shaped piece of wood called a "rack" on which is a affixed a spring loaded quill (as in bird feather quill). The jack lifts the damper, the quill contacts the string, moves it upward, then bends, and releases the string for that delightful "plucked" sound. At the uppermost extent of its travel, the jack bumps into the felt-lined wooden action cover. On the way back down, not only must the quill be tucked in out of the way of the string, but the felt damping pad must also be made to come in contact with the string to stop the sound. Wood, felt, feather quills, and glue do not age well. Take it from me; if you come across a harpsichord with problems, it may be best to run for the hills.

Percussion Instruments

Pianos

The collection of levers, wooden rods, felt pads, metal pivot pins, and strips of leather that make up the piano's hammer mechanism, or "action." at first appears primitive until you take into consideration the outstanding job it does. It moves the damper onto and off of the string and causes the hammer to: strike and completely lower or stop at an intermediate position ready to strike again, all with a dynamic range that is minutely variable. In addition, every key must feel identical, and there should be no hammer bounce, which might cause double notes to sound. This mechanism must reset itself quickly enough to allow for a rapid succession of trill notes, yet be ready to dampen the strings instantly when the key is released.

A piano's soundboard is huge and the steel frame that supports it is massive. This combination allows for the use of very thick steel strings under high tensioning. This is the reason for the piano's high output level.

I remember one physicist claiming that since the hammer is not actually connected to the key, the only variation in the dynamics of play available to musicians was the velocity of the key movement. In this particular experiment, they were not measuring the response caused by a player, but that of objects made to fall on the keys. In fact, the piano string partials not only decay at different rates when struck with varying force, but musicians also have touch, timing of hits, damper release, and a sustain pedal, all of which provide them with a fairly wide dynamic range.

WARNING: *As an audio recordist, you'd have to be out of your mind to muck about with piano, clavichord, or especially harpsichord mechanisms.*

The radiation pattern of an acoustic grand from 10 to 11 feet out is almost omnidirectional. However, the lower frequencies will usually be around 15 to 25% louder if the lid is partially or fully closed, as this attenuates the higher frequencies more than the lows. You can help balance this out by positioning your high-end mic a little closer in. I prefer a close-in positioning at the hammers, as it renders more of that percussive, hammer-hitting, string-attack sound which, when added to an overall or "room" mic, makes for a truer approximation of the actual sound, and at the same time gives you something that will cut through most any mix.

Blocks

Blocks made of wood, plastic, or metal when struck bend slightly, and their elasticity causes them not only to spring back to their original position but also to continue on, overshoot it, and then rebound back. This displacement is a perpendicular movement and is, therefore, called a "transverse" vibration. The duration of this vibration, while short (only fractions of a second), still produces an audible tone. Xylophones are an example of using only this one simple method of tone production.

Marimbas and Xylophones

These two instruments are almost identical except for two important distinctions. The underside of the sound blocks, or bars, on the marimba is undercut to a greater degree, and the mallets used to play this instrument are softer than those of a xylophone. This equates to the xylophone having a brighter, more attack-oriented sound than the mellower marimba. They both utilize tubular resonators located underneath each bar to help amplify their sound. These resonator tubes are closed at the bottom end. Each is tuned to match the bar (their length being one quarter of the block's wavelength), but moving the tube closer in lowers the resonant frequency.

Some ethnic and tribal percussion instruments use gourds that are shaped like bottles to amplify percussion sounds in this same manner.

Vibraphones

These fit right into the same category as marimbas and xylophones. Their aluminum bars are undercut at the bottom, and their first overtone is double that of the fundamental. Aside from the longer decay time afforded due to the use of aluminum bars, their resonators also differ in that there is a motor driven disk fitted into each tube. The vibrato caused by the opening and closing of the resonator affects both the amplitude and, to a lesser degree, the frequency of the resonance. Vibes can be played with soft or hard mallets depending on what the piece calls for.

Sound Actuators

Rubber mallets can effectuate the same kind of grab and release action as bows if their handles are flexible enough so as not to impede the vibration of the head. Almost every conga player will wet his fingers for that distinctive vibratory drag across the skin just before an accentuated pop. I know a couple of percussionists who apply a little rosin to the top of their heads and one who keeps a well resin-impregnated cloth at the ready.

While everyone is familiar with the Latin "guiro" or ridged gourd, which is played by scraping it with a stick, there are also serrated sticks, which can be used on all kinds of objects to produce the same type of percussive sound. When it comes to mallets, the selection is practically endless. The three most important parameters are weight, hardness, and shape. The choice can make a huge difference on the sound produced. A hard mallet, when used on a marimba, will excite more overtones, which enhances its "woody" sound, while a soft mallet excites only the lower harmonics, which augments this instruments mellowness.

Mallets act differently according to their weight. A mallet that equals the weight of the struck object will transfer the most energy. Light mallets bounce off of the object after a short contact time, while heavier mallets remain in contact longer. This longer contact period has the effect of cutting off much of the higher harmonics. While this would certainly not be ideal for vibes and xylophones, it results in louder drum sounds with shorter decay times and less ringing. Try giving the drummer a heavy set of sticks if you've got a set in your recording bag of tricks. I always carried several different sets; but then again I'm a big fan of percussion.

Membrane Radiators

Some instruments' sounds are produced by a vibrating membrane that has been stretched to increase its tension, while others use stretched membranes just to help radiate their sound. Of these, the banjo is the most notable Western instrument that utilizes a membrane resonator, but it is more common in other cultures.

Membrane sound actuators are very effective in doing their job. They are all fairly large and are capable of moving a good deal of air, and since this is done directly, there is no loss due to a mismatched interface.

Drums have three distinct components that determine their sound: the membrane, the body, and the size of the air mass within that body. When the head and air cavity are tuned to the same resonant frequency, the drum will "sing" much clearer (have a more defined pitch) and have a longer rate of decay.

Timpani or Kettle Drums

Striking timpani or kettle drum-like heads at a point midway between the center and the side produces a sound that is rich in both the fundamental and upper harmonics. In fact, you end up with a duller sound when you strike timpani at the dead center point. Yet the sound also changes if the kettle is made of fiberglass instead of copper. Don't try to talk a classical percussionist into changing their calfskin heads to synthetic ones. They will not be interested in loosing the warm tone that this drumhead material provides. With this instrument, the body or the kettle itself baffles the sound of the membrane and increases the decay time of upper harmonics.

Bass Drums

Orchestral bass drums are quite a bit larger than those commonly found in the average drum kit. Instead of 22- to 26-inch diameters, here we are dealing with drums with diameters of 30 to 40 inches. These big boys put out a lot of sound power. To avoid distortion, make sure you use a microphone that can handle high SPLs.

Drum Heads

Single-headed drums produce a defined (more singular) pitch, but when two heads are used, normally each one is adjusted to a different tension. From my experience most classical players tune the striking head to have more tension (for better mallet rebound) than the resonating (non-struck) head. This also accentuates the attack sound. On the other hand, drums with both heads tuned to equal tensioning have a longer decay. A longer resonance or decay rate can also be gained from using thin drum bodies, as they are excited and vibrate more easily. Malleable risers (or legs) under floor-standing drums will also lengthen the decay time, as does the use of lighter and thinner heads made from a rigid material like plastic as opposed to soft goatskin. Thicker cowhide is responsible in great part for the wide bandwidth and high frequency overtones of congas, while thin pigskin or goatskin and hair-on drum heads under looser tension will accentuate the fundamental.

Instead of droning on and on about all the different types of drum heads (Kelvar, Mylar, rubberized canvas, etc.), I'll end with a few final tips.

If the player wishes to utilize a "grab and release" style of play, it's easier when the surface of the skin is roughed up a bit. I know of several percussionists who carry tools, such as sharpened nails, shortened knifepoints, and even quill-type ink pens, specifically for this purpose. It's the same for brushwork on a snare, which is facilitated by the use of a rough-skinned drumhead. Synthetic skins tend to be more uniform in thickness, have even tone production, are more stable with changes in temperature and humidity, and are also more affordable. Natural skins can dry out and crack. In this case, a small amount of natural oil applied to the skin will improve the sound and stop the head from splitting. There are many theories as to the best oil to use, but I've found olive oil to work well, and no matter where you are, it's easy to come by.

Drumhead modification can include adding damping or weight to the head. At one point I tried to get real scientific about this. Many test results giving the vibrations of circular membranes are available, which show the locations of harmonic modes. But I quickly gave up on this because the type of head, stick, and playing style used can radically affect mode location, making it practically impossible to nail down a single method of damping for all situations.

However you will have a good deal of success by using standard drum dampers and by weighting the center of drumheads. Drum manufactures continue to use those adjustable, spring-loaded dampers with thick felt tops located at the perimeter of the drum shell because they work very well at reducing overtones and ringing. If you really want to augment the fundamental for that low thunder drum sound, try taping coins to the inside or outside surface of the drumhead. My favorite method was to gaffer tape a ring of quarters (half-dollars were too heavy) under the head around the circumference of a Black Dot. However I've also gotten good results with washers, screws, and even magnets taped to the top of a head. I know of one engineer who carried a wallet exclusively for this purpose; it worked rather well at adding low-frequency sustain while reducing overtones. I've seen conga players sanding down the outside perimeter of their heads to achieve this effect, but melted wax on the center is easier to modify. Taking a closer look at a tabla will reveal that a black sticky putty-like substance at the center of the head will help give this instrument its well-defined (fewer overtones) pitch and long decay time. One tabla player told me that lead was included in the mixture.

Drum rims and hoops are used for wrapping and holding the heads to the top of the shell (or body) of the drum. They also facilitate gunfire-like rim shots. Usually there is some form of tensioning mechanism involved, and for the sake of an even sound around the whole perimeter of the drum, an even tensioning is your best bet. Lightly tapping the drumhead at the area in from of each tensioning adjustment while you tune each to the same tone usually does the trick. If you're dealing with laced-on skins, make sure the cord used is not stretchy, or your tuning work will not hold for long. Fibrous rope like jute works well, but some nylons can be too elastic.

Drum Bodies or Shells

Aside from a surface for attaching a membrane, which is just about all that frame drums and tambourines provide, the drum body or shell defines the size of the air chamber, which has a lot to do with the resonant frequency of a drum. Here, the body and air chamber influence the way the drumhead vibrates and can augment the fundamental tones, depending on how the head is tuned. The contact area between the skin and the shell must be level to avoid uneven tensioning (or worse yet, wrinkles in the drumhead). In general, the larger the shell, the lower the resonant sound. Narrowing the opening at the bottom of the body lowers the resonance and decreases the bandwidth.

Drum Tuning

The idea is to tension the drumhead to get the best sound you can out of the drum. The rigidity of the drum shell helps determine the drum's resonant frequency, and the head can be thought of as part of this shell. You adjust the head tension looking for the point where it and the air chamber enhance each other so that the drum sings with a fuller resonant tone. This is not a hit or miss attempt we're talking about, but a gradual altering. Head modes are very complex, and shells seldom have a defined fundamental resonance. Some shell resonances are so low that to match them by detuning the head would make for unresponsive playing (little stick bounce). The trick is to find a point were the skin is playable, and the drum emits a song enhancing pitch.

Luckily the air chambers inside of drum shells do not have a defined resonance but cover a large band of frequencies. Therefore, the tuning process does not require one to go after a specific pitch, but you can try. It's worth the effort, because when drums sound notes that match the key being played, they become more a part of the song. Also a shell's edges are often rounded to avoid excessive strain on the head. A trick I learned from a conga player that works on all drums is to apply a lubricant, like wax, powder, or oil (with synthetic plastic heads such as mylar), to the drum's rim. This facilitates the tuning process, as the head will be less likely to catch, adhere to, or bind on the rim.

Drum Setup

The heads and shell openings of each drum should remain unblocked so that their sound is free to radiate. Reflective surfaces can aid or hinder the sound depending on their distance from the drum and, especially, the microphone (think delay, phase response, and comb filtering). The three to one rule can help negate this problem.

If the drum has a mount, (feet or a stand), cushioning the bottom enhances the sound. The acoustics of the room will have a definite affect on the sound of drums, so try moving them around the room to find the location that enhances the sound the most.

Percussion instruments, like marimbas, xylophones, and vibes, which have an added resonator for each note, differ from something like a thumb piano, which utilizes a single resonator for all notes, but in both cases, each note sounds separately with little interplay between them. You'll find that there is only a slight difference in the levels and timbre quality of notes across the whole instrument.

Bowls can be tuned and rung like bells when hit, or they can even be bowed to lengthen the duration of the note. To aid the understanding of how the surrounding material influences the sound within a cavity, picture air enclosed in a bottle. When you open the bottle up, it acts like a bell. Further expansion gives it the shape of a bowl, a gong, then finally a cymbal, and a flat resonator.

Pellet Drums

These are like maracas, but they have dual membranes instead of a round enclosure. The pellets inside strike the membranes when the instrument is shaken. For more thunder, try dropping marbles on a kick drum with its head side up, or for the sound of rain, drop some rice on a snare drum with the snares disengaged. Have some fun.

Tenor Drums

These marching drums do not have a set pitch. Instead their shells, which are shortened on one side, resonate less and provide a dry sound that is also deep. They were originally used by the military and evolved into a staple of marching bands. I am surprised that they have not gained much popularity with rock drummers, since they are loud, dead, and come with arch-shaped mounts. In marching bands, they wrap around the player in a semicircle, while a "side" drum (usually a snare) hangs by a strap from the player's shoulder.

Snare Drums

The snare itself lies across the bottom resonating (non-struck) head. Snares were originally made of guitar gut strings with several strands laid across the bottom head, and some players still prefer their warmer sound. Today most snares, however, are made of coiled wire that looks like a spring that's been stretched out.

When the top head is hit, it induces sympathetic vibrations in the bottom head. Mechanical adjustment allows for tensioning the snare so that you can vary the rattle. Snares add a high frequency white-noise-like sound with a sharp attack. They also increase the output level of the drum and help dampen any ringing tones. Using a heavy top and a light bottom head can augment the snap and resonance of a snare drum.

Tube Drums

These often come in tuned sets with each drum having a clearly defined pitch. This is possible because the tubes can be adjusted for specific resonant frequencies. Their lengths can range from a few inches to several feet while their diameters may range from a couple of inches up to eight inches. I've seen the shells made from bamboo, acrylic, and metal tubes. I'll bet hard wood tubing would sound even better. The sound of these drums is quite remarkable in that because of the shape of their bodies, they can be tuned by adjusting their length so that the air columns within them have a specific resonant frequency with a narrow bandwidth. Because the fundamental sound is from the shell, the drumhead can be tuned for a snap-like attack and good stick response.

Talking Drums

With these drums, the heads are not firmly attached, but laced to a drum shell, which has an hourglass shape. This allows the lacing to be squeezed into the body (usually with a headlock hold by the left arm), while the right hand hits the head with a stick that has a right angle curve to its striking end. Many of these drums can be varied in pitch by up to an octave. By raising and lowering the pitch, good players can mimic speech inflections. The resulting sound is not only delightful, but can also become comical at the player's will.

Tambourines

In order to record these instruments correctly, you must first know what you are dealing with. The womp of a membrane-covered tambourine can put out quite a high SPL. I've seen older models with snares positioned against a bottom head. Most modern plastic head-less instruments are merely frames to hold the five or more pairs of flattened circular "jingles" mounted in slots cut out of the shell. Since they are hand-held, the positioning is usually constantly changing. In general, a good omnidirectional mic will work out well with those from the test and measurement field, providing an undistorted, full-frequency sound. Just make sure the player's motion won't cause the instrument to collide with the mic.

Castanets

These seem like they would be the most straight ahead micing job you can run into, but they can easily fool you. While shaped like a pair of seashells, their name, which is derived from the Latin word for chestnut, more aptly describes their cup-shaped bodies. So what? Well clap your flattened hands together in front of your face. You may feel a slight puff of air. Now cup your palms and slap them together in front of your face. Not only is the sound lower in pitch, but also that breath of air is much more pronounced. Your measurement mic may handle the SPL output, and correct positioning may keep these handheld instruments from colliding with the mic and causing distortion, but you'll be better off starting out with a windscreen on the microphone.

Friction Drums

With these drums, catch and release vibration is induced directly or indirectly onto the drumhead. Like talking drums, good players can make these instruments sing, wail, cry, laugh, *and* talk. The most common form of these is the "cuica" from Brazil. In this case the catch-release friction is induced onto a stick that is either attached to or held against the skin. The friction is augmented by the player's wet or rosin coated fingers. If an independent (non-attached) stick is used, varying its pressure against the head changes the pitch. I've witnessed players momentarily touch or grab the stick with their other hand to temporarily cut off the sound for enhanced articulation. A cord, either attached to or passed through a skin can also be used to induce vibration, and as mentioned before, wet fingers and rubber mallets can also be used for this affect. I'm waiting for someone to play a talking drum with an hourglass body, laced skins using a rubber ball on a flexible mallet, or their rosin/wetted fingers to make the drum sound like it is carrying on a conversation.

Steel Drums

Most people are familiar with these as singular one-piece instruments, but they often come in sets of three to six drums. When it's Carnival time, marching bands may be playing dozens of "pans" to the delight of everyone within earshot. While hand hammered, they are quite precise in their tuning and note layout. Medium sized instruments may span only an octave and a half, but it's remarkable how the notes are laid out so that when playing a run or a crescendo, the position of the notes is such that the beats go from the left hand, to the right hand, back to the left hand, etc.

Cymbals

These instruments sound brighter when well polished, but adding tape and dampening to their surfaces can reduce their decay time. However, just as with drumhead modes, it is nearly impossible to take a scientific approach to adding damping because cymbal modes are difficult to identify. However, I strongly urge you to seek out holograms of vibrating cymbals, as they do give you clues as to where to add tape and weights. In addition, they are quite entertaining to look at.

HINT: *By drilling a round hole at its end point, a crack in a cymbal can be stopped dead in its tracks (for a while).*

Some Drummer Stories

Up to this point I've tried to stay away from reminiscing about the past, reciting lengthy "war stories," and what might be termed useless fill. There was a limit set on the length of this book, and there was just too much that had to be covered to allow it to turn into some sort of a "story book" instead of the functional how-to manual it was intended to be. However, I've chosen to digress a bit because I want to make a specific point about audio engineering in general and drummers in particular.

Drummers and percussionists are a much-maligned group. They're the brunt of jokes that place their intelligence in question. They are often portrayed as Neanderthal "semi-musicians." But I've had the good fortune to work with a number of very impressive drummers over the years—some while I was employed in a rehearsal studio, a place where you learn a lot about drums very quickly because of the continual adjustment and other abuses that are inflicted on the kit, and others, while acting as a second engineer during recording sessions. In quite a few instances, I was called upon to execute repair work while the kit was in use. The physical position I'd have to put myself in while making these adjustments often afforded me an unbelievable view of the drummer's performance. Here are a few of those occasions that stick out most in my mind.

When the points on the ends of the kick's front two legs become dull, it can slide away from the drummer. Once while attaching rope between the bass drum and the drum stool to prevent this, I was down low on my knees tying knots and watching Lenny White's right foot in action. I distinctly remember this because I was awestruck by this gentleman's dexterity. What he was playing on the kick with a single drum pedal most people would have difficulty performing with their hands. When you stomp on a bass drum pedal from a standing position, the force of the beater contacting the skin is increased to the point of it being a noticeably sharp wallop. When you hold the beater against the skin momentarily after it contacts the skin, the sound of the hit is sharply cut off, and this adds finality to the note that it accents. I witnessed Lenny White do all of these things while in a sitting position and playing accented paradiddles, using nothing but his right foot. He needed no additional pedal attached to a second beater by a cable.

Holding the beater against the skin to accent the attack and reduce the ringing overtones seemed like the specialty of a drummer named Yogi Horton because he was also able to do it with his drumsticks. I remember standing in front of the kit and re-positioning the mics on the toms the way the engineer liked them and smiling as I packed up my tape, padding, and dampening materials because they would not be needed. This guy was big, had the physique of a body builder; yet his strength did not interfere with his dexterity. He was able to womp a tom and momentarily continue pressing the stick into the skin, even during a tom fill! It sure grabbed my attention quick, and when I mentioned it to the engineer, he said something to the effect of, "any mic, placed anywhere on the kit, no Eq, no gates, no compression, it flat out sounds good." I'm a witness!

Speaking of drummers and dexterity brings the name Steve Gadd to mind. Everyone should be familiar with his famous double-stick-in-each-hand flammed tom sound as well as his intricate rolls and fills. Well, when someone like this books time in a rehearsal studio, they get the royal treatment. The kit he was using had been given two floor toms. Hey, I'd never deny any drummer the right to dual "coffee tables," but on this occasion because a second kit was needed for the other drummer in his band, it meant that the drummer in studio B down the hall had no floor tom. I told him not to worry; "I'll just steal one of Gadd's." This freaked him out no end, but I had something else in mind. The studio had another floor tom available, but it was shy one leg and its mounting hardware. Since the two floor toms Mr. Gadd was using were positioned right next to each other, I could add a piece of dampening material between them, join them together with wire, rope, tape, or whatever it took, and the two would easily be supported by the five remaining legs.

I wasn't worried about disturbing the rehearsal because I'd been climbing all over music instrument setups since I was a kid and could swap out a guitar speaker, a whole combo amp, or a high hat while they were being played, all without a hitch. I got myself positioned low so as not to block anyone's eye contact and commenced tying the two floor toms together. Two short lengths of wire and a single clamp later left me with just the leg hardware to deal with. I had been listening to what Gadd had been playing, and I now took the opportunity to view it as well. Yeah, he could flam, fill, and roll just like you hear him do it on recordings, but witnessing him go from one method to the other, while playing the drum set like a music instrument was something to behold. I walked out of that room carrying the floor tom leg and mounting hardware, while shaking my head over the amount of body control this individual possessed. OK, I was straight out in awe.

The last time a drummer had caused me to walk out of a rehearsal room like that was close to a decade earlier. I was the weekend guy, which meant I worked all the sessions from Friday evening until Sunday night. This left me with just enough time to sleep and eat, but I thought it was great because I had the rest of the week off.

During this period of the week, the studio was almost constantly booked, and the drum kit would take a real beating. This particular weekend was no different. Saturday night started with a disco band, and the snare was tightened to sound like a gunshot. The next band played soul music, and their drummer liked to emphasize the ring in the three mounted and single floor toms. The next and last band for the evening were Grateful Dead enthusiasts whose drummer mainly used the high hat, an occasional snare rim shot, and kick, which he adjusted to be bright. Then another band showed up at 2:00 a.m.! Someone had booked them and not told me. But I had worked with them before; they were all right guys, and they promised to be out by four.

They were Led Zeppelin fans, and besides cranking the amps all the way up, the drummer attempted to lower the pitch of two of the toms by loosening their skins until they sagged. They got out on time, which was okay with me, as I was able to get home for some sleep before my first Sunday morning booking at 10:00 a.m.

I was a tad late on Sunday morning so when I arrived, I told the band, a Latin group whose drummer always tuned two of the three mounted toms to sound like timbales, that they could have an extra half hour's worth of makeup time at the end of their four hours. At 2:00 p.m. on the dot, a guy shows up with an upright bass. I hadn't even looked to see what kind of band was booked in. While explaining the merits of using a Fender twin over an Acoustic head with an 18" driver in a folded horn cabinet on an upright bass with a bridge-mounted pickup, in terms of feedback in a small 25' × 30' rehearsal room, I surmised that I was dealing with a pro, who was also a nice guy and part of a traditional four piece Jazz band. When the Latin band starts to load out and the other Jazz cats are showing up and moving into the studio, I finally take a peek at the scheduling book and what do I see but Elvin Jones no less! I was mortified.

All I could do was gather up my drum keys, which included a couple of wrenches for the common Ludwig type square lugs that had different handles, a "Z" shaped speed wrench, a "T" shaped wrench with a wing nut adjustment cut out at the end, plus a slotted lug wrench for the Hayman eight lug snare and toms. I brought these, a roll of masking tape, and some paper towels up to the drum riser. I told Mr. Jones that I normally would have at least evened out the tuning between all the drums a little between sessions, but that I didn't get the chance to. I apologized for the set having a disco gunshot snare, a heavy metal floor tom, two tomes that sounded like timbales, a bass drum tuned overly high, and a third mounted tom with unknown tuning. He just smiled, put all the keys off to the side, and lightly tapped each drum with his stick. This *really* made all the different tunings stick out like a handful of sore thumbs. I was disgraced to the point where all I could do was hang my head and walk away in shame.

Again the depth of the studio was only 25 feet, so it couldn't have taken me even a full minute to reach the door from the drum riser, but as I walked, I heard him stop tapping the skins and just up and start to play. Within two steps of the door, I froze, turned around, and with my jaw on the ground witnessed Elvin Jones make a completely out-of-tune drum set sound in tune just by where and how he hit each skin. I still felt bad but this rendered me completely speechless!

I remember once showing up for an 11:00 a.m. recording session a little early and seeing an older gentleman standing in front of the building with two plastic milk crate-like boxes (about 18 cubic inches in size) filled with handwritten music charts. Since the studio was two flights down, I offered him a hand, and as we descended, he told me all about the band that he was with.

It was a big band, and since I had been a fan of this style of music for a long time, we talked a little bit about some of the old bands. He told me he was the drummer so I told him my favorite Count Basie band drummer story. Then he went into his rehearsal room, and I started setting up the recording studio for a date that promised to be interesting, as the band was supposed to be a blend of R&B, Funk, and Disco with a little Progressive Jazz mixed in. I was just the assistant on this date, and during a short break, the studio owner told me that the other drummer had mentioned my helping him and was thankful. He then went on to explain that the gentleman's name was Jim Chapin, that it was *his* big band, and that *he* was the one who had written out all those charts. Furthermore, this guy had written the definitive book on drumming. It had sections of transparent pages, which were used to add more and more beats to the basic rudiments printed on the paper page below them.

When I told the drummer in the recording studio that Jim Chapin was rehearsing his big band in the other room, the cat was floored. He knew all about the guy, had his book, and went off explaining it all to me. He mentioned some difficulties he was having understanding one part so I offered to ask Mr. Chapin about it. The drummer was hesitant, but I asked anyway and witnessed Jim Chapin take the time during a break to sit down with this guy and answer his questions by tapping his sticks on the table in the lounge. That's what I call a real nice guy. Oh yeah, that favorite Count Basie Band drummer story?

Okay, NYC Village Gate circa 1974. I'm in the audience and have nothing to do with the show, just enjoying my third row, bird's-eye, and side-stage view. The band is smokin' and everyone takes a solo. Comes the drummer's turn, the whole band goes quiet. The guy's playing like a banshee, bouncing all around, sweat pouring off him, while he jive sings along with his playing. The whole place is caught up in this guy's groove. The punch line is that he never touched a drum. That's right, the cat was drumming his butt off, but he was able to stop each stick about 3/4 of an inch above each skin. When the rest of the band came back in, signaling the end of his solo everyone in the audience *jumped* out of their seats to give this guy a standing "O" that lasted more than a few minutes. I don't know the name of the drummer that soloed without ever making a sound, but man, he sure was playing those drums.

Well at least Mr. Chapin liked the story.

The point? Gadd's control can't be touched, but he probably can't write horn charts as well as Jim Chapin. Nobody's got a foot like Lenny White's, but Yogi Horton was the only drummer I've ever run across who needed no adjustments when the kit was close miced, and Elvin Jones is on a whole different level. The next drummer you work with might not be able to read charts and may have difficulty keeping a steady beat. Part of your job is to figure out ways to help musicians out when they are recording. You can't do this unless you take the time to learn about the instruments they play, or if you question their worth. From my experience, the process of "nudging" beats around a computer screen and trying to come up with a drum track takes a lot more time and often doesn't sound nearly as good as when the tracks are laid down by a real live drummer.

It's also part of the audio engineer's responsibility to take an interest in every musician booked into the studio, not just drummers. As an example of what I'm taking about, we'll take this pair of musicians. Most everyone knows Dolly Parton to be a top singer/songwriter, musician, actress, performer, and celebrity superstar, but her vocal ability tells me she has, if not perfect, at least near-perfect pitch. That's because I've heard her position her voice in between two background vocalists to bring all three into perfect harmony. You may also know Bonnie Raitt to be another perennial singer/songwriter guitarist who's always at the top of the charts but she's also an excellent slide guitarist and a master at playing the National steel guitar. I say this because I've heard her move a song from the verse through a bridge and into the chorus by strumming a chord and gently gliding the pitch up or down. She often uses this effect in her own especially downplayed manner to move a song from one section to the next just as if she were lifting a newborn baby out of its bassinet and into its waiting mother's arms.

Knowing this you'd be sure to feed the cue system of someone like Dolly Parton an even blend of all the vocalists and make certain that the guitar amp intended for Bonnie Raitt was capable of producing a nice clean sound along with a touch of raunch.

But how can you know the nuances and expertise of every musician you're going to work with? You can't, but the answer is so simple many overlook it. Whenever a noted artist is booked in to your studio, buy $50 worth of their CDs along with one or two of their published charts

Fifty bucks? That's right, that is all it will cost you to break down the barriers to an easy flowing conversation, make a session run more smoothly, and get answers to questions like "Hey, how'd you guys get that great rhythm guitar sound on such and such a track?" especially if that same producer is present. That's all you have to do once you develop your listening ability to the point where you can perceive what's going on down underneath in the pocket of a rhythm section. Do you think I have some kind of special insider's knowledge about these two musicians? Believe me when I say a little time and effort and about fifty bucks should do it.

What It All Boils Down To

Knowing how instruments operate and the way musicians play them points the record-ist to the best means of capturing their sound. As an example, just watch accordion players, who are also front men, sing and play. They will aim the bass box at the vocal microphone when they want to accentuate the rhythm part and get the band going. Then the treble box will be aimed into the mic to emphasize the melody. But when it comes time for the solo, they will then hoist the instrument up higher on their chest so as to put the vocal mic in the line of fire from both cases.

If possible, try giving these players a break by using separate mics for each box and another for the vocal. If that's not practical because of stage bleed, those cases are almost perfect resonators and trigger contact pick-ups very well. In the studio, I've had good results with distant micing techniques, or when the player was willing to wear them, a set of headphone style binaural mics. It should be obvious that you will get a better performance from someone who doesn't have to worry about moving their instrument around while they're playing.

Microphone Placement

The sound radiation pattern of an organ pipe with a closed end is fairly basic because the sound only emanates from the single lip opening. While the opening is relatively small and the whole pipe radiates the lower frequencies in an almost omnidirectional pattern, the resonant frequency radiation is concentrated at the opening and spreads out more on the vertical plane than the horizontal. When the end of the pipe is open, things get a little more complicated, as now the sound is projected from both this area as well as the lip opening. These two sounds are in phase for odd harmonics, but out of phase when even harmonics are produced. While the opening of the pipe ends is much larger than the lip opening, because of the jet-steam effect from the lip, the sound emitted from both openings are usually about equal in level. Equal levels and varying phase relationships? You guessed it; different location points within the overall radiation pattern will range between a doubling in level to almost completely canceling it out. You'd never notice this when listening to a pipe organ being played because the multiple propagation angles combine together to give the audience a well-balanced sound. But close micing these instruments should to be attempted only after extensive experimentation. So just start off with the mic positioned a little further out and gradually bring it in closer, while listening for anomalies.

Multimicing Instruments

No matter what you're recording, there will always be some conflict between the need to close mic the instrument to gain separation and the desire to back the mic placement off to capture the instrument's full frequency response. During overdubs this presents less of a problem, as multiple mics can be used in close along with "room" mics backed off to capture the full sound of the instrument. I've found that in most cases it is actually necessary to at least double mic an instrument up close to get its full bandwidth, and if it's an overdub session, I'll always add a room mic as well. Music instruments have fairly complicated multidirectional sound projection patterns. This method takes into consideration the way an instrument propagates sound as well as the way this sound fills or integrates with the acoustics of the room.

Many engineers take exception to this when it comes to brass instruments (trumpets and bones) because most of their sound is radiated on axis. But I've found a meaty sound to come from the side of the horn and that a single cardioid mic in a deadened acoustical atmosphere just cannot capture the full frequency content of these instruments.

You can try to compensate for this by boosting or cutting the bandwidth at certain points or adding reverb and other effects, but if a partial or some harmonic content is not recorded, there'll be no way to get it back or "fix it in the mix." Since you're better off getting it right from the start, I'll give you some examples to set you off in the right direction. However, good old common sense will usually serve you well as a guide once you take the way an instrument produces sound into consideration.

Violins and other bowed string instruments have an intricate sound radiation pattern, but it changes with frequency. Luckily we can nail down areas of sound propagation, which are fairly consistent. The lows come from all around the instrument but are usually concentrated at the "F" holes. The bottom or back plate is a good place to aim a mic if you need a little more low-mids. Mids come from the left-hand side of the top plate and high-mids from the right. If you need to accentuate the sound of bowed strings, aim the mic where the neck meets the body. So a single mic placed at the bridge will lack lows and highs, while a mic at the neck will provide an overly bright bow on string sound. Therefore, you may want to try the neck position to get the sound of the bow and also aim a mic up from the floor at the bottom plate for those dark low-mids.

For comparison let's look at another member of the violin family, the cello. While this instrument also radiates lows almost omnidirectionally, because of the player's position micing the back plate is often not available. Here you may be better off picking up the low end at the bottom of the instrument or right off the sound hole. Low-mids can be had from below the bridge, high-mids at the upper part of the top plate, and the highs at the neck. Don't forget to try micing an instrument from the player's perspective, as long as this position is not intrusive. People generally don't like mics positioned in real close to their heads.

Woodwinds radiate sound through their ports as opposed to just the bell. However, most of the reed sound is projected from the bell, so when only a single mic is placed in this position, the horn is going to sound raspy. A mic positioned at the player's right side will face the low ports, but the sound will be partially blocked by the pads. Positioning the mic at the player's right side but more toward the front and aimed at the slot-like pad openings gets you a fuller sound. Like I said, it's just common sense.

When placing microphones, the difference of a couple of inches can be critical especially with membrane (drumhead) resonators. That's why I believe that if you really want to be good at this job, you have to understand a bit about the physics involved in the sound production of the music instrument you are planning to record. Always remember that high frequencies are very directional and generally beam, while the lows cannot be corralled or channeled, as they go right around obstructions. This is the reason a close mic position generally provides more highs. While room mics will capture more low-mid frequency content they will often lack the clear attack sounds of the instrument.

Also, never disregard the affect the room has on the sound of an instrument, so don't be afraid to move an instrument's position around to find the "sweet spot" when necessary.

Floors have a pronounced effect on the sound of upright basses and cello, while side walls play an important part in the sound from woodwinds and violins. Studios with carpeting, hung "acoustic" ceilings, and foam panels have a tendency to overly absorb the high frequencies. In rooms like this, you may have to place a mic in close to get more of the instrument's high-end output. Again a single mic will tend to pick up only part of the frequency content, depending on the radiation pattern of the instrument.

While using only the low-frequency proximity boost inherent in ribbon and single D cardioid mics or using a built-in high-frequency roll-off may seem like a good remedy, in reality you may be eliminating part of the signal content. Trying to add it back with equalization or reverberation will not always work. Even in a situations where it's appropriate to back off on the mic positioning so that the room can augment the low-end sound, an engineer still needs to know the way the instrument radiates sound, the effect the acoustical environment will have on the sound, as well as the polar patterns, sensitivity, and frequency response of the microphones being used. If multiple mics are used, what will be the effect of placing a microphone in close at various locations around the instrument? This may mean that you'll need to watch a phase meter or a scope while recording. Don't forget about phase relationships and use the 3 to 1 rule on larger instruments.

The 3 to 1 rule states that when placing multiple microphones on a single source, the second mic (B) should be positioned away from the initial mic (A) at least three times the distance that the (A) mic is from the source or instrument. I have seen folks get all hung up on this using tape measurers and rulers. But when speedy setup is important (and it usually is), exactness is not that necessary. Let's use an acoustic guitar as an example. If you place a mic 6 inches out from where the neck meets the body and then want to add another mic on the sound hole, it should be close to a foot and a half back to avoid phase cancellation.

If you solo the two microphones and hear a "wooshing" sound in the mid-range, a hole in the bottom end or screechy highs, reposition the second mic. To save yourself multiple trips between the studio and control room, set up a low-level cue feed of the microphones before you start. Positioning microphones while monitoring through headphones can be *very* instructive.

Music instruments are designed to output sound at a consistent level. Therefore, if a portion of the sound is insufficient at one microphone position, it may be overly magnified at another. If the mic position is overly limited in frequency content, you may end up emphasizing a sound with a narrow bandwidth. A single close micing position can seldom provide the full tone or "timbre" of a music instrument. This is because they normally radiate the fundamental frequency and overtones or harmonics from different places and often in different directions. Music instruments are designed to sound best at the listener's position. This works because, outside of anechoic chambers, the sound will combine at some distance to provide the listener with a full frequency content.

Examples of Microphone Placement

Again it's really just common sense. Take micing a dual-concentric speaker. If you place a mic in close, at the center, you'll get more high end whereas you'll get more lows if it's placed at the side. These dual speakers work well because the high-frequency content that normally comes from the dome section of the rear speaker is replaced by the more efficient high-frequency driver positioned in front of it. This radiation pattern holds true for single driver configurations as well. A mic at the side will provide more low-frequency content and more highs when at the center dome. However, like music instruments, the exact radiation pattern differs slightly from speaker to speaker and frequency to frequency.

Therefore, there are no set rules in this game. Often the recordist must attempt to find a single mic position that is far enough out to pick up the full tonal content of the instrument, while avoiding any bleed from other instruments and adverse sounds from room reflections. Remember that comb or notch filtering caused by reflections will be more pronounced at the microphone than it is at the listener's ears. This is due to the small size of the diaphragm and its being closer to the instrument than the normal listening position. Also remember that while omnidirectional measurement microphones are very sensitive, have a flat frequency response, pick up sound from all directions, and have virtually no proximity effect, they are not connected to a brain. By the time the sound gets to the speaker, it may be too late for the listener's brain to "discern" some of the instrument's subtle nuances. It's much better when you get it all from the start. Here's a couple of examples of what I'm talking about.

Acoustic Guitar

A mic in close at the bridge provides low-mids, lows at the sound hole, and highs at the neck. Back it off a couple of feet, and you will be in business. But here you may have to deal with bleed, so instead try aiming a mic down at the bridge from the player's ear position. You'll find that the sound is quite natural. If there's not too much ruckus going on (as when overdubbing) you may like the sound from a single PZM on the floor in front of the instrument. Use a small throw rug under foot tappers.

Acoustic Grand Piano

Can't take the lid off and hang a priceless antique tube mic six feet above it in the center in an acoustically perfect room? Remember you may want more of the attack sound, so locating a mic closer to where the hammers strike the strings may get you what you're after. Want more high end? Then place the mic (or a PZM) at the curve near the high strings. More bottom will come from the left side by the bass strings or underneath the piano from the soundboard. You may also be pleased by the sound of a PZM taped to the center of the lid, at the front over the hammers with the lid on the tall stick position.

Where Do You Go from Here?

Starting with your local library is easy even if it does not turn out to be all that productive a site. Nearby colleges with music degree programs may be a better source of information, if you can somehow gain access to their libraries. This is a bit tougher today than it was thirty or forty years ago. Try making friends with some of the musicians on campus. You're a recordist, they're musicians; the two usually go well together, and you might pick up some not-so-well-paying work and a ticket into the school's library stacks in the bargain. These places will have a great deal of information on music instrument making. This stuff is only so-so as far as recording is concerned. I mean how important is it to mic placement whether the violin tailpiece button is made of knotted steel cable or nylon? Yet you'll find reams of arguments in favor of one or the other. They'll also have a lot of periodicals like *Luthier* and *The Strad*, which are devoted to a narrow subject area (acoustic guitars and violins). You will also find publications with broader based subject areas, such as *The Catgut Acoustical Society Journal* as well as publications that cover orchestra and marching band interests. Simply checking out the ads in these rags ends up being highly enlightening for a recordist. If you want to really get into string instrument design theory, I'd suggest spending some time checking out the Catgut Acoustical Society. Started in 1963, this organization is made up of musicians, instrument makers, and scientists interested in the acoustical makeup of the string family.

However if it's only music instrument acoustics, their radiation patterns, and the test results of in-depth experimentation that you're after, then head to your local technical college. In addition to all the books covering the physics of music instruments, you may find that they keep back issues of journals from The Acoustic Society of America, The Audio Engineer Society, and those that cover many other far ranging subjects, such as HVAC systems and noise control.

The job may seem daunting, but that's only because it is. Hey, I've got over forty years of this stuff under my belt, yet when I enter a physics or music section of a library, I still get a little overwhelmed by the depth of information available. Wadding through all of this stuff can be wearisome, but at the same time for an enthusiastic recordist, it can be Candy Land!

You don't do this job because you like or enjoy it. No, you have no choice. You need to do it because you love it. You want to solder all your own connectors just to make sure the job's done right. You take the time to align your equipment so that you get the best possible S/N ratio. You use an oscilloscope to insure phase linearity. Why? Because it's *all* about the sound, and every one of these things will enhance it.

When you get to this point, you'll find yourself peeking over the rail into the orchestra pit before the start of the "Carnival of the Animals" by the New York City Ballet. You'll spy a 30+ inch bass drum, a xylophone, a French horn with its player's right hand in its bell, a gorgeous 12-foot acoustic grand piano, two contra basses, flutes, several violins, and cello as well as an actual antique glass harmonium.

This will send you back to your seat grinning ear to ear, anticipating the sound of these instruments, which are all in top condition and played by professionals who have near-perfect intonation. You'll be in heaven right here on earth, my friend.

Believe me, I know. 'cause I've been there!

Enjoy.

Sound Absorption Coefficients for General Building Materials and Furnishings

Complete tables of coefficients of the various materials that normally constitute the interior finish of rooms can be found in the various books on architectural acoustics. The following list of materials gives approximate values, which will be useful in making simple calculations of the reverberation in rooms.

	Coefficients					
Materials	125 Hz		500 Hz	1000 Hz	2000 Hz	4000 Hz
Brick, unglazed	.03	.03	.03	.04	.05	.07
Brick, unglazed painted	.01	.01	.02	.02	.02	.03
Carpet						
⅛ " pile height	.05	.05	.10	.20	.30	.40
¼ pile height	.05	.10	.15	.30	.50	.55
³⁄₁₆ " combined pile and foam	.05	.10	.10	.30	.40	.50
⁵⁄₁₆ " combined pile and foam	.05	.15	.30	.40	.50	.60
Concrete block, painted	.10	.05	.06	.07	.09	.08
Fabrics						
Light velour, 10 oz. per sq. yd., hung straight, in contact with wall	.03	.04	.11	.17	.24	.35
Medium velour, 14 oz. per sq. yd., draped to half area	.07	.31	.49	.75	.70	.60
Heavy velour, 18 oz. per sq. yd., draped to half area	.14	.35	.55	.72	.70	.65
Floors						
Concrete or terrazzo	.01	.01	.01	.02	.02	.02
Linoleum, asphalt, rubber or cork tile on concrete	.02	.03	.03	.03	.03	.02
Wood	.15	.11	.10	.07	.06	.07
Wood parquet in asphalt on concrete	.04	.04	.07	.06	.06	.07
Glass						
¼ ", sealed, large panes	.05	.03	.02	.02	.03	.02
24 oz., operable windows (in closed condition)	.10	.05	.04	.03	.03	.03
Gypsum board, ½ " nailed to 2 × 4's 16" o.c., painted	.10	.08	.05	.03	.03	.03
Marble or glazed tile	.01	.01	.01	.01	.02	.02
Plaster, gypsum or lime, rough finish on lathe	.02	.03	.04	.05	.04	.03
Same, with smooth finish	.02	.02	.03	.04	.04	.03
Hardwood plywood paneling ¼ " thick, wood frame	.48	.22	.07	.04	.03	.07
Water surface, as in a swimming pool	.01	.01	.01	.01	.02	0.2
Wood roof decking, tongue-and-groove cedar	.24	.19	.14	.08	.13	.10
Air, sabins per 1000 cubic feet at 50%RH				.9	2.3	7.2

Reprinted from ACOUSTICAL CEILINGS—USE AND PRACTICE
by permission of the Ceilings & Interior Systems
Contractors Association.

APPENDIX B

Sound Absorption Coefficients of Owens-Corning Products

OWENS-CORNING FIBERGLAS CORPORATION
Absorption Coefficients For Type 703 Fiberglas
Semi-rigid Boards (3 lbs/cu. ft. density) and Allied Materials

Facing Material	Core Material	OCTAVE BAND CENTER FREQUENCIES, Hz						NRC
		125	250	500	1000	2000	4000	
None[1]	1″ 703	.06	.20	.65	.90	.95	.98	.70
¼″ Pegboard[2]	1″ 703	.08	.32	.99	.76	.34	.12	.60
⅛″ Pegboard[3]	1″ 703	.09	.35	.99	.58	.24	.10	.55
None	1″ TIW Type I	.11	.33	.70	.80	.86	.85	.65
¼″ Pegboard	1″ TIW Type I	.08	.41	.99	.82	.26	.32	.60
1″ Nubby Glass Cloth Board	None	.04	.21	.73	.99	.99	.90	.75
1″ Textured Glass Cloth Board[5]	None	.05	.22	.67	.93	.99	.85	.70
1″ Painted Linear Glass Cloth Board	None	.03	.17	.63	.87	.96	.96	.65

Facing Material	Core Material	125	250	500	1000	2000	4000	NRC
None	2″ 703	.18	.76	.99	.99	.99	.99	.95
¼″ Pegboard	2″ 703	.26	.97	.99	.66	.34	.14	.75
Perforated Metals[4]	2″ 703	.18	.73	.99	.99	.97	.93	.95
1″ Painted Linear Glass Cloth Board	1″ 703	.18	.71	.99	.99	.99	.99	.90
1″ Nubby Glass Cloth Board	1″ 703	.25	.76	.99	.99	.99	.97	.95
None	2″ TIW Type I	.25	.75	.99	.99	.99	.99	.95
¼″ Pegboard	2″ TIW Type I	.26	.89	.99	.58	.26	.17	.70
Perforated Metal	2″ TIW Type I	.25	.64	.99	.97	.88	.92	.90
1″ Linear Glass Cloth Board	1″ TIW Type I	.23	.72	.99	.99	.99	.99	.90
1″ Nubby Glass Cloth Board	1″ TIW Type I	.26	.75	.99	.99	.99	.99	.95
1″ Linear Glass Cloth Board	1″ Air space	.04	.26	.78	.99	.99	.98	.75

NRC = Noise Reduction Coefficient. It is the average of absorption
Coefficients for 250, 500, 1000, and 2000 Hz rounded to the
nearest multiple of 0.05.

Facing Material	Core Material	125	250	500	1000	2000	4000	NRC
None	3 " 703	.53	.99	.99	.99	.99	.99	.95
¼ " Pegboard	3 " 703	.49	.99	.99	.69	.37	.15	.75
1 " Painted Linear Glass Cloth Board	2 " 703	.59	.99	.99	.99	.99	.99	.95
1 " Nubby Glass Cloth Board	2 " 703	.50	.99	.99	.99	.99	.97	.95
None	3 " TIW Type I	.46	.99	.99	.99	.99	.99	.95
¼ Pegboard	3 " TIW Type I	.53	.99	.97	.51	.32	.16	.70
1 " Painted Linear Glass Cloth Board	2 " TIW Type I	.48	.99	.99	.99	.99	.99	.95
1 " Nubby Glass Cloth Board	2 " TIW Type I	.51	.99	.99	.99	.97	.95	.95
1 " Painted Linear Glass Cloth Board	2 " Air space	.17	.40	.94	.99	.97	.99	.85

Facing Material	Core Material	125	250	500	1000	2000	4000	NRC
None	4" 703	.99	.99	.99	.99	.98	.98	.95
¼ Pegboard	4 " 703	.80	.99	.99	.71	.38	.13	.75
1 " Painted Linear Glass Cloth Board	3 " 703	.88	.99	.99	.99	.93	.98	.95
1 " Nubby Glass Cloth Board	3 " 703	.75	.99	.99	.99	.99	.97	.95
None	4 " TIW Type I	.57	.99	.99	.99	.99	.99	.95
¼ " Pegboard	4 " TIW Type I	.70	.99	.94	.58	.37	.19	.70
1 " Painted Linear Glass Cloth Board	3 " TIW Type I	.77	.99	.99	.99	.99	.99	.95
1 " Nubby Glass Cloth Board	3 " TIW Type I	.71	.99	.99	.99	.99	.92	.95
1 " Painted Linear Glass Cloth Board	3 " Air space	.19	.53	.99	.99	.92	.99	.85

[1]Absorption values would be unchanged for open facings such as wire mesh metal lathe, or light fabric.
[2]Perforated ¼ " holes, 1 " o.c.
[3]Perforated ⅛ " holes, 1 " o.c.
[4]24 gauge, ³⁄₃₂ " holes, 13% open area.
[5]Absorption values of textured glass cloth may be interpolated to lie between linear and nubby glass cloth for all other thickness of wall treatment.
TIW = Thermal insulating wool

Facing Material	Core Material	125	250	500	1000	2000	4000	NRC
None	5 " 703	.95	.99	.99	.99	.99	.99	.95
¼ " Pegboard	5 " 703	.98	.99	.99	.71	.40	.20	.75
1 " Painted Linear Glass Cloth Board	4 " 703	.87	.99	.99	.99	.99	.99	.95
1 " Nubby Glass Cloth Board	4 " 703	.88	.99	.99	.99	.99	.96	.95
None	5 " TIW Type I	.83	.99	.99	.99	.99	.99	.95
¼ " Pegboard	5 " TIW Type I	.78	.99	.89	.63	.34	.14	.70
1 " Painted Linear Glass Cloth Board	4 " TIW Type I	.77	.99	.99	.99	.99	.99	.95
1 " Nubby Glass Cloth Board	4 " TIW Type I	.79	.99	.99	.99	.99	.98	.95

Facing Material	Core Material	125	250	500	1000	2000	4000
None	6″ 703	.99	.99	.99	.99	.99	.99
¼″ Pegboard	6″ 703	.95	.99	.98	.69	.36	.18
1″ Painted Linear Glass Cloth Board	5″ 703	.99	.99	.99	.99	.99	.99
1″ Nubby Glass Cloth Board	5″ 703	.92	.99	.99	.99	.99	.99
None	6″ TIW Type I	.93	.99	.99	.99	.99	.99
¼″ Pegboard	6″ TIW Type I	.95	.99	.88	.64	.36	.17
1″ Painted Linear Glass Cloth Board	5″ TIW Type I	.87	.99	.99	.99	.99	.99
1″ Nubby Glass Cloth Board	5″ TIW Type I	.92	.99	.99	.99	.99	.93
1″ Painted Linear Glass Cloth Board	5″ Air space	.41	.73	.99	.98	.94	.97

Facing Material	Core Material	125	250	500	1000	2000	4000
1″ Painted Linear Glass Cloth Board	6″ 703	.86	.99	.99	.99	.99	.99
1″ Nubby Glass Cloth Board	6″ 703	.85	.99	.99	.99	.99	.99
1″ Painted Linear Glass Cloth Board	6″ TIW Type I	.95	.99	.99	.99	.99	.99
1″ Nubby Glass Cloth Board	6″ TIW Type I	.95	.99	.99	.99	.99	.94

All material combinations installed and tested against a solid wall (i.e., # 4) mounting) TIW = Thermal Insulating Wool

16″ O.C.

Wood Studs	Facing	Insulation	125	250	500	1000	2000	4000	NRC
2 × 2's	None	2¼″ Fiberglas	.30	.69	.94	.92	.92	.98	.85
2 × 4's	None	3½″ Fiberglas	.34	.80	.99	.97	.97	.92	.95
2 × 4's	1″ Painted Linear Glass Cloth Board	3½″ Fiberglas	.66	.99	.99	.99	.99	.97	.95
2 × 4's	1″ Nubby Glass Cloth Board	3½″ Fiberglas	.67	.99	.99	.99	.99	.90	.95
2 × 4's	¼″ Pegboard	3½″ Fiberglas Paper faced*	.45	.99	.87	.41	.30	.14	.70
2 × 4's	None	3½″ Fiberglas	.38	.96	.99	.68	.47	.35	.80
2× 4's	1″ Painted Linear Glass Cloth Board	3½″ Fiberglas	.66	.99	.99	.96	.99	.99	.95
2 × 4's	1″ Nubby Glass Cloth Board	Paper faced* 3 ½″ Fiberglas	66	.99	.99	.98	.99	.95	.95
2 × 4's	¼″ Pegboard	Paper faced* 3 ½″ Fiberglas	.50	.99	.70	.41	.38	.27	.60
2 × 6's	None	6″ Fiberglas	.67	.99	.99	.99	.99	.98	.95
2 × 6's	1″ Painted Linear Glass Cloth Board	6″ Fiberglas	.89	.99	.99	.99	.99	.99	.95

Reprinted by permission from "Industrial Noise Control," publication No. 5-BMG-8277 (1978), Owens-Corning Fiberglas Corporation, Fiberglas Tower, Toledo, Ohio 43659

Sound Absorption Coefficients for Tectum*

Absorption Coefficients Hz

Thick,	Mfg.	125	250	500	1000	2000	4000
	2	.08	.14	.27	.57	.59	.63
	4	.07	.12	.24	.44	.70	.54
1″	7	.45	.43	.31	.42	.56	.79
	8	.18	.53	.96	.90	.71	.90
	2	.09	.16	.36	.70	.49	.78
1½″	4	.09	.14	.31	.65	.66	.64
	7	.44	.43	.33	.49	.66	.77
	2	.13	.20	.50	.70	.58	.72
2″	4	.12	.20	.48	.80	.62	.94
	7	.48	.46	.36	.55	.74	.79
	2	.14	.29	.77	.67	.80	.85
2½″	4	.18	.28	.63	.87	.62	.80
	2	.14	.32	.78	.60	.84	.91
3″	4	.18	.35	.82	.76	.75	.88

*Gold Bond Building Products

References

1. Everest, F. Alton, *Acoustic Techniques for Home and Studio*, 2nd edition, TAB Books Nl. 1696 (1984), especially Chapter 5 "Resonances in Rooms," and Chapter 6, "Standing Waves in Listening Rooms and Small Studios."
2. Ibid. Fig. 10-4, page 183.
3. Ibid. Pages 164–169.
4. Ibid. Chapter 13, "Adjustable Acoustics."
5. Ibid. Chapter 9, "Acoustical Materials and Sturctures," especially pp. 169–173.
6. Gilford, C.L.S., *The Acoustic Design of Talk Studios and Listening Rooms*, Proc. I.E.E., Vol. 106, Part B, No. 27, May, 1959, pp. 245–258.
7. Sepmeyer, L.W., *Computer Frequency and Angular Distribution of the Normal Modes of Vibration in Rectangular Rooms*. Jour. Acous. Soc. Am., Vol. 37, No. 3, March, 1965, pp. 413–423.
8. Louden, M.M., *Dimension–Rations of Rectangular Rooms with Good Distribution of Eigentone*, Acustic, Vol. 24 (1971) pp. 101–104.
9. Bolt, Richard H. and Manfred Schroeder, Personal communication from Dr. Bolt.
10. Springs, N.F., and K.E. Randall, *Permissible Bass Rise in Talk Studios*, BBC Engineering, No. 83, July, 1970, pp. 29–34.
11. Davis, Don, Editor, *Syn-Aud-Con Newsletter*, Vol. 5., No. 4, July, 1978, page 25.
12. Souther, Howard, *Improved Monitoring with Headphones*, dB The Sound Engineering Magazine, Vol. 3, No. 2, Feb., 1969, pp. 28–29 and Vol. 3, No. 3, March, 1969, pp. 17–20.
13. Runstein, Robert E., *Modern Recording Techniques*, Howard W. Sams & Co., (1974).
14. Eargle, John M., *Sound Recording*, (1976), Van Nostrand Reinhold.
15. Woram, John, *The Recording Studio Handbook*, Sagamore Publishing Co., Plainview, NY (1976).
16. Rettinger, Michael, *Instrument Isolation for Multiple Track Music Recording*,Preprint No. 1119 (J-3), presented at the 54th Audio Engineering Society Convention, May, 1976.
17. Rettinger, Michael, *Sound Insulation for Rock Music Studios*, dB The Sound Engineering Magazine, Vol. 5, No. 5, May (1971), page 30.
18. Rettinger, Michael, *Recording Studio Acoustics*, dB The Sound Engineering Magazine, Part 1, Vol. 8, No. 8, Aug. (1974) pp. 34–47
    Part 2, Vol. 8, No. 10, Oct. (1974) pp. 38–41.
    Part 3, Vol. 8, No. 12, Dec. (1974) pp. 31–33.
    Part 4, Vol. 9, No. 2, Feb. (1975) pp. 34–46.
    Part 5, Vol. 9, No. 4, Apr. (1975) pp. 40–42.
    Part 6, Vol. 9, No. 6, June (1975) pp. 42–44.
19. Rettinger, Michael, *Noise Level Limits in Recording Studios*, dB The Sound Engineering Magazine, Vol. 12, No. 4, April (1978) pp. 41–43.
20. Rettinger, Michael, *Studio Rumbles*, dB The Sound Engineering Magazine, Vol. 7, No. 9, Sept. (1973) pp. 46–48.
21. Cooper, Jeff, *Building a Recording Studio*, Recording Institute of America, New York, (1978).
22. Matel, Juval, *Advanced Room Acoustics*, Preprint No. 1312, presented at the 59th Audio Engineering Society Convention, Feb.–Mar. 1978.
23. Hansen, Robert, *Studio Acoustics*, dB The Sound Engineering Magazine, Vol. 5, No. 5, May (1971) pp. 16–24.
24. Bruce, Robert H., *How to Construct Your Own Studio in One Easy Lesson*, Preprint No. 1245, presented at the 57th Audio Engineering Society Convention, May (1977).
25. Storyk, John and Robert Wolsch, *Solutions to 3 Commonly Encountered Architectural and Acoustic Studio Design Problems*, Recording Engineer/Producer, Vol. 7, No. 1, Feb. (1975) pp. 11–18.
26. Brown, Sandy, *Recording Studios for Popular Music*, 5th International Congress on Acoustics, Liege, 1965, paper G036.
27. Olson, N., *Survey of Motor Vehicle Noise*, Jour. Acous. Soc. Am., Vol. 52, No. 5 (part 1) (1972) pp. 1291–1306.

28. Hudson, R.R. and K.A. Mulholland, *The Measurement of High Transmission Loss (The Brick Wall Experiment)*, Acustica, Vol. 24, (1971) pp. 251–261.

29. Burroughs, Lou, *Microphones: Design and Application*, Sagamore Publishing Co., Plainview, NY (1974).

30. —*Performance Data–Architectural Acoustical Materials*, Acoustical and Insulating Materials Association, Bulletin No. 31, 1971–72.

31. —*Performance Data–Acoustical Materials*.

32. Harris, Cyril M., editor, *Handbook of Noise Control*, McGraw-Hill, (1957) pp. 9–10.

33. Randall, K.E. and F.L. Ward, *Diffusion of Sound in Small Rooms*, Proc. Inst. of Elect. Engrs., Vol. 107-B, pp. 349–350.

34. Wente, E.C., *The Characteristics of Sound Transmission in Rooms*, Jour. Acous. Soc. Am., Vol. 7, Oct. (1935) pp. 123–126.

35. Audio Visual Source Directory, Spring/Summer 1987, published by Motion Picture Enterprises Publications, Inc., Tarrytown, NY 10591, (212) 245-0969.

36. Bolt, R.H. and R.W. Roop, *Frequency Response Fluctuations in Rooms*, Jour. Acous. Soc. Am., Vol. 22, No. 2, March (1950) pp. 280–289.

37. Volkmann, John E., *Polycylindrical Diffusers in Room Acoustic Design*, Jour. Acous. Soc. Am., Vol. 13, Jan. (1942) pp. 234–243, especially Fig. 2.

38. Somerville, T. and F.L. Ward, *Investigation of Sound Diffusion in Rooms by Means of a Model*, Acustica, Vol. 1, No. 1 (1951) pp. 40–48.

39. Head, J.W., *The Effect of Wall Shape on the Decay of Sound in an Enclosure*, Acustica, Vol. 3 (1953) pp. 174–180.

40. Mankovsky, V.S., *Acoustics of Studios and Auditoria*, Hastings House Publishers, New York (1971).

41. Shiraishi, Y., K. Okumura, and F. Fujimoto, *Innovations in Studio Design and Recording in the Victor Record Studios*, Jour. Audio Engr. Soc., Vol. 1, No. 5, May (1971) pp. 405–40.

42. Sabine, Paul E. and L.G. Ramer, *Absorption-Frequency Characteristics of Plywood Panels*, Jour. Acous. Soc. Am., Vol. 20, No. 3, May (1948) pp. 267–270.

43. Young, Robert W., *Sabine Reverberation Equation and Sound Power Calculations*, J. Acous. Soc. Am., Vol. 31, No. 12, Dec. (1959) p. 1681.

44. Schroeder, M.R., *Diffuse Sound Reflection by Maximum-Length Sequences*, Jour. Acous. Soc. Am., Vol. 57, No. 1 (1975) pp. 149–150.

45. Schroeder, M.R., *Binaural Dissimilarity and Optimum Ceilings for Concert Halls: More Lateral Sound Diffusion*, Jour. Acous. Soc. Am., Vol. 65, No. 4, April (1979) pp. 958–963.

46. D'Antonio, Peter and John H. Konnert, *The Reflection Phase Grating Diffusor: Design Theory and Application*, Jour. Audio Engr. Soc., Vol. 32, No. 4, April (1984) pp. 228–238.

47. D'Antonio, Peter and John H. Konnert, *The RPG Reflection Phase Grating Acoustical Diffusor: Applications*, 76th Convention of the Audio Engineering Society, Oct. 1984, preprint 2156.

48. D'Antonio, Peter and John H. Konnert, *The RFZ/RPG Approach to Control Room Monitoring*, 76th Convention of the Audio Engineering Society, Oct. 1984, preprint 2157.

49. D'Antonio, Peter and John H. Konnert, *The RPG Reflection Phase Grating Acoustical Diffusor: Experimental Measurements*, 76th Convention of the Audio Engineering Society, Oct. 1984, preprint 2158.

50. D'Antonio, Peter and John H. Konnert, *The Role of Reflection Phase Grating Diffusors in Critical Listening and Performing Environments*, 78th Convention of the Audio Engineering Society, May 1985, preprint 2255.

51. D'Antonio, Peter and John H. Konnert, *The Acoustical Properties of Sound Diffusing Surfaces: The Time, Frequency, and Directivity Energy Response*, 79th Convention of the Audio Engineering Society, Oct. 1985, preprint 2295.

52. D'Antonio, Peter, John Konnert, and Bill Peterson, *Incorporating Phase Grating Diffusors in Worship Spaces*, 81st Convention of the Audio Engineering Society, Nov. 1986, preprint 2364.

53. D'Antonio, Peter, *New Acoustical Materials and Designs Improve Room Acoustics*, 81st Convention of the Audio Engineering Society, Nov. 1986, preprint 2365.

54. London, A., *Transmission of Reverberant Sound through Double Walls*, Jour. Acous. Soc. Am., Vol. 22, No. 2, (1950), pp. 270–279.

Index

CPSIA information can be obtained
at www.ICGtesting.com
Printed in the USA
LVHW102220120620
657965LV00006B/148